Lecture Notes in Electrical Engineering

Volume 298

For further volumes:
http://www.springer.com/series/7818

About this Series

"Lecture Notes in Electrical Engineering (LNEE)" is a book series which reports the latest research and developments in Electrical Engineering, namely:

- Communication, Networks, and Information Theory
- Computer Engineering
- Signal, Image, Speech and Information Processing
- Circuits and Systems
- Bioengineering

LNEE publishes authored monographs and contributed volumes which present cutting edge research information as well as new perspectives on classical fields, while maintaining Springer's high standards of academic excellence. Also considered for publication are lecture materials, proceedings, and other related materials of exceptionally high quality and interest. The subject matter should be original and timely, reporting the latest research and developments in all areas of electrical engineering.

The audience for the books in LNEE consists of advanced level students, researchers, and industry professionals working at the forefront of their fields. Much like Springer's other Lecture Notes series, LNEE will be distributed through Springer's print and electronic publishing channels.

Sabnam Sengupta · Kunal Das
Gitosree Khan
Editors

Emerging Trends in Computing and Communication

ETCC 2014, March 22–23, 2014

Springer

Editors
Sabnam Sengupta
Kunal Das
Gitosree Khan
Department of Information Technology
B.P. Poddar Institute of Management
and Technology
Kolkata, West Bengal
India

ISSN 1876-1100 ISSN 1876-1119 (electronic)
ISBN 978-81-322-1816-6 ISBN 978-81-322-1817-3 (eBook)
DOI 10.1007/978-81-322-1817-3
Springer New Delhi Heidelberg New York Dordrecht London

Library of Congress Control Number: 2014931361

Printed on acid-free paper

Springer is part of Springer Science+Business Media (www.springer.com)

Preface

It is a great pleasure with which we are releasing the Proceedings of the National Conference on Emerging Trends in Computing & Communication (ETCC 2014), which is organized by Department of Information Technology, B.P. Poddar Institute of Management and Technology, Kolkata, India, on March 22–23, 2014. The conference is being technically sponsored by IEEE Kolkata Section and Computer Society of India (CSI), Kolkata Chapter.

The objective of ETCC 2014 is to intensify the information exchange of the results in theoretical research and practical developments in the emerging domains. This Conference seeks to bring together national researchers, industrial experts, and academicians to present papers and generate discussions on current research and development in state-of-the-art technologies in computing and communication.

We hope that ETCC 2014 will provide an excellent platform for researchers and practitioners in exchanging new ideas and exploring future directions of research.

We want to thank all the authors for submitting their manuscripts and to the reviewers, whose effort and hard work has helped in ensuring high quality of submissions. We are thankful to the management of BPPIMT for their wholehearted support and trust on the organizers. We are also thankful to the funding agencies; their support helped us to take this forward. Sincere thanks to Mr. Aninda Bose, Editor, Springer (India) Pvt. Ltd., for extending his cooperation in shaping these Proceedings. A special thanks to all the faculty, student, and staff members of B.P. Poddar Institute of Management and Technology, who have been tirelessly working and collaborating round the clock to make sure that these Proceedings, sees the light of the day.

Sabnam Sengupta
Kunal Das
Gitosree Khan

About the Institute

B.P. Poddar Institute of Management and Technology (BPPIMT) was established in the year 1999, under the aegis of B.P. Poddar Foundation for Education, a Trust dedicated to enriching the quality of technical education in the country. The Institute was set up as a tribute to the memory of Late Badri Prasad Poddar, philanthropist and educationist, and the founder of the B.P. Poddar Group.

The B.P. Poddar Institute of Management and Technology is affiliated to the West Bengal University of Technology, Kolkata. The courses offered are B.Tech in the disciplines of Computer Science and Engineering, Electronics and Communication Engineering, Electrical Engineering, and Information Technology. The Institute blends a dynamic and progressive approach to education with high quality, innovative, and result-oriented programs which have been approved by the All India Council for Technical Education (AICTE), New Delhi.

The mission of the institute include creation of conducive learning atmosphere to inculcate value-based education, development of state-of-the-art technology, and making education responsive to changes, creation of effective interface with industry and strengthening Industry–Institutional interaction and Entrepreneurship Development, promoting research activities, and establishing collaboration with National and International Institutes of repute.

The institute's objective is to develop an Institution of Excellence for advancement of Science and Technology and for creation of professionals with high sense of ethical values and commitment to the Society who are capable of functioning unfettered by the shackles of caste, creed, political dogmas, and religious parochialism, and who can effectively contribute to and lead in the shaping of the nation.

ETCC 2014 Committee

President

Shri Arun Poddar

Chief Patrons

Sri Ayush Poddar, Vice Chairman, BPPIMT
Dr. Subir Choudhury, Director, BPPIMT

Chairperson

Prof. Dr. Sutapa Mukherjee, Principal, BPPIMT

Advisory Committee

Prof. Rajib Mall, IIT Kharagpur
Prof. Bhargab B. Bhattacharya, ISI Kolkata
Prof. Swapan Bhattacharya, NIT Surathkal
Prof. Aditya Bagchi, ISI Kolkata
Prof. Indranil Sengupta, IIT Kharagpur
Prof. K. P. Ghatak, National Institute of Technology, Agartala
Prof. Bhabani P. Sinha, ISI Kolkata
Prof. Santanu Chaudhury, IIT, Delhi
Prof. B. Yegnanarayana, IIIT, Hyderabad
Prof. Dr. M. Bhattacharya, Advisor, BPPIMT
Prof. Dr. B. N. Chatterji, Dean Academics, BPPIMT
Prof. Sandip Ghosh, Registrar, BPPIMT

Convener

Dr. Sabnam Sengupta

Program Committee

Prof. K. N. Dey, University of Calcutta
Prof. Nabanita Das, ISI Kolkata
Prof. Partha Pratim Das, IIT Kharagpur
Prof. Sanghamitra Bandyopadhyay, ISI Kolkata
Prof. Hafizur Rahman, BESU, Shibpur
Prof. Sankhayan Choudhury, University of Calcutta
Prof. Satchidananda Dehuri, Fakir Mohan University, Orissa
Prof. Nabendu Chaki, University of Calcutta
Mr. Abhik Sengupta, Cognizant Technology Solutions
Dr. Mallika De, Kalyani University
Prof. Debashis De, West Bengal University of Technology
Prof. G. P. Biswas, ISM, Dhanbad
Prof. J. P. Singh, NIT Patna
Dr. Ram Sarkar, Jadavpur University
Dr. Sourav Chattopadhyay, Amity University, Delhi
Dr. Usha Banerjee, College of Engineering Roorkee
Prof. Shila Ghosh, BPPIMT
Prof. Ananya Kanjilal, BPPIMT
Prof. Arijit Saha, BPPIMT
Mr. Kunal Das, BPPIMT
Ms. Gitosree Khan, BPPIMT

Organizing Committee

Mr. Sabyasachi Chakraborty
Ms. Lipi Begum
Mr. Sudarsan Biswas
Mr. Asim Kr. Panda
Mr. Ashis Bhnuia

Finance Committee

Ms. Swagata (Gayen) Kundu
Mr. Balaram Ghosal
Mr. Joy Roy

Contents

Reviewers (ETCC 2014)

1. Abhijit Chandra, Bengal Engineering and Science University, Howrah, WB
2. Abhik Sengupta, Cognizant Technology Solution, Kolkata, WB
3. Amit Phadikar, MCKV
4. Ananya Kanjilal, B.P. Poddar Institute of Management and Technology, Kolkata, WB
5. Animesh Dutta, NIT Durgapur, Durgapur, WB
6. Anirban Sarkar, NIT Durgapur, Durgapur, WB
7. Arijit Saha, B.P. Poddar Institute of Management and Technology, Kolkata, WB
8. Ayatullah Faruk Mollah, Jadavpur University, Kolkata, WB
9. Amlan Chakrabarti, Calcutta University, Kolkata, WB
10. B. N. Chatterji, B.P. Poddar Institute of Management and Technology, Kolkata, WB
11. Bhaskar Som, B.P. Poddar Institute of Management and Technology, Kolkata, WB
12. Bibhash Sen, NIT Durgapur, Durgapur, WB
13. Chandan Kr. Bhattacharya, Techno India SaltLake, Kolkata, WB
14. Debashis De, West Bengal University of Technology, Kolkata, WB
15. Debasis Chanda, Cognizant Technology Solution, Kolkata, WB
16. Debasish Jana, West Bengal University of Technology, Kolkata, WB
17. Debatosh Guha, University of Calcutta, Kolkata, WB
18. Dipankar Bandyopadhyay, IIT Guwahati, Assam, Guwahati
19. Dipankar Majumdar, RCC, Kolkata, WB
20. Diptiman Roychoudhury, University of Calcutta, Kolkata, WB
21. G. P. Biswas Biswas, ISM, Dhanbad, Jharkhand
22. Gitosree Khan, B.P. Poddar Institute of Management and Technology, Kolkata, WB
23. Hafizur Rahman Rahman, Bengal Engineering and Science University, Howrah, WB
24. Himadri Dutta, Kalyani Government Engineering College, Kalyani, WB
25. Indranil Sengupta, IIT Kharagpur, Kharagpur, WB
26. Indra Kanta Maitra, B.P. Poddar Institute of Management and Technology, Kolkata, WB

27. Iti Saha Misra, Jadavpur University, Kolkata, WB
28. Ivy Majumdar, B.P. Poddar Institute of Management and Technology, Kolkata, WB
29. J. P. Singh, NIT Patna, Bihar
30. Jayeeta Chanda, B.P. Poddar Institute of Management and Technology, Kolkata, WB
31. Kamakhya Prasad Ghatak, NIT Agartala, Tripura
32. Kaushik Roy, West Bengal State University, Kolkata, WB
33. Kunal Das, B.P. Poddar Institute of Management and Technology, Kolkata, WB
34. Mallika De, Kalyani University, Kalyani, WB
35. Nabanita Das, ISI Kolkata, Kolkata, WB
36. Nabarun Das, Hindustan Aeronautical Limited
37. Nabendu Chaki, University of Calcutta, Kolkata, WB
38. Nandini Sidnal Sidnal, KLE DR MSS CET Belgaum Karnataka
39. Paramartha Dutta Dutta, Visa varati
40. Partha Pratim Das, IIT Kharagpur, Kharagpur, WB
41. Poulami Das, Heritage Institute of Technology, Kolkata, WB
42. Prakash Biswas, IIT Roorkee, Roorkee, Uttarakhand
43. Rajib Mall, IIT Kharagpur, Kharagpur, WB
44. Ram Sarkar, Jadavpur University, Kolkata, WB
45. Ranjan Sen, Exide Industries Limited
46. Ravi Shankar, Hindustan Aeronautics Limited
47. Rituparna Chaki, University of Calcutta, Kolkata, WB
48. Sabnam Sengupta, B.P. Poddar Institute of Management and Technology, Kolkata, WB
49. Saket Srivastava, University of Lincoln, UK
50. Sankhayan Choudhury, University of Calcutta, Kolkata, WB
51. Satadal Saha, MCKV
52. Shila Ghosh, B.P. Poddar Institute of Management and Technology, Kolkata, WB
53. Soharab Hossain, University of Calcutta, Kolkata, WB
54. Soumya Sen, University of Calcutta, Kolkata, WB
55. Sourav Chattopadhyay, Amity University, Delhi
56. Subir Das, IIT Benaras Hindu University, Varanasi, UP
57. Surajit Mandal, B.P. Poddar Institute of Management and Technology, Kolkata, WB
58. Sutapa Mukherjee, B.P. Poddar Institute of Management and Technology, Kolkata, WB
59. Swagata (Gayen) Kundu, B.P. Poddar Institute of Management and Technology, Kolkata, WB
60. Tanay Chattopadhyay, Kolaghat Thermal Power Station, Amalhara, WB
61. Usha Banerjee, College of Engineering, Roorkee, Uttarakhand
62. V. S. Malemath Malemath, KLES College of Engineering and Technology, Belgaum, Karnataka

Part I
Communication and Wireless Network

Chapter 1
Analysis and Design of Power Line Filter Using Transmission Parameters

Shashwatee Paul and Dipankar Dan

Abstract A conventional method of designing power line filter using a unique combination of constant-k and m-derived composite section has been introduced. The Composite LPF is implemented in the classical design technique of power line filter. The filter is designed for both the Common mode (CM) and the Differential mode (DM) of the filter. The final circuit of the filter is designed by using the Cascaded structure of the CM and DM section of the filter in an Equivalent Circuit. The entire Equivalent circuit is further analyzed by the two-port Network parameters using Transmission. The design of the filter circuit is simulated using Advanced Design software (ADS) based on S-parameters. A comparison on the Filter response, before and after tuning, of the components in the design, is presented in the last section of the paper.

Keywords Conducted emission (CE) · Common mode (CM) · Differential mode (DM) · Composite low pass filter · Transmission parameter

1.1 Introduction

In Communication systems, distributed power systems are widely used. These power modules are often cited as one of the main source of conducted electromagnetic interference (EMI). Widely used *Switched mode power supplies* (SMPS) for powering today's electronics loads is a most common example [1].

S. Paul (✉)
Gitam University, Gandhi Nagar, Rushikonda, Visakhapatnam, Andhra Pradesh 530 045, India
e-mail: shashwatee.paul@gmail.com

D. Dan (✉)
SAMEER-Kolkata Centre, Kolkata, India
e-mail: dipankardan@rediffmail.com

S. Sengupta et al. (eds.), *Emerging Trends in Computing and Communication*, Lecture Notes in Electrical Engineering 298, DOI: 10.1007/978-81-322-1817-3_1,

Unfortunately, all the power control techniques deliberately distort sinusoidal wave form of power frequency and generate unwanted interfering signals. They all are often cited as one of the main source of conducted electromagnetic interference (EMI). These interferences are generally divided into two components: the common mode (CM) interference and the differential mode (DM) interference [2]. Therefore, for the design of EMI filters, separate common mode and differential mode section are needed in order to filter out the switching noise and eliminate electromagnetic interference to other equipments.

1.1.1 EMI Issues of Power Line Filter

These are electromagnetic disturbances carried by the electrical power supply lines, are classified into two categories, common-mode current/voltages and differential-mode (or normal mode) current/voltages (Fig. 1.1).

Examples of CM and DM Interferences [3].

1.1.2 Characterization of Conduction Currents/Voltages

In single-phase applications the mains cable of the EUT, which is connected to the LISN, consists of three parallel wires: Phase, neutral and ground. Considering the currents flowing through the phase and neutral conductors with phase current (I_p) and neutral current (I_n) respectively. These currents can be decomposed into two auxiliary currents, which are referred to as the common mode current (I_{cm}) and differential mode current (I_{dm}) [4]. The CM and DM noise components are generally not specified in the Conducted noise compliance tests. However, they can be determined from the measurement data of phase line and neutral line.

$$I_{cm} = \frac{I_p + I_n}{2} \quad I_{dm} = \frac{I_p - I_n}{2} \tag{1.1}$$

The CM and DM Voltage can be obtained across the 50 Ω impedance of the LISN port:

$$V_{dm} = 50\, I_{dm} = 50\left(\frac{I_p - I_n}{2}\right) = \frac{V_p - V_n}{2} \tag{1.2}$$

$$V_{cm} = 50\, I_{cm} = 50\left(\frac{I_p + I_n}{2}\right) = \frac{V_p + V_n}{2} \tag{1.3}$$

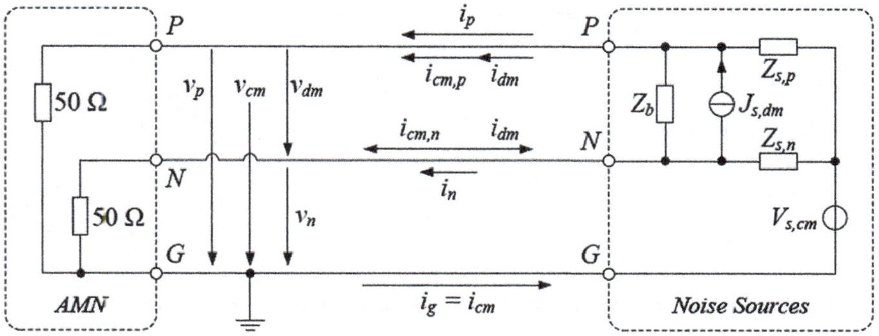

Fig. 1.1 Conducted emission measurement equivalent circuit

1.2 Design Process

1.2.1 Differential Mode Filter Design Process Using K-Constant, M-Derived Composite LPF

A design methodology to realize low pass filter is presented by using a unique combination of constant-k, m-derived section. Cascading both k-constant and m-derived section gives a composite LPF that gives both sharp cut-off and good impedance matching at input and out sides. And a terminating π-section is connected at the source and the load side to increase the matching impedance. The circuit is designed using vendor components. So the component values are chosen nearest to the calculated values.

Constant-k Section—The nominal characteristic impedance of constant-k section is made constant as 50 Ω for the assigned frequency that is given by [5] the value of Capacitance and Inductance is determined as

$$L = \frac{Z_o}{\pi f_c}; \quad C = \frac{1}{\pi f_c Z_o} \tag{1.4}$$

The cut-off frequency (f_c) and the infinite attenuation occurred at frequency (f_∞) is defined as

$$f_\infty = \frac{f_c}{\sqrt{1 - m^2}}; \quad f_c = \frac{1}{\pi\sqrt{LC}} \tag{1.5}$$

M-derived Section—The constant-k section is followed by the m-derived section, in order to improve the impedance match of the filter. The ‘m’ is fixed at 0.6.The independent element values of m-derived section is obtained as

$$L^{'} = \frac{ml}{2}; \quad C^{'} = mc; \quad L^{''} = \frac{1 - m^2}{4m} \tag{1.6}$$

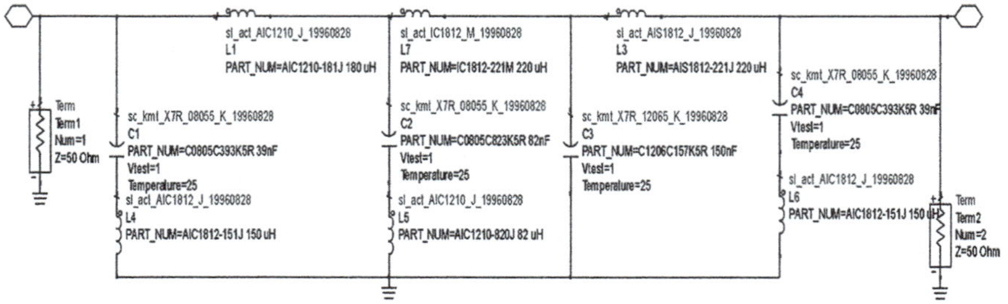

Fig. 1.2 Circuit of composite LPF for differential-mode using vendor components

By the combination of *cascaded* structure of Constant-k, m-derived section and the matching section, the circuit of the Composite filter for the DM section is realized within the frequency range of 150 kHz–1 MHz and cut-off frequency (f_c): 50 kHz (Fig. 1.2).

In the similar manner, the composite filter can be designed for the common mode (CM) section of the filter. The cut-off frequency (f_c) is taken at 333.33 kHz to provide attenuation to the common mode noise signal within the frequency range of 1–30 MHz (Fig. 1.3).

1.3 Insertion Loss of the Power Line Filter

The entire power line filter can be coupled into two section of network. One is the common mode (CM) network section and the other is the differential mode (DM) as shown in the Fig. 1.4.

1.3.1 Basic Definition

The Insertion loss (IL) as a function of frequency is the most fundamental characteristic of a filter. It is defined as [6]

$$\mathrm{IL(dB)} = 20\, log_{10} \frac{V_1}{V_2} \tag{1.7}$$

where,

V_1 the output voltage of the signal source without the filter connected to the circuit

V_2 the output voltage of the signal source at the output terminal of the filter with the filter inserted in the circuit.

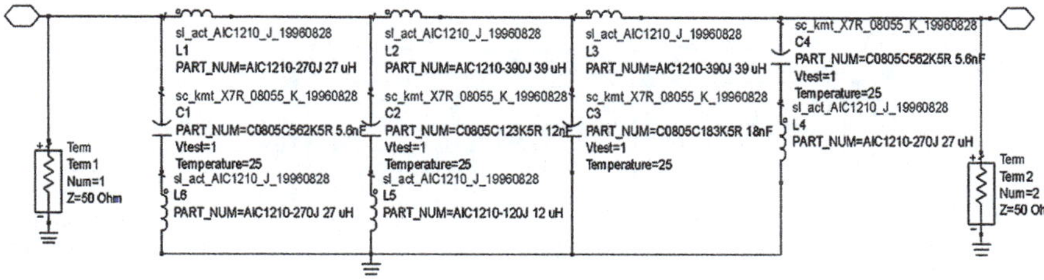

Fig. 1.3 Circuit of composite LPF for common-mode using vendor components

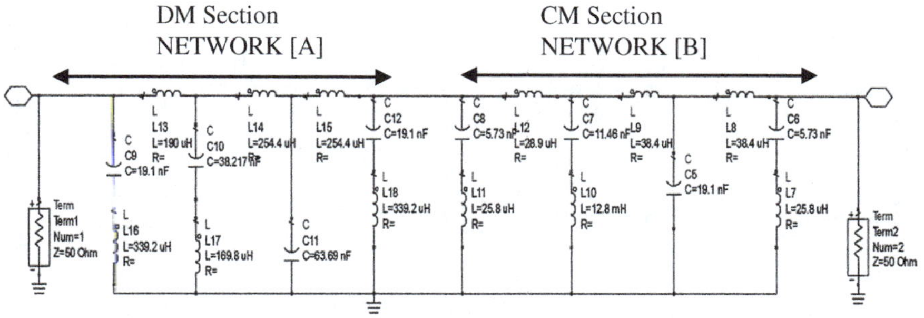

Fig. 1.4 Design structure of power line filter

1.3.2 IL in Terms of Transmission Parameters

The network analysis of the filter is carried out, by initially considering the differential mode (DM) section of the filter. And all the passive elements are denoted by their equivalent impedance. The vendor passive components are now replaced with their respective original calculated values for the determination of IL (Fig. 1.5).

Applying Nodal equations, to the network:

$$\mathrm{I}_1 - \frac{V_1}{Z_1} - \frac{V_1 - V_2}{z_2} = 0 \tag{1.8}$$

$$\frac{V_1 - V_2}{Z_2} - \frac{V_2}{Z_3} - \frac{V_2 - V_3}{Z_4} = 0 \tag{1.9}$$

$$\frac{V_2 - V_3}{Z_4} - \frac{V_3}{Z_5} - \frac{V_3 - V_4}{Z_6} = 0 \tag{1.10}$$

$$\frac{V_3 - V_4}{Z_6} - \frac{V_4}{Z_7} + I_o = 0 \tag{1.11}$$

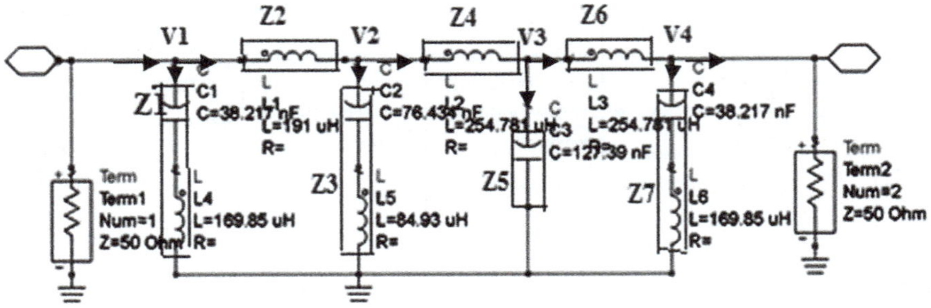

Fig. 1.5 Filter network

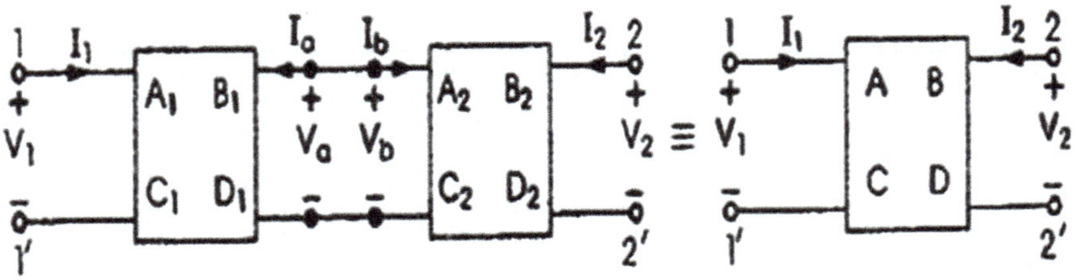

Fig. 1.6 Cascaded structure of the two-port DM and CM filter section

Solving the above equations, the chain parameters A_1, B_1, C_1 and D_1 of the network [A] and the network [B] can be determined for the DM network [A] and the CM network [B] as shown in the Fig. 1.6, that can be written as [6]

$$\begin{bmatrix} V_1 \\ I_1 \end{bmatrix} = \begin{bmatrix} A1 & B1 \\ C1 & D1 \end{bmatrix} \begin{bmatrix} V_a \\ -I_a \end{bmatrix} \tag{1.12}$$

$$\begin{bmatrix} V_b \\ I_b \end{bmatrix} = \begin{bmatrix} A_2 & B_2 \\ C_2 & D_2 \end{bmatrix} \begin{bmatrix} V_o \\ -I_o \end{bmatrix} \tag{1.13}$$

Thus the two filters Section (CM and DM) can be cascaded and the parameters of the final Equivalent network can be obtained with their matrix multiplication. Thus, the transmission or ABCD parameters of the overall two port-network can be written as the product of the matrix of differential mode and the common mode filter as shown in the Fig. 1.7.

1.4 Simulation Result

The equivalent circuit of the filter is simulated using *advanced design software* (ADS) as shown in the first section of the figure. This frequency response is quiet impractical to achieve and hence further modifications are done in the circuit,

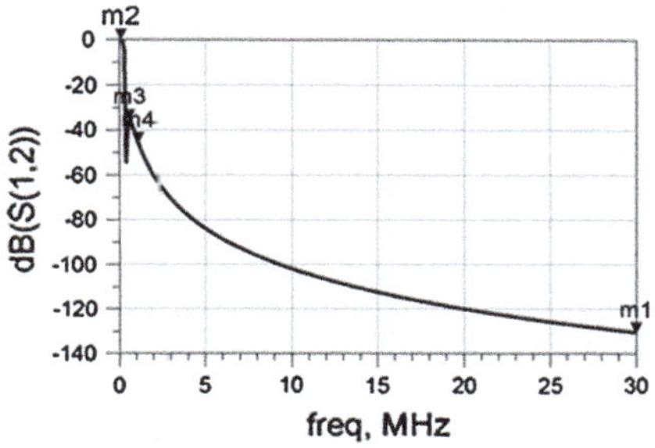

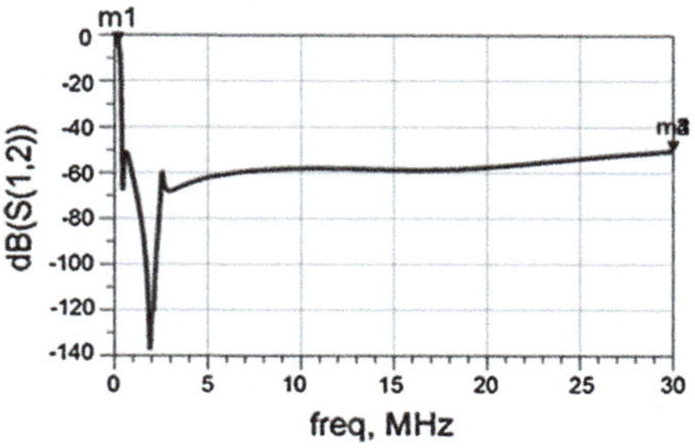

Fig. 1.7 Simulation result of composite low pass power line filter before and after tuning

considering the parasitic effect of the passive components. Now the attenuation level of the circuit has reached to a certain level of 60 dB for the conducted emission frequency range.

1.5 Conclusion

- Design of Power line filter using constant-k and m-derived composite LPF has been proposed.
- The study of CM and DM interference in the designed power line filter section has been attempted to show with the help of Network parameters to present a better understanding of the propagation phenomena of CM and DM section.
- The implementation of Transmission parameters in the cascaded structure of filter in the form of Matrix multiplication has been presented.

References

1. Tarateeraseth V, See KY, Canavero FG, Chang RW (2010) Systematic electromagnetic interference filter design based on information from in-circuit impedance measurements. IEEE Trans Electromagn Compat 52(3):588–598
2. Paul CR (1992) Electromagnetic compatibility. Wiley, New York
3. Wang S (2010) Investigation of the transformation between DM and CM noise in an EMI filter due to unbalance. IEEE Trans EMC 52(3):578–587
4. Kostov K (2009) Design and characterization of single-phase power filter. Helsinki University of Technology, Department of Electrical Engineering, Helsinki
5. Pozer DM (2009) Microwave engineering. Wiley, New York
6. Kostov K, Kyyra J, Suntio T (2003) Analysis and design of EMI filters for DC–DC converters. In: 10th European conference on power electronics and applications, Toulouse

Chapter 2
Studies and Implementation of Subband Coder and Decoder of Speech Signal Using Rayleigh Distribution

Sangita Roy, Dola B. Gupta, Sheli Sinha Chaudhuri and P. K. Banerjee

Abstract In the last 40 years a number of coding techniques for analog sources (speech and images) has been employed. Subband coding, a kind of transform coding, splits analog speech signal into a number of different smaller frequency bands. By subbanding data rate has been reduced to 12.13804 Kbps [Sangita et al. Studies and implementation of subband coder and decoder of speech signal, Proceedings of national conference on electronics, communication and signal processing, 8–16, 1] on 64 Kbps telephone line. In this paper a method has been proposed by which data rate has been reduced to 9.4875 Kbps using Rayleigh distribution where data rate can be reduced to 9.6 Kbps [Crochiere et al. Digital coding of speech in subbands, The BELL System Technical Journal, 2]. Proposed method can save data rate which in turn saves bandwidth as well as spectrum. Moreover this proposed method provides acceptable probability of error and quantization noise i.e. SNR.

Keywords DM · PCM · SNR · Subband · Probability of bit error

S. Roy (✉) · D. B. Gupta (✉)
ECE Department, Narula Institute of Technology, WBUT, Kolkata, India
e-mail: roysangita@gmail.com

D. B. Gupta
e-mail: dola.ju@gmail.com

S. S. Chaudhuri (✉) · P. K. Banerjee (✉)
ETCE Department, Jadavpur University, Kolkata, India
e-mail: shelism@rediffmail.com

P. K. Banerjee
e-mail: Pkb.ju65@rediffmail.compkb.ju68@gmail.com

S. Sengupta et al. (eds.), *Emerging Trends in Computing and Communication*,
Lecture Notes in Electrical Engineering 298, DOI: 10.1007/978-81-322-1817-3_2,

2.1 Introduction

Sub-band coding (SBC) is a kind of transform coding [3, 4]. A signal is divided into a number of different frequency bands and encodes each one independently. It enables a data compression by discarding information about frequencies which are masked. The result differs from the original signal, but if the discarded information is chosen carefully, the difference will not be noticeable, or more importantly, objectionable [5–7]. A paper—"**A low-complexity audio data compression technique using subband coding (SBC) and a recursively indexed quantizer (RIQ)**" compared SBC and RIQ to conventional coding techniques with SNR 2–5 dB higher than that of other coders of similar computational complexity of wideband audio signals [8]. The basic concept of "**Frequency Domain Coding of Speech**" methods is to divide the speech into frequency components by a filter bank (sub-band coding), or by a suitable transform (transform coding), and then encode them using adaptive PCM. Recent developments and examples of the "Vocoder-driven" adaptive transform coder for low bit-rate applications is also discussed [9]. In "**Subband Coding of Speech Signals Using Decimation and Interpolation**"—**a** structure of a two-channel quadrature mirror filter with low pass filter, high pass filter, decimators and interpolators, is proposed to perform subband coding of speech signals in the digital domain. The results show that the proposed structure significantly reduces the error and achieves considerable performance improvement compared to delta-modulation encoding systems [10]. Subband coder reduces and controls quantization noise. Here bit allocation on each subband is done on perceptual criterion. So that quality of the coded signal is improved over the full spectrum coding. Computer simulated data provides 16 and 9.6 Kbps over 64 Kbps data rate [2].

2.2 Basic Idea of the System

Figure 2.1a is an example of Power Spectral Density (PSD) of an Voice Signal. Here, Voice signal has been considered to be restricted to 3.5 kHz only. Power Spectral Density to be in watt/HZ or dB. In this figure frequency axis is divided into number of subbands (say 0–f1, f1–f2, f2–f3, f3–f4, etc.). The frequency band (0–f_1) is base band signal whereas (f_1–f_2), (f_2–f_3), (f_3–f_4), etc. are band pass signals. Each band will translated to baseband by multiplying with lowest frequency component of the said subband. Here seven subbands have been considered (Fig. 2.1b).

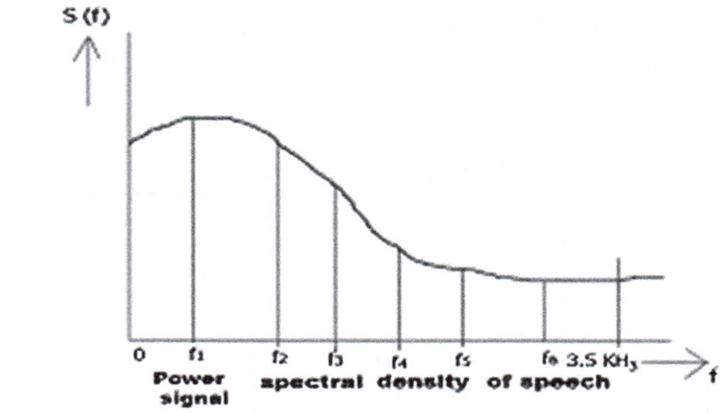

Fig. 2.1a Power spectral densities versus frequency of speech signal using subband

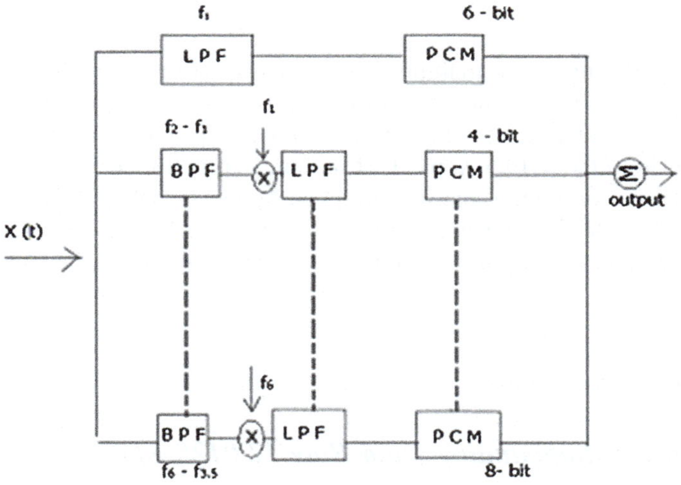

Fig. 2.1b Block diagram of subband coding transmitter

Transmitter consists of one LPF and six BPFs. All BPFs outputs are multiplied by the lowest frequency component of those bands at the multiplier block. Then outputs are PCM and then added by summer. Finally the summed output is put into channel.

At the receiver signals are decoded by seven decoders. Then each signal is passed through LPF of cut-off frequency f_1, $f_2–f_1$, $f_3–f_2$ etc. From second to

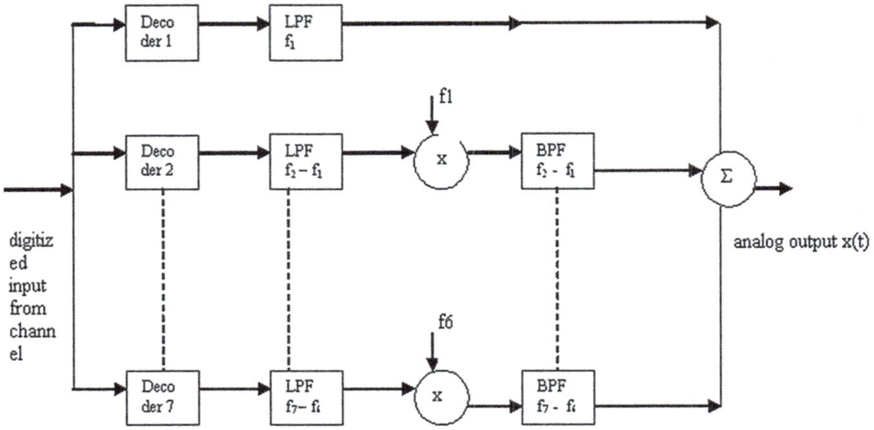

Fig. 2.1c Block diagram of receiver

seventh signal outputs are multiplied by their respective lowest frequency components and then passed through BPFs of f_2–f_1, f_3–f_2 etc. Then the outputs are summed up to get the replica of the original signal. Data rate from the above signal is reduced from 64 to 19.5 Kbps [1]. Data rate can be further reduced, if Power spectral density of voice signal is multiplied by the probabilities of occurrences which is 12.13804 Kbps lower than 19.5 Kbps by using MATLAB Simulation [1]. Figure 2.1c shows SNR, quantization noise produced out of subbanding.

2.3 Proposed Scheme

2.3.1 Case I: Subbanding Data Rate 12.0128 Kbps

Matlab simulation of subband data rate [1] has been reduced to 12.0128 Kbps where existing telephone line data rate 64 Kbps. Figures 2.2a, 2.2b, 2.2c, 2.2d have been shown as MATLAB simulation of data rate, cumulative data rate, SNR and probability of bit error of subbands respectively [11].

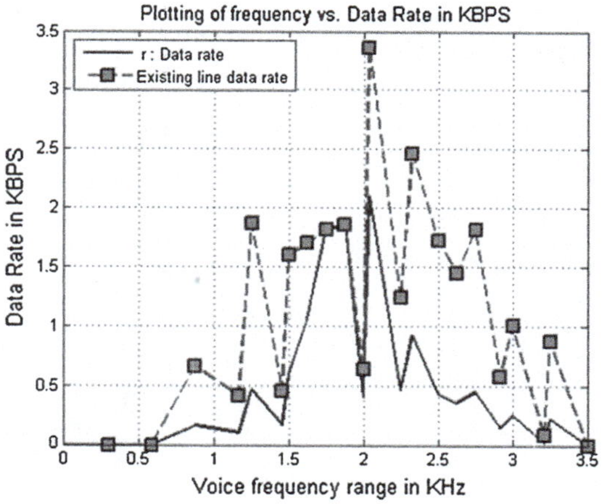

Fig. 2.2a Matlab simulation output of frequency versus data rate

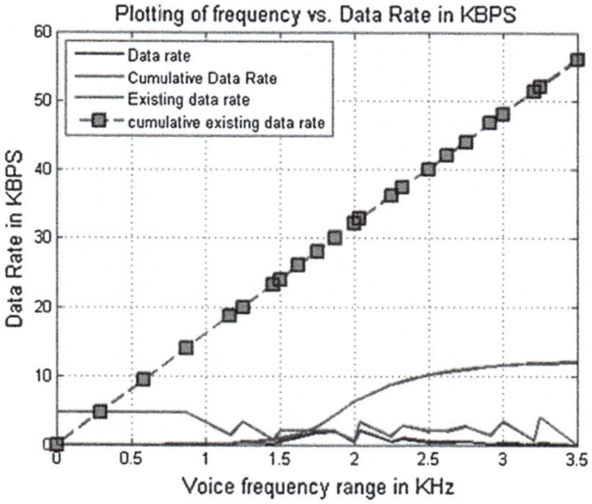

Fig. 2.2b Matlab simulation output of frequency versus cumulative data rate

2.3.2 Case II: Subbanding Data Rate 11.0959 Kbps

By changing the bit allocation by perceptual criterion data rate has been reduced to 11.0959 Kbps from 12.0128 Kbps. Figure 2.3a, 2.3b, 2.3c, 2.3d have been shown as MATLAB simulation of data rate, cumulative data rate, SNR and probability of bit error of subbands respectively.

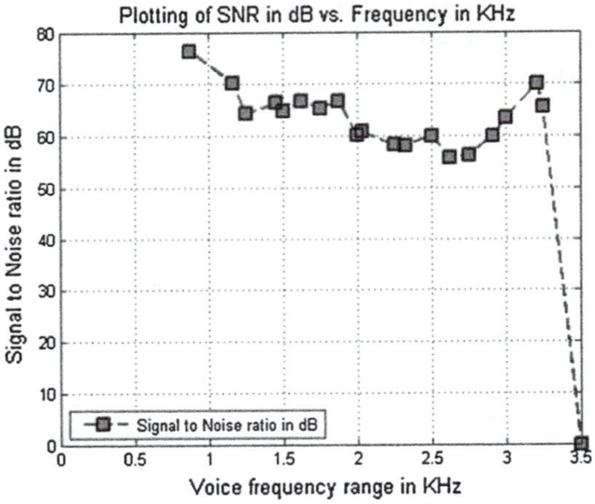

Fig. 2.2c Matlab simulation output of frequency versus SNR

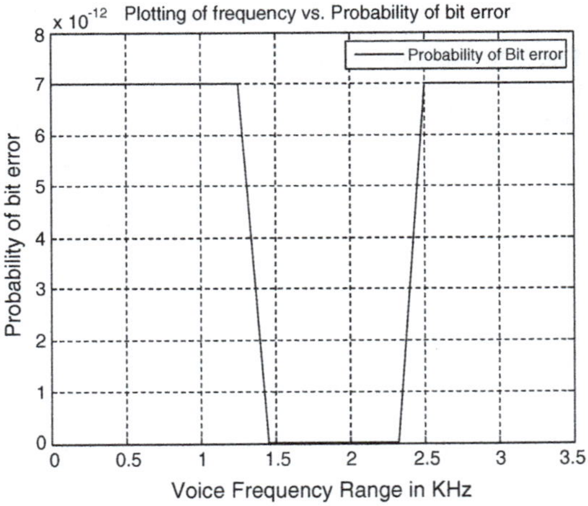

Fig. 2.2d Probability of bit error

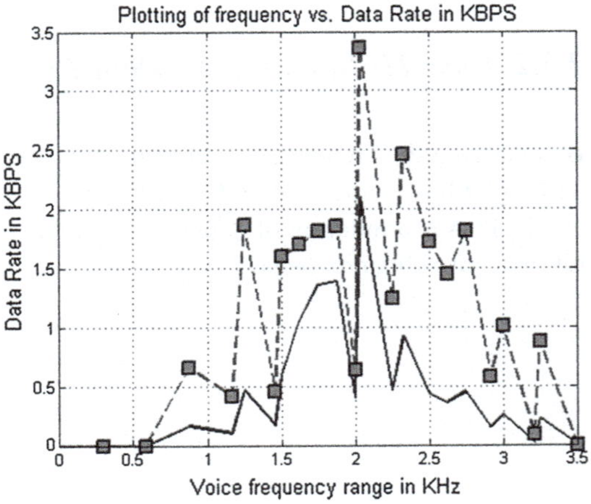

Fig. 2.3a Matlab simulation output of frequency versus data rate

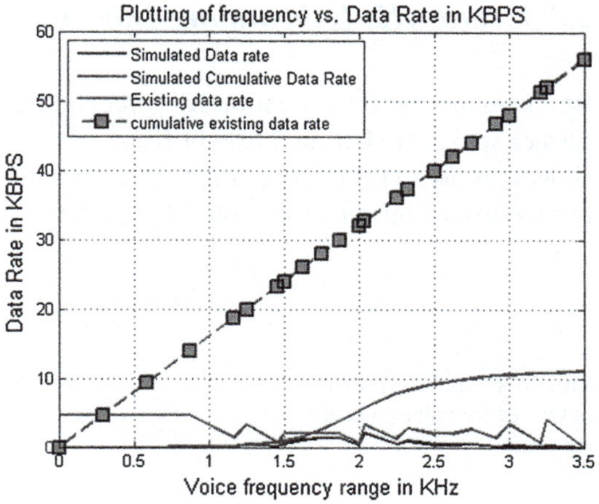

Fig. 2.3b Matlab simulation output of frequency versus cumulative data rate

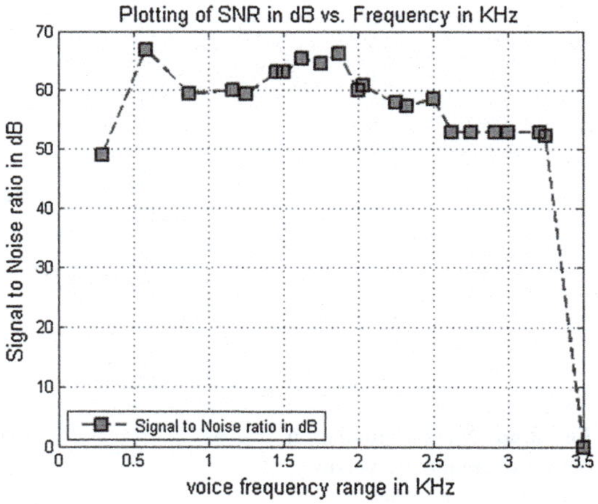

Fig. 2.3c Matlab simulation output of frequency versus SNR

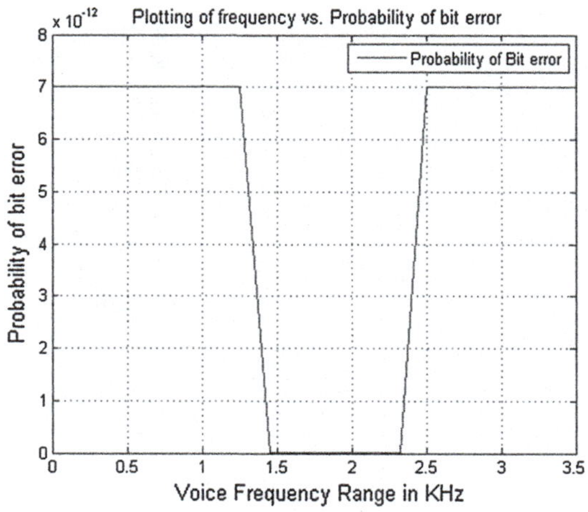

Fig. 2.3d Probability of bit error

2.3.3 Case III: Subbanding Data Rate 10.0494 Kbps

By changing the bit allocation by perceptual criterion data rate has been reduced further to 10.9494 from 11.0959 Kbps. Figures 2.4a, 2.4b, 2.4c, 2.4d have been shown as MATLAB simulation of data rate, cumulative data rate, SNR and probability of bit error of subbands respectively.

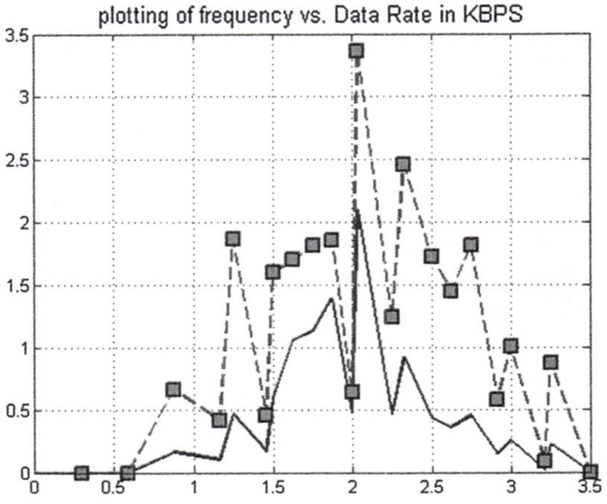

Fig. 2.4a Matlab simulation output of frequency versus data rate

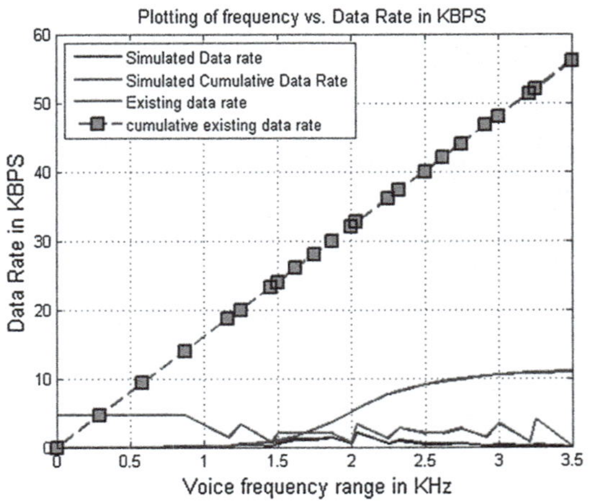

Fig. 2.4b Matlab simulation output of frequency versus cumulative data rate

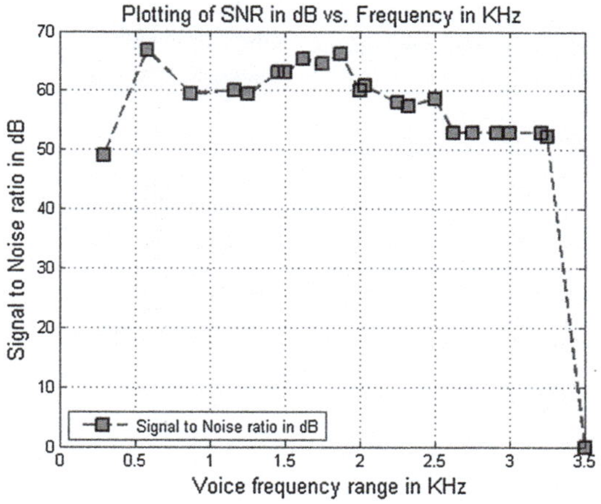

Fig. 2.4c Matlab simulation output of frequency versus SNR

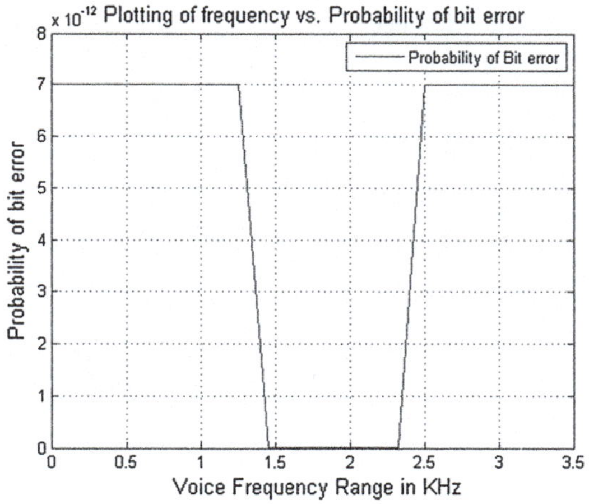

Fig. 2.4d Probability of bit error

2.3.4 *Case IV: Subbanding Data Rate 10.7175 Kbps*

By changing the bit allocation by perceptual criterion data rate has been reduced to 10.7175 Kbps from 10.9494 Kbps. Figures 2.5a, 2.5b, 2.5c, 2.5d have been shown as MATLAB simulation of data rate, cumulative data rate, SNR and probability of bit error respectively of subbands.

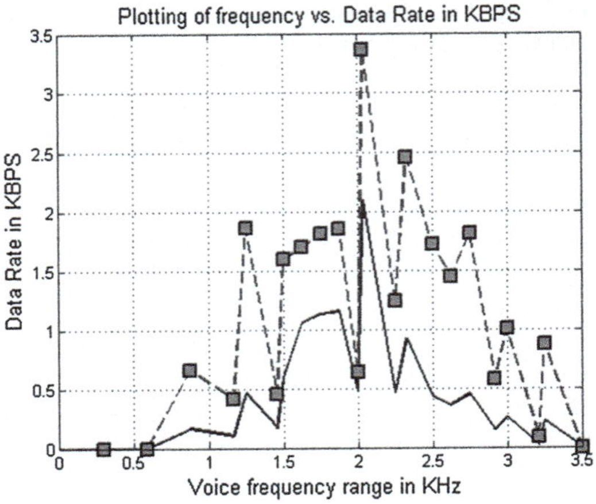

Fig. 2.5a Matlab simulation output of frequency versus data rate

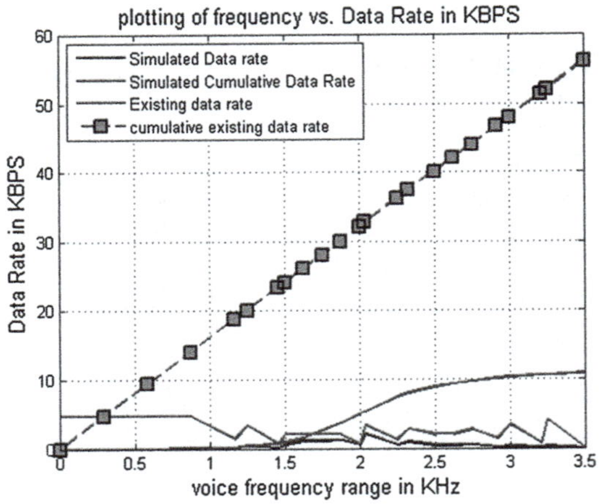

Fig. 2.5b Matlab simulation output of frequency versus cumulative data rate

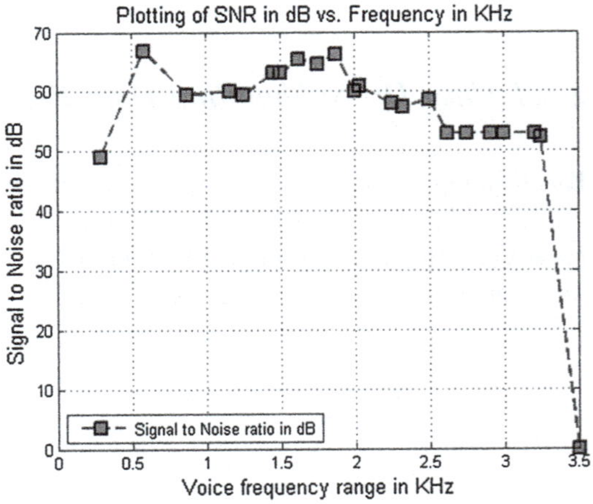

Fig. 2.5c Matlab simulation output of frequency versus SNR

Fig. 2.5d Probability of bit error

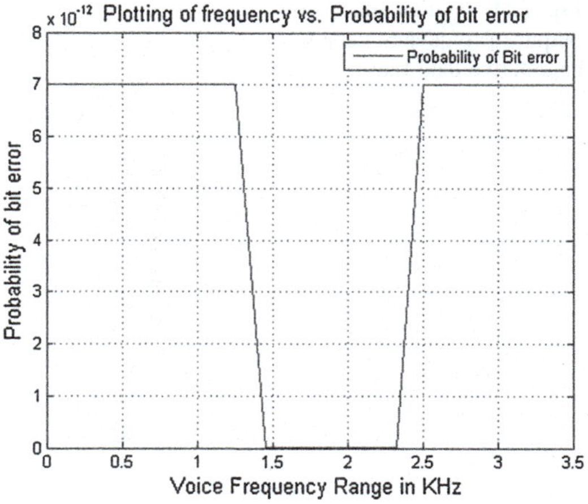

2.3.5 *Case V: Subbanding Data Rate 10.4867 Kbps*

By changing the bit allocation by perceptual criterion data rate has been reduced to 10.4867 Kbps from 10.7175 Kbps. Figures 2.6a, 2.6b, 2.6c, 2.6d have been shown as MATLAB simulation of data rate, cumulative data rate, SNR and probability of bit error respectively of subbands.

Fig. 2.6a Matlab simulation output of data rate versus frequency

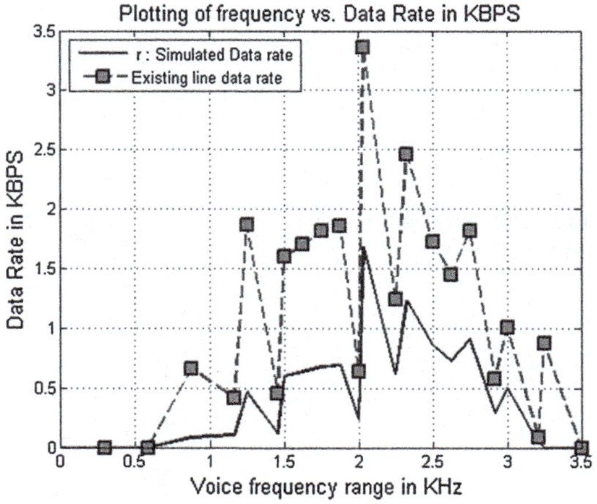

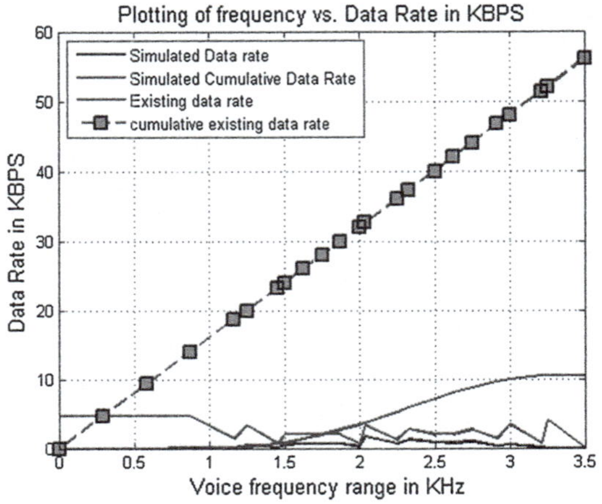

Fig. 2.6b Matlab simulation output of frequency versus cumulative data rate

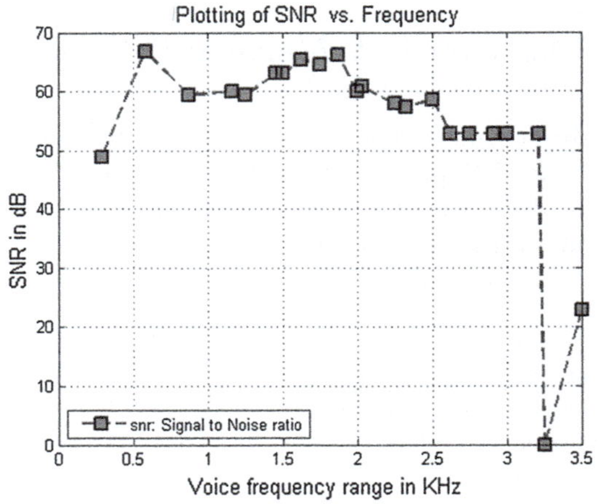

Fig. 2.6c Matlab simulation output of SNR versus frequency

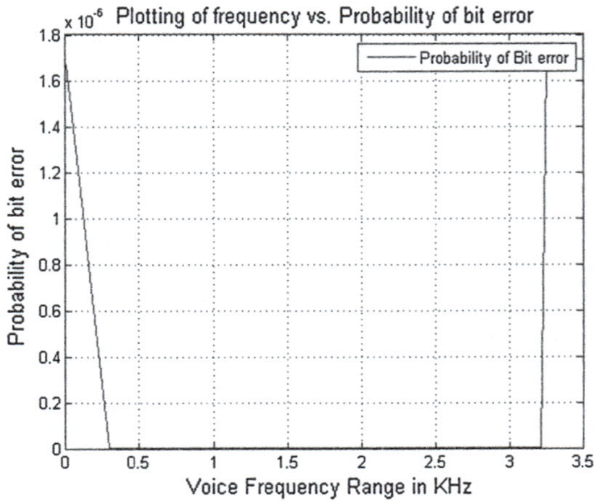

Fig. 2.6d Probability of bit error

2.3.6 *Case VI: Subbanding Data Rate 9.4875 Kbps Using Rayleigh Distribution*

Changing the bit allocation by perceptual criterion and Rayleigh distribution Power Spectral Density, data rate has been reduced to 9.4875 Kbps from 10.4867 Kbps which is lowest in this study [12]. Figures 2.7a, 2.7b, 2.7c, 2.7d have been shown as MATLAB simulation of data rate, cumulative data rate, SNR and probability of bit error respectively of subbands.

The above studied cases are tabulated in Table 2.1.

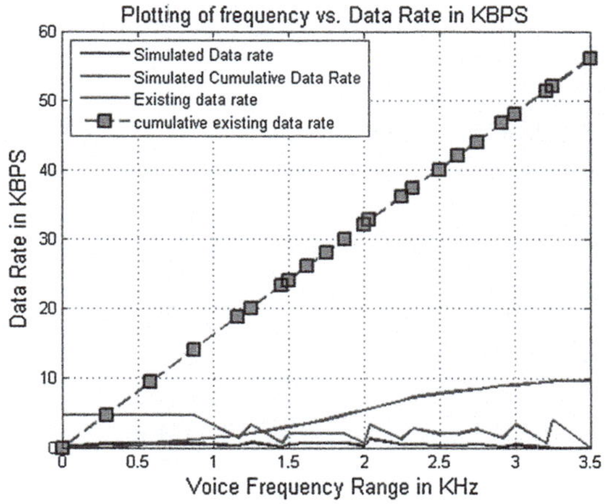

Fig. 2.7a Matlab simulation output of frequency versus cumulative data rate

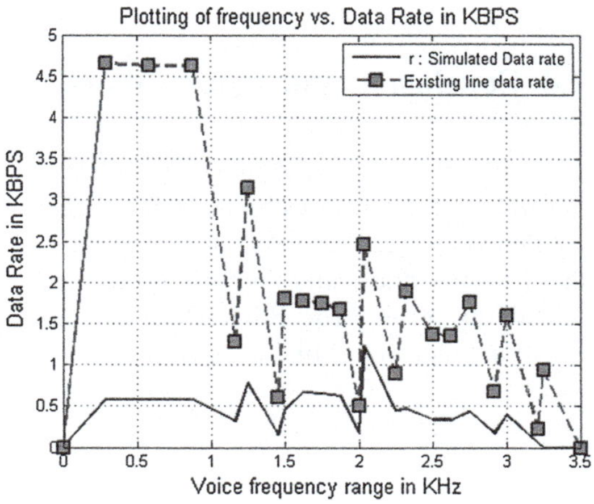

Fig. 2.7b Matlab simulation output of frequency versus data rate

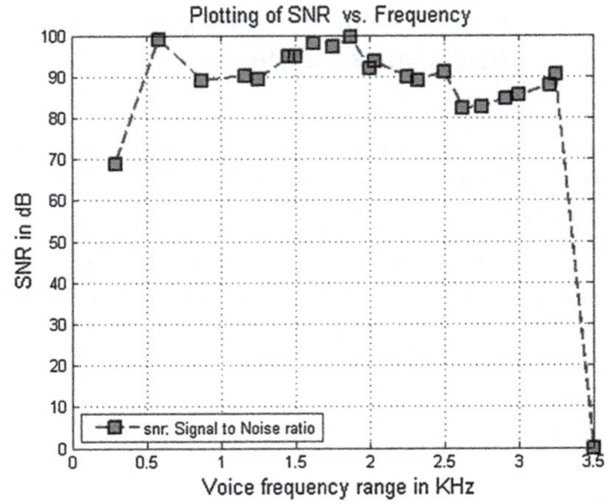

Fig. 2.7c Matlab simulation output of frequency versus SNR

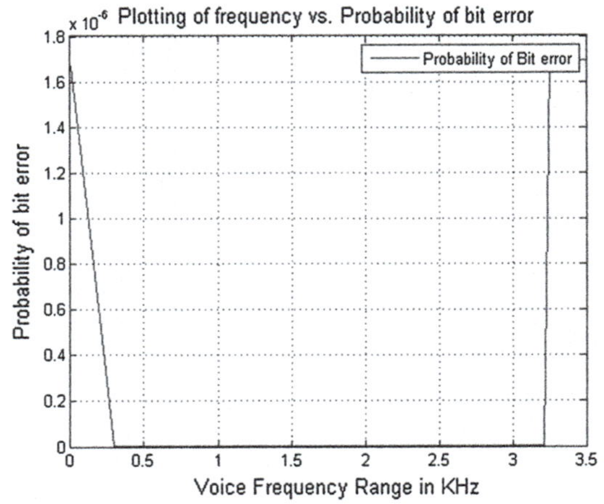

Fig. 2.7d Probability of bit error

2.4 Conclusion and Future Work

The new proposed method reduces data rate by using Rayleigh distribution. Several bit combinations by perceptual criterion have been taken. Result of those combinations have been tabulated in Table 2.1. It is clear that all the cases reduces the data rate drastically without losing signal to noise ratio criterion as well as probability of bit error. Rayleigh distribution outperforms all the other cases. In future authors would like to optimize those schemes by different algorithms.

Table 2.1 Subband coding data rate comparative table

Exixting data rate in Kbps	Simulated data rate in Kbps [1]	Simulated data rate in Kbps	Simulated data rate in Kbps	Simulated data rate in Kbps	Simulated data rate in Kbps	Simulated data rate in Kbps	Result in Kbps [2]
	Case I	Case II	Case III	Case IV	Case V	Case IV	
64	12.0128	11.0959	10.9494	10.7175	10.4867	9.4875	9.6

References

1. Sangita R, Gupta DB, Banerjee PK (2012) Studies and implementation of subband coder and decoder of speech signal. In: Proceedings of national conference on electronics, communication and signal processing (NCECS 2012), 19 Sept 2012, pp 8–16
2. Crochiere RE, Webber SA, Flanagan JN (1976) Digital coding of speech in subbands. The BELL Syst Technical Journal, October 1976
3. Knutson FG,Ramaswamy K, Richardson JW (2001) Subband ADPCM voice encoding and decoding. PCT/US2000/034410, July 2001
4. Szczutkowski CF (1986) Subband encoding method and apparatus. EP 0178608 A2, April 1986
5. Proakis JG (2001) Digital communications, 4th edn. Mcgraw-Hill, New York
6. Leon W, Couch II (1995) Modern communication systems principles and applications. Prentice-Hall, New Jersey
7. Taub H, Schilling DL (1986) Principles of communication systems, 2nd edn. Tata Mcgraw-Hill, Noida
8. Chen Y-J, Maher RC (1995) Sub-band coding of audio using recursively indexed quantization, Department of Electrical Engineering and Center for Communication and Information Science, University of Nebraska–Lincoln
9. Tribolet JM, Crochiere RE (1979) Frequency domain coding of speech. IEEE Trans Acoust Speech Signal Process 27(5):512
10. Aziz AM (2009) Subband coding of speech signals using decimation and interpolation. In: 13th international conference on aerospace sciences and aviation technology, ASAT-13, 26–28 May 2009, Military Technical College, Kobry Elkobbah
11. Roy S (2011) Studies and implementation of subband coder and decoder of speech signal. M-Tech (Communication Engineering) Thesis, WBUT
12. Rivet B, Girin L, Jutten C (2007) Log-Rayleigh distribution: a simple and efficient statistical representation of log-spectral coefficients. IEEE Trans Audio Speech Lang Process 15(3):796

Chapter 3
Microstrip Hairpin Bandpass Filter with Improved Out of Band Performance

Dibakar Yadav, Tamasi Moyra and Kaushik Debbarma

Abstract This paper presents an microstrip bandpass filter using folded hairpin resonators with improved out of band performance. The 5 pole hairpin filter has a centre frequency of 2 GHz with a Fractional Bandwidth (FBW) of 0.2. The filter performance has been improved using a combination of Defected Ground Structure (DGS) and Defected Microstrip Structure (DMS) in the feed line to remove the spurious passband extending from 3.1 to 4 GHz. The combination of DGS and DMS provides a relatively wide stopband and at the same time has a steep rising and falling edge skirts so as to minimize the distortion in the actual passband. The structure has been designed using RT/Duroid 6006 substrate with a height 1.27 mm and dielectric constant 6.15. The proposed structure has been simulated by Method of Moment (MoM) based IE3D electromagnetic simulator. Finally, the filter exhibits a good bandpass response suitable for use in modern microwave and millimeter wave communication.

Keywords Hairpin resonators · DGS · DMS · Bandpass filter · Stopband

3.1 Introduction

Bandpass filters are important components in a microwave communication system [1]. Bandpass filters (BPF) with improved out of band performance, reduced size and high spurious band rejection are required in modern RF and microwave

D. Yadav (✉) · T. Moyra (✉) · K. Debbarma (✉)
National Institute of Technology, Agartala, India
e-mail: ecedibakar@gmail.com

T. Moyra
e-mail: tamasi_moyra@yahoo.co.in

K. Debbarma
e-mail: kausharmika@gmail.com

S. Sengupta et al. (eds.), *Emerging Trends in Computing and Communication*,
Lecture Notes in Electrical Engineering 298, DOI: 10.1007/978-81-322-1817-3_3,

communication systems. BPF designed using split ring resonators has relatively large circuit losses, while BPF using parallel coupled lines has large size and poor selectivity [4]. Hairpin-line band pass filters are compact structures [2]. These are widely used at lower microwave frequencies where compactness is desirable for coupled lines. The design of Hairpin-line band pass filters has been investigated widely in literature [3, 4]. In [4] the out-of –band performance has been improved by using DGS and open stub in the feed lines. In this paper a 5 pole hairpin filter is presented. The filter has passband extending from 1.8 to 2.2 GHz with a centre frequency of 2 GHz. The response of this proposed hairpin filter shows a spurious passband starting around 3.2 GHz and extending up to 4 GHz. To enhance the filter performance it is necessary to suppress this spurious passband. The requirement is of a structure that can introduce a stopband in this spurious passband region. This is done by introducing a DGS and DMS in the feed line. The structures to be used have to satisfy the constraint of providing relatively wide stopband and at the same time to have sharp rising and falling skirts so that they don't introduce any distortion in the actual passband of the filter. Here a square-shaped Dumbbell has been used in the ground plane with the lower square split in two equal halves with a gap between them. The DMS has the shape of a rectangular ring. The DGS and DMS combined provide the required stopband without affecting the passband significantly. The final response has a considerably attenuated spurious passband.

3.2 Design and Simulation

3.2.1 Design of Hairpin Line Bandpass Filter

Hairpin filters are compact structure designed using U-shaped resonators also called hairpin resonators. The use of compact structure results in a reduction of length thereby resulting in reduced coupling than conventional parallel coupled band pass filters. Also, if the two arms of each hairpin resonator are closely spaced, they function as a pair of coupled line themselves, which can have an effect on the coupling as well. To design this kind of filter efficiently full wave EM simulation has been employed.

The lowpass prototype chosen is a five-pole Chebyshev filter with a passband ripple of 0.1 dB. The lowpass prototype parameter for the normalized cutoff frequency $(\Omega = 1)$ are $g_0 = g_6 = 1.0$, $g_1 = g_5 = 1.1468$, $g_2 = g_4 = 1.3712$, $g_5 = 1.9750$. Using the lowpass prototype parameters, the parameters for the band pass filter can be obtained [2]. The coupling between the adjacent resonators is determined by the spacing between them and the spacing required for the given value of the coupling co-efficient can be determined using the formula $M = \frac{f_2^2 - f_1^2}{f_2^2 + f_1^1}$ and carrying out full wave simulation where f_2, f_1 are two peak resonant frequencies obtained from the simulation of S_{21} for the adjacent resonators.

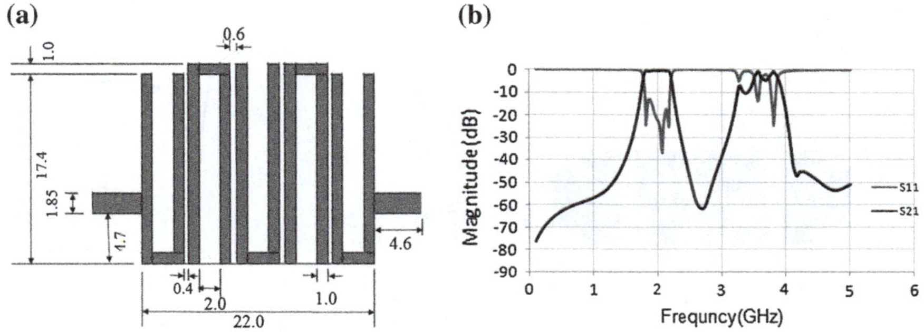

Fig. 3.1 **a** Optimized proposed hairpin structure. **b** S parameters for hairpin filter

The filter is designed to have tapped input and output. The characteristic impedance of the tapped line is obtained such that it matches to the 50 Ω terminating impedance. For the proposed filter characteristic impedance of tapped line is 68.4 Ω and the width is 1.85 mm. The hairpin line filter and its response are shown in Fig. 3.1a, b

3.2.2 Suppression of Spurious Passband

The designed hairpin line filter has a spurious passband extending from 3.2 to 4 GHZ. A combination of DGS and DMS has been used in the feed line for suppression of unwanted harmonic to improve the out-of-band performance.

DESIGN OF PROPOSED DGS. A defected structure created in the ground plane by etching out a slot alters the current distribution in the line thereby changing the inductance and capacitance associated with the transmission line. The resonant properties of the line can be altered by changing the size and shape of this slots [5].

The utilization of microstrip Hi-Lo with a DGS in the ground plane for generation of a stopband has been investigated in [6, 7]. Figure 3.2a shows the proposed DGS structure with a Hi-Lo line for out of band harmonic suppression. A rectangular DGS has been used on either side of the feed line in the ground plane with the lower DGS module being split in two sections with a gap of 1.5 mm between them. The dimension for the DGS structure has been optimized to provide a stopband at the required frequency band with the ratio of length of rectangular head (L) to its width (W) being maintained constant at 2.5:1. The location of attenuation pole has been shown in Fig. 3.2b and tabulated in Table 3.1 as the length and width of the rectangular head is varied. The optimized DGS structure has been shown in Fig. 3.2a, optimized to provide a stopband in the spurious passband region.

DESIGN OF PROPOSED DMS STRUCTURE. For improved out of band suppression a Defected Microstrip Structure (DMS) has been used by etching out slots in the in the feed line as DMS is also known to provide a stopband [8, 9].

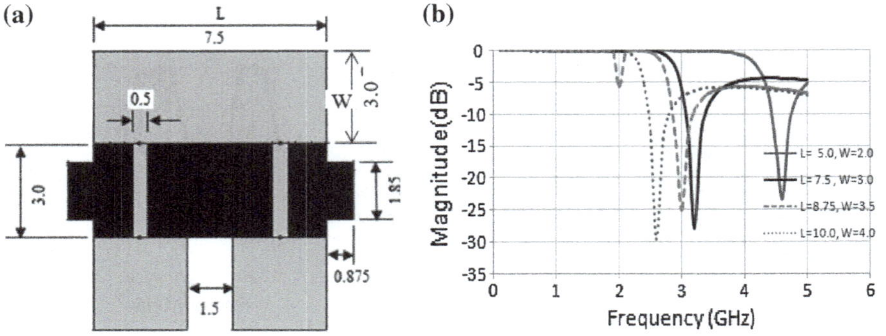

Fig. 3.2 **a** Feed line with proposed DGS structure. **b** Simulated S_{21} parameters (*Black color* Microstrip line; *Gray color* DGS section)

Table 3.1 Variation of the attenuation pole with dimension of DGS structure

Length of rect. head (L) (mm)	Width of rect. head (W) (mm)	Attenuation Pole (GHz)
10.0	4.0	2.60
8.75	3.5	2.90
7.5	3.0	3.20
5.0	2.0	4.70

The length of the slot has been optimized using simulation with the slot width being maintained at 0.2 mm and the total length of the slot has been optimized using simulation to obtain a stopband in the desired frequency band. Figure 3.3a shows the optimized dimensions for the DMS structure. The location of resonant frequency as a function of length of the slot (L1), including vertical portions is shown in Fig. 3.3b and is tabulated in Table 3.2.

3.3 Final Filter Structure and Results

The final filter structure is obtained by incorporating the DGS and DMS structure into the feed line of the hairpin structure to suppress the out of band response. The use of DGS and DMS has facilitated the removal of the first spurious passband thereby enhancing the out of band performance. The final structure and its response are shown in Fig. 3.4a, b respectively.

As evident from Fig. 3.4b the spurious passband has been attenuated due to introduction of DGS and DMS in the feed line. The total structure has a size of $(0.82 \times \lambda_g) \times (0.25 \times \lambda_g)$ for RT/Duroid 6006 substrate including the input–output feed lines. The final structure provides a good sharpness factor of in the rising and falling edges of 42 and 140 dB/GHz respectively.

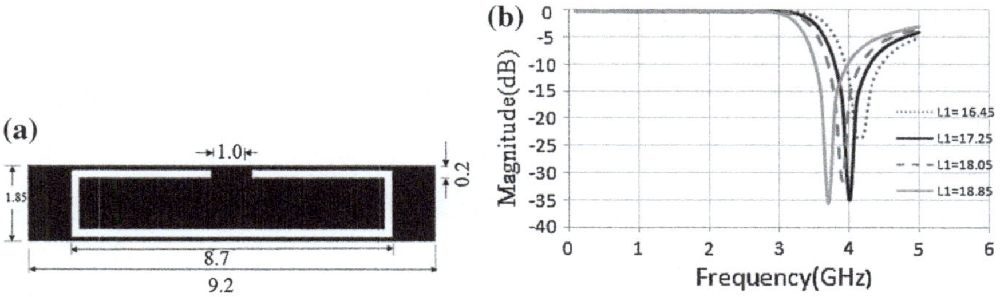

Fig. 3.3 **a** Optimized DMS structure. **b** Simulated S_{21} Parameters

Table 3.2 Variations of attenuation pole with the total length of DMS slot

Total length of slot (mm)	Attenuation pole (GHz)
16.45	4.15
17.25	4.0
18.05	3.90
18.85	3.70

Fig. 3.4a Modified hairpin BPF with DGS and DMS in feed line

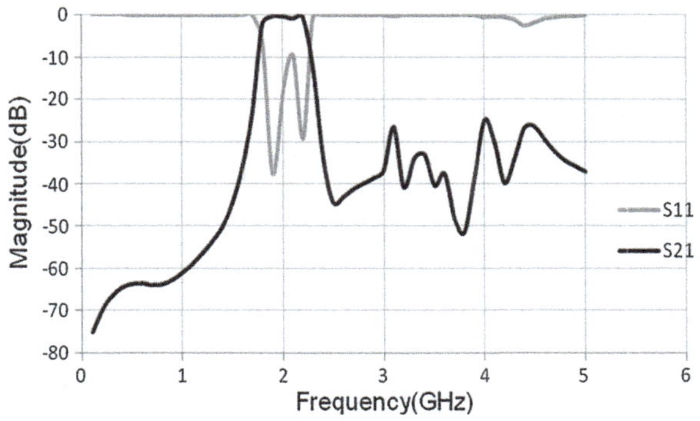

Fig. 3.4b Simulated scattering parameters

3.4 Discussion and Conclusion

In this paper a 5 pole band pass filter has been designed by using hairpin resonators. One of the key features of this design is an improved out of band rejection with attenuation greater than 20 dB in the first spurious passband. Thus a filter with centre frequency of 2 GHz with 10 dB FBW = 25 % has been obtained with improved out of band response. The proposed bandpass filter has lots of application in modern microwave and millimeter wave communication systems.

References

1. Matthaei GL, Young L, Jones EMT (1980) Microwave filters, impedance-matching networks, and coupling structures. Artech House, Norwood, MA
2. Hong J-S, Lancaster MJ (2001) Microstrip filter for RF/microwave application. Wiley, New York
3. Sulaiman AA, Ain MF, Hassan SIS (2008) Design of hairpin bandpass filters for K-Band application. In: 2008 IEEE international RF and microwave conference
4. Vidhya K, Jayanthy T (2011) Design of microstrip hairpin band pass filter using defected ground structure and open stubs. In: 2011 international conference on information and electronics engineering IPCSIT, vol 6
5. Ahn D, Park JS, Kim CS, Kim J, Qian Y, Itoh T (2002) A design of the low-pass filter using the novel microstrip defected ground structure. IEEE Trans Microwave Theory Tech 2001 49(1):86–93
6. Abdel-Rahman AB, Verma AK, Boutejdar A, Omar AS (2004) Control of bandstop response of Hi-Lo microstrip lowpass filter using slot in ground plane. IEEE Trans Microwave Theory Tech 52(3):1008–1013
7. Moyra T, Parui SK, Das S (2010) Application of a defected ground structure and alternative transmission line for designing a quasi-elliptic lowpass filter and reduction of insertion loss. Int J RF Microwave Comput Aided Eng 20(6):882–888
8. La D, Lu Y, Sun S (2010) Novel bandstop filter using dual-U shape defected microstrip structure. In: International symposium on signal, systems and electronics, vol 1, Nanjing, China, pp 1–3
9. Kazerooni M, Rad GR, Cheldavi A (2009) Behavior study of simultaneously defected microstrip and ground structure (DMGS) in planar circuits. In: PIERS proceedings, vol 895(900), Beijing, China

Chapter 4
Size Reduction of 4 × 4 Butler Matrix Using Defected Microstrip Structure

Kaushik Debbarma, Tamasi Moyra and Dibakar Yadav

Abstract The paper presents a design of a 4 × 4 planar Butler matrix for an operating frequency of 2.4 GHz which can be used for WLAN applications. Defected Microstrip Structures (DMSs) have been introduced with conventional microstrip line for the design of compact phase shifter. The parametric study of different size DMSs for the desired length of microstrip line has been done. Then the optimized dimension of microstrip line with DMS is calculated and simulated to obtain the required phase shift at the operating frequency. The proposed Butler matrix has been designed using glass epoxy (FR4) substrate with height 1.59 mm and dielectric constant 4.4 and simulated by method of moment (MoM) based IE3D electromagnetic simulator. The errors in phase distribution are between 2° and 7°, couplings are in the range −6.53 to −7.84 dB and various losses of the matrix are within tolerable range.

Keywords Butler matrix · Crossover · Coupler · DMS · Phase shifter · Slow wave factor

4.1 Introduction

Smart antenna is one of the most promising technologies that will enable a higher capacity in wireless effectively reducing multipath and co-channel interference [1, 2]. Smart antennas employ a set of radiating elements arranged in the form of

K. Debbarma (✉) · T. Moyra (✉) · D. Yadav (✉)
National Institute of Technology, Agartala, India
e-mail: kausharmika@gmail.com

T. Moyra
e-mail: tamasi_moyra@yahoo.co.in

D. Yadav
e-mail: ecedibakar@gmail.com

S. Sengupta et al. (eds.), *Emerging Trends in Computing and Communication*, Lecture Notes in Electrical Engineering 298, DOI: 10.1007/978-81-322-1817-3_4,

an array. The signals from these elements are combined to form a movable or switchable beam pattern that follows the desired user. The process of combining the signals and then focusing the radiation in a particular direction is often referred to as digital beam forming [3]. Smart antennas are most often realized with either switched-beam or fully adaptive array antennas. An array consists of two or more antennas (the elements of the array) spatially arranged and electrically interconnected to produce a directional radiation pattern. Switched-beam antenna systems are the simplest form of smart antenna. The Butler matrix has been used extensively over the years in radar and satellite systems for this purpose [4]. The Butler matrix is a versatile device. For N $\times$ N Butler matrix has N input ports and N output ports. As a beam-forming network, it is used to drive an array of N antenna elements. It can serve as a beam-forming network permitting volumetric beams to be generated that are orthogonal and independent, and each port will have the gain of the full array. Being passive and reciprocal, they can be used for both reception and transmission in an antenna array. The beams may be deployed simultaneously or sequentially depending on the application. Adaptive antennas can also be used for direction finding, with applications including emergency services and vehicular traffic monitoring. The transmission quality can be improved by increasing desired signal power and reducing interference.

4.2 Analysis and Design

Before designing the Butler matrix, the constituent components i.e. the 90° coupler, the crossover and the phase shifter are designed and simulated.

4.2.1 Hybrid Coupler

The 90° hybrid coupler or 3 dB Branchline coupler has been designed by conventional branch-line technique. Basically the 90° hybrid is made by two main transmission lines shunt connected by the two secondary branch lines [5].

It has two 50 Ω and two 35.4 Ω transmission lines with length $\lambda/4$. So the perimeter of the square is approximately equal to one wavelength. The Fig. 4.1b shows the phase of 90° coupler which has been simulated on IE3D software. From the figure it is clear that there is 93.45° phase shift maintained around the desired 2.4 GHz frequency. The Fig. 4.1c shows the coupling and direct coupling in dB. As it is evident from the figure that both the coupling and direct are as desired ($S_{13} = -3.15$ dB & $S_{12} = -3.11$ dB) with tolerable errors. Isolation and return losses are also less ($S_{11} = -23.37$ dB & $S_{14} = -23.74$ dB) (Fig. 4.1a, 4.1d).

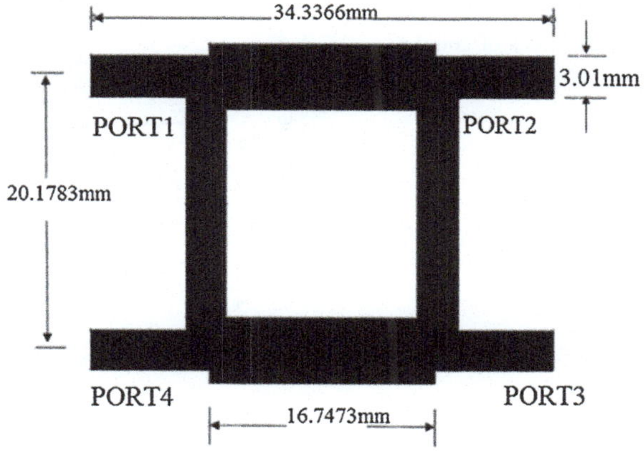

Fig. 4.1a 3 dB branch line coupler

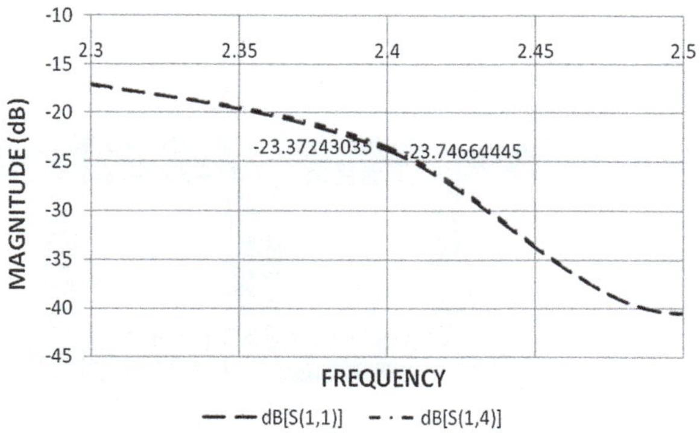

Fig. 4.1b Isolation and return loss

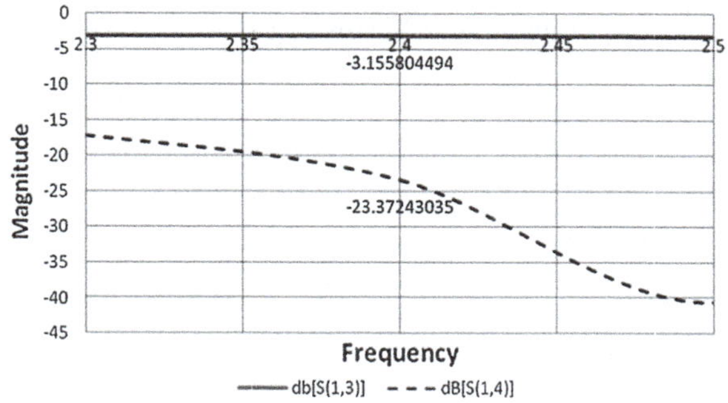

Fig. 4.1c Simulated phase response of coupler

4.2.2 Crossover

In the butler matrix array the signal path has to physically crossover while maintaining the high isolation. In this paper the crossover is implemented by cascade of the two 90° hybrids with slight modification on line widths. The geometry of the crossover is given in Fig. 4.2a (Fig. 4.2b, 4.2c).

The simulation results of the above geometry are given below:

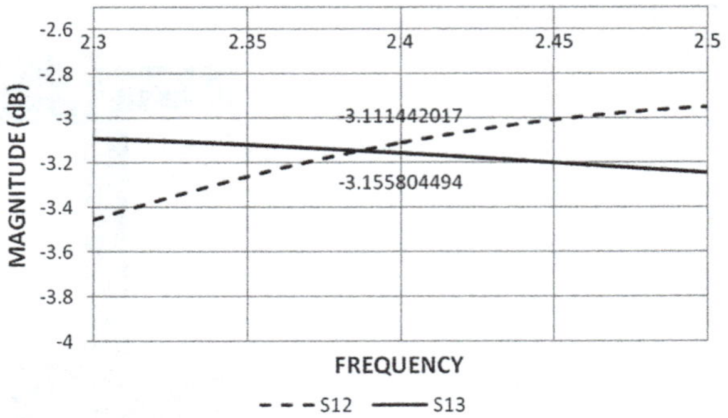

Fig. 4.1d Coupling and direct values of coupler

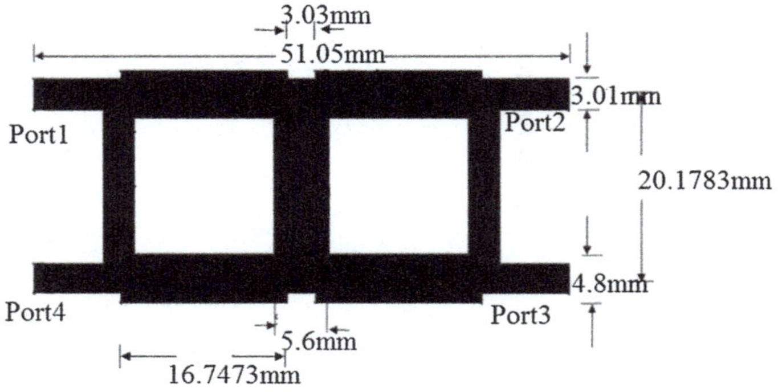

Fig. 4.2a Geometry of 0 dB crossover

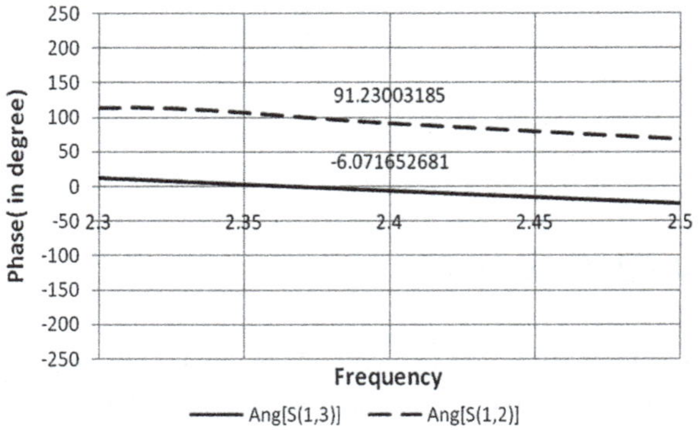

Fig. 4.2b Phase difference between ports

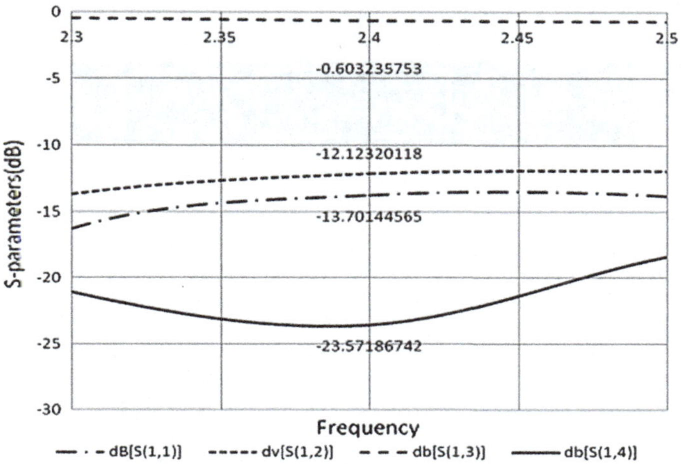

Fig. 4.2c S-parameters of the crossover

4.2.3 Phase Shifter

The required 45° phase shifter is designed using defected microstrip structure (DMS). It has been widely discussed that the size of a microstrip line can be reduced by induction of a defect on the microstrip signal line as well as the ground plane. The problem with defect on the ground plane (known as DGS) is that on making any defect on the ground plane the radiation become high. The main aim of the defect (DMS) is to decrease the phase constant and phase velocity and hereby increasing the slow wave factor (SWF). The dimension of the phase shifter is exactly the gap between two branch-line couplers. A parametric study has been done to acquire the exact dimension of the DMS so as to get the required phase shift.

The length of the DMS has been varied by keeping other parameters fixed. Graph of the parametric study done is given in Fig. 4.3b (Fig. 4.3a, 4.3c).

As it can be seen from the simulation result that the phase shifter is working properly giving a phase shift of 134.73° at 2.4 GHz frequency which is within the tolerable range. The phase shifter designed gives the following response.

4.3 Final Butler Structure and Results

The required dimensions of the individual components of proposed Butler matrix which is operating at 2.4 GHz have been determined using IE3D full wave EM simulator in the previous discussion at Sect. 4.2.3. It is evident that all the individual components are designed and simulated. The simulation results of the components are satisfactory. Now all the individual components are combined on

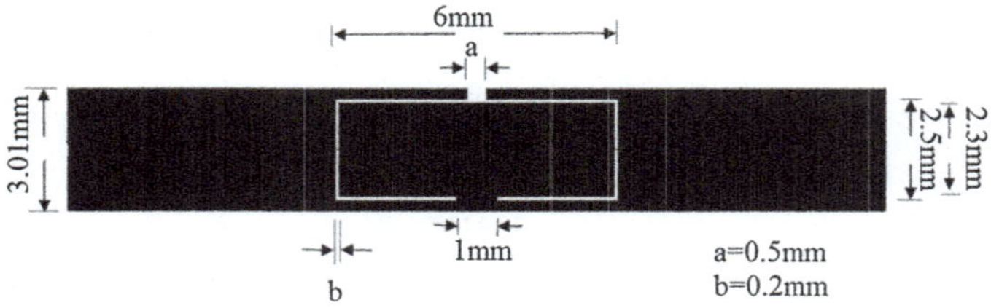

Fig. 4.3a Geometry of phase shifter using DMS

Fig. 4.3b Parametric study of DMS structure

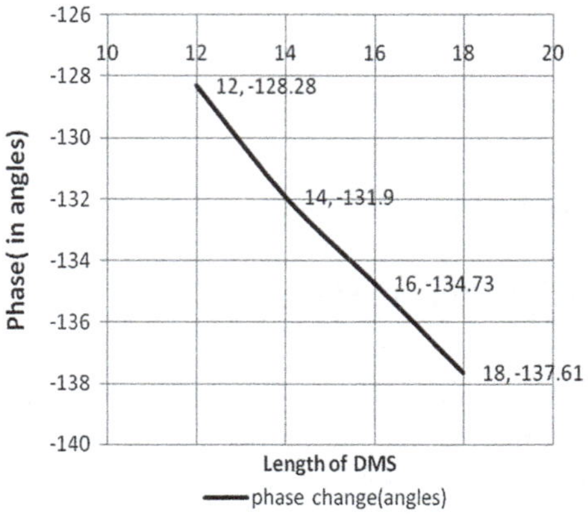

Fig. 4.3c Simulated response of the phase shifter

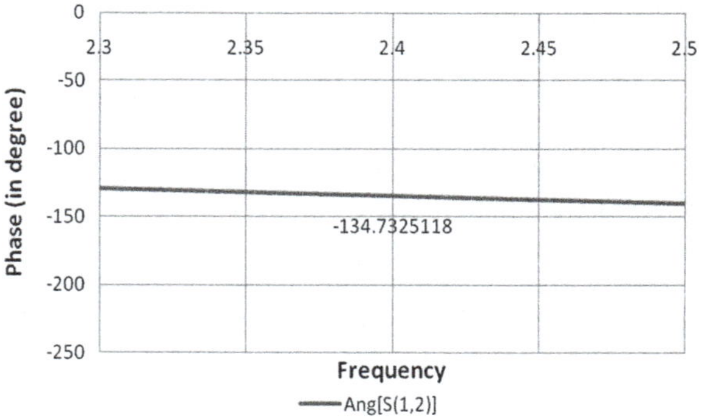

a single substrate FR4 to implement the Butler matrix. The theoretically calculated and simulated phase responses of the Butler matrix are given as tabular form in Table 4.1. It can be observed from the table that output beams are separated by either a phase difference of 45° or integer multiple of 45°. There is a deviation of maximum 7° from the calculated value. The geometry of the final design of the

Table 4.1 Calculated value of different phases

	Port5		Port6		Port7		Port8	
	Calculated	Simulated	Calculated	Simulated	Calculated	Simulated	Calculated	Simulated
Port1	−135	−128.04	135	132.87	−90	−94.9	180	173.61
Port2	135	127.52	45	44.77	0	−6.00	−90	−96.18
Port3	−90	−95.27	0	−6.15	45	48.06	135	129.29
Port4	180	−173.91	−90	−96.96	135	129.45	−135	−131.71

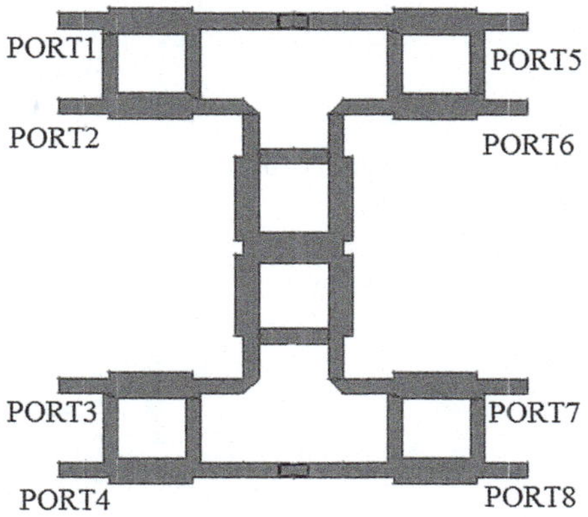

Fig. 4.4a Geometry of the Butler matrix

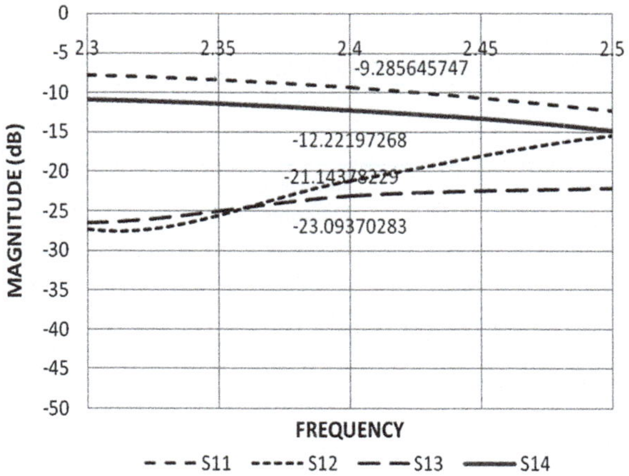

Fig. 4.4b Various losses of the Butler matrix

Butler matrix is given in Fig. 4.4a. Figure 4.4c represents the phase at 2.4 GHz for port 1 to all other output ports, i.e. port 5, 6, 7 and 8. The Fig. 4.4d shows the coupling of port1 with all other output ports. Figure 4.4b represents various losses of the proposed Butler matrix. Table 4.2 represents the magnitude of various parameters i.e. of both losses and couplings of the proposed Butler matrix in dB.

Fig. 4.4c Phase response of the four ports

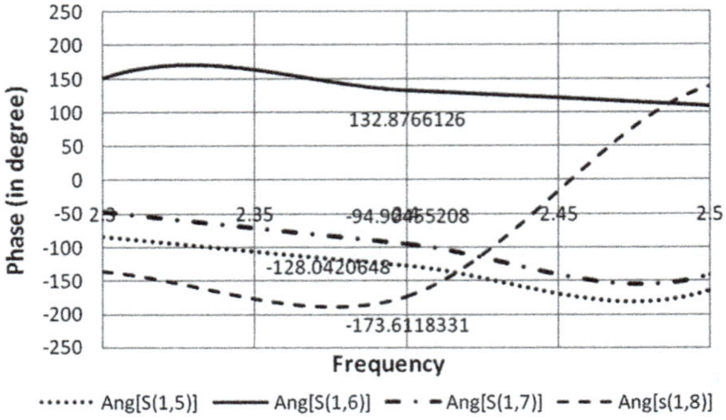

Fig. 4.4d S-parameters of the 4 output ports

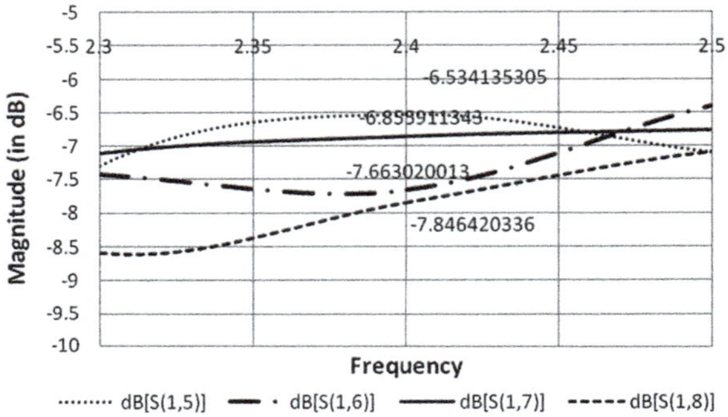

Table 4.2 Magnitude of Butler matrix

s-parameter	Parameter	Magnitude	
		Theoretical	Simulated (dB)
S_{11}	Return loss	−infinite	−9.28
S_{12}	Isolation	−infinite	−21.14
S_{13}	Isolation	−infinite	−23.09
S_{14}	Isolation	−infinite	−12.22
S_{15}	Coupling	6 dB	−6.53
S_{16}	Coupling	6 dB	−7.66
S_{17}	Coupling	6 dB	−6.89
S_{18}	Coupling	6 dB	−7.84

4.4 Conclusion

Finally, the proposed Butler matrix has been designed using glass epoxy (FR4) substrate with height 1.59 mm and dielectric constant 4.4 with negligible loss tangent and simulated by method of moment (MoM) based IE3D electromagnetic simulator. The errors in phase distribution are between 2° and 7°, couplings are in the range −6.53 to −7.84 dB and various losses of the matrix are within tolerable range. Thus, the obtained simulated result shows that the proposed Butler matrix is able to work at 2.4 GHz with desired response. This circuit is suitable for the application in GSM planar antenna arrays, RADAR, satellites etc. in our modern RF, microwave and millimetre wave communication systems.

References

1. Winters J (1998) Smart antennas for wireless systems. IEEE Pers Commun 5:23–27
2. Balanis CA (1997) Antenna theory analysis and design, 2nd edn. Willey, London
3. Yadav AK, Upmanu V, Yadav SK (2012) Design and analysis of a beam forming network for WLAN application. Int J Sci Res Eng Technol (IJSRET) 1(6):004–009
4. Bhowmik W, Srivastava S (2010) Optimum design of a 4 × 4 planar butler matrix array for WLAN application. J Telecommun 2(1):68–74
5. Pozar DM (2005) Microwave engineering, 3rd edn. Wiley, London

Chapter 5
Interference Mitigation in Overlay Cognitive Radio Using Orthogonal Polarization

Sandip Karar and Abhirup Das Barman

Abstract Interference mitigation is one of the significant challenges in overlay cognitive radio. This paper proposes two schemes to avoid the interference in an overlay cognitive radio scenario by exploiting the polarization diversity. In the first scheme both the primary and the secondary user use dual polarized antenna at the receiver end which combines the received signals in a proper way to cancel out the interference. In the second scheme a dual polarized antenna is incorporated at the secondary transmitter one of which acts as a relay of the primary user's signal and cancels out interference via space–time coding. Analytical results about the error performance and the outage probability for the two schemes are shown.

Keywords Decode-and-forward relay · Interference mitigation · Overlay cognitive radio · Polarization diversity

5.1 Introduction

With rapid growth of wireless applications, the problem of spectrum utilization has become much more critical. As a promising solution cognitive radio (CR) technology can reuse the licensed spectrum in three ways [1] namely, *underlay*,

---

Both the authors are affiliated with *ITRA project* "Mobile Broadband Service Support over Cognitive Radio Networks".

S. Karar (✉) · A. D. Barman (✉)
Department of Radiophysics and Electronics, University of Calcutta, Kolkata, India
e-mail: sk7632@gmail.com

A. D. Barman
e-mail: abhirup_rp@yahoo.com

S. Sengupta et al. (eds.), *Emerging Trends in Computing and Communication*,
Lecture Notes in Electrical Engineering 298, DOI: 10.1007/978-81-322-1817-3_5,

overlay and *interweave* method. In *underlay* and *overlay* method both the primary and the secondary users can simultaneously use the same spectrum whereas in *interweave* method the secondary user continually searches for the spectrum holes and use those spectrum holes for communication. Different interference mitigation techniques in *overlay* cognitive radio network have been proposed such as dirty paper coding [2], beamforming [3, 4] etc. In most of the overlay techniques it has been assumed that the secondary transmitter has non-causal information about the primary user's message which is quite impractical.

In this paper we exploit the polarization diversity to mitigate the interference in an overlay spectrum sharing scheme. Using the polarization of the signal is not new in wireless communication but only a few papers have addressed the issue [5–9]. Here we introduce two schemes using the polarization diversity to mitigate the interference in an overlay cognitive radio. In the first scheme both the primary and the secondary user use dual polarized antenna at the receiver end which combines the received signals in a proper way to cancel out the interference. One minor drawback in this scheme is that the accurate knowledge of fractional cross-polar leakage power α is necessary. We also propose a second scheme which does not require the knowledge of α. It incorporates a dual polarized antenna at the secondary transmitter one of which acts as a relay of the primary and cancels out interference via space–time coding. A scheme similar to this has been previously proposed in [10] where the secondary transmitter uses two separate antennas. One of the main disadvantages of this scheme is that the antennas must be separated large enough resulting in larger size and higher cost of the secondary user making it quite ineffective which one is replaced here by a dual-polarized antenna as a space and cost effective alternative. Also larger number of channels need to be estimated beforehand in [10] compared to the dual-polarized scheme used here.

The rest of the paper is organized as follows: Sect. 5.2 gives the analysis of a 2×2 cognitive radio model where the primary and the secondary user share the spectrum using polarization multiplexing. In Sect. 5.3, we propose a method of overlay spectrum sharing scheme which uses dual polarized antenna at both the receivers and in Sect. 5.4, we propose a second scheme with dual polarized antenna at the secondary transmitter one of which is used for relay cooperation. Simulation results are given in Sect. 5.5 and conclusions are drawn in Sect. 5.6.

5.2 System Model

We consider a 2×2 overlay cognitive radio model shown in Fig. 5.1 where the primary users use vertically polarized antennas and the secondary users use horizontally polarized antennas in both transmitter and receiver sides. The symbols $h_{pp}, h_{ps}, h_{sp},$ and h_{ss} denote the flat Rayleigh fading channel gains each with parameter σ_h. In ideal scenario the cross polar transmission should be zero. But actually this is not the case as there is some cross-polar leakage due to the rotation of the polarization plane when propagating through the atmosphere.

Fig. 5.1 A 2 × 2 cognitive radio model using polarization multiplexing

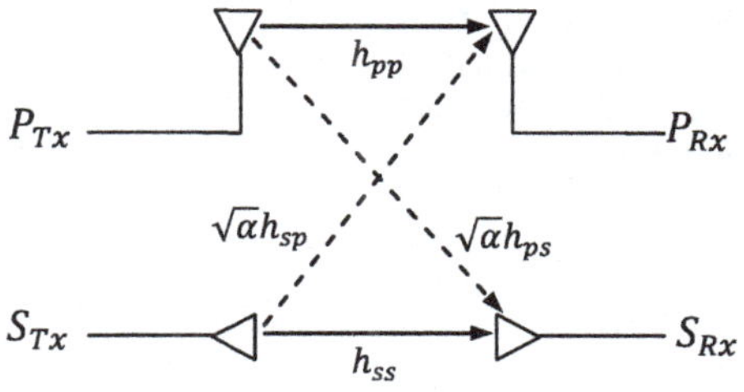

It has been shown in [9] that the leakage from vertical to horizontal polarized transmission and horizontal to vertical polarized transmission have the same power on average. Let α be the fraction of the cross-polar leakage power where $0 < \alpha \leq 1$. To incorporate the cross polar leakage into the model the channel gain between different polarized antennas should be multiplied by $\sqrt{\alpha}$. Let the primary transmitter transmits the symbol x_p with power P_p and the secondary transmitter transmits the symbol x_s with power P_s. We assume $E(|x_p|^2) = 1$ and $E(|x_s|^2) = 1$. The received signals at the primary and secondary receivers are given by:

$$y_p = \sqrt{P_p}\, h_{pp} x_p + \sqrt{\alpha P_s}\, h_{sp} x_s + n_p \tag{5.1}$$

$$y_s = \sqrt{P_s}\, h_{ss} x_s + \sqrt{\alpha P_p}\, h_{ps} x_p + n_s \tag{5.2}$$

Here n_p and n_s are the additive white Gaussian noise with zero mean and variance σ_n^2 at the primary and secondary receiver respectively. In ideal scenario $\alpha = 0$, the term x_s vanishes in the expression of y_p and so there would be no interference at each receiver. But in actual case $\alpha \neq 0$ due to cross-polar leakage and hence this causes some interference both to the primary and secondary receiver. In the next two sections we will show two of our proposed methods to mitigate the interference in overlay cognitive radio.

5.3 Interference Mitigation Using Dual Polarized Antennas at the Receivers

It is evident that only orthogonal polarization between two users cannot remove interference completely. Instead of single polarized antennas at the receivers as in the previous case each of the receivers are equipped with dual polarized antenna one vertical and one horizontal as shown in Fig. 5.2. The receivers need to know the value of α and only their corresponding channel state information (CSI) i.e. primary receiver need to know only h_{pp} and secondary receiver only h_{ss}.

The received signals at the vertical and horizontal polarized antennas of the primary receiver are respectively given by

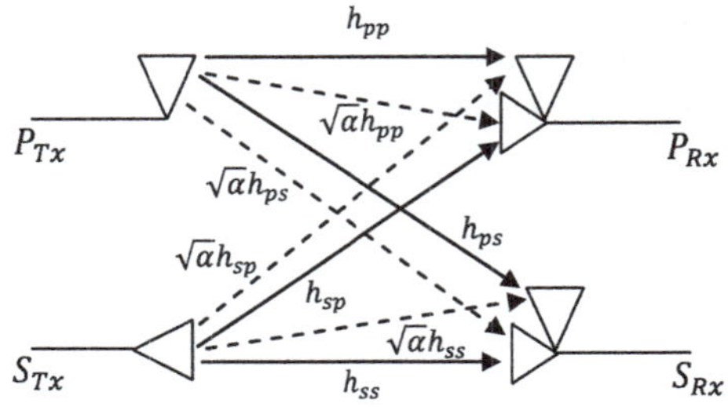

Fig. 5.2 A 2 × 2 cognitive radio model with dual polarized antennas at receivers

$$y_{pv} = \sqrt{P_p}\, h_{pp} x_p + \sqrt{\alpha P_s}\, h_{sp} x_s + n_{pv} \tag{5.3}$$

$$y_{ph} = \sqrt{\alpha P_p}\, h_{pp} x_p + \sqrt{P_s}\, h_{sp} x_s + n_{ph} \tag{5.4}$$

and the received signals at the vertical and horizontal polarized antennas of the secondary receiver are respectively given by

$$y_{sv} = \sqrt{P_p}\, h_{ps} x_p + \sqrt{\alpha P_s}\, h_{ss} x_s + n_{sv} \tag{5.5}$$

$$y_{sh} = \sqrt{\alpha P_p}\, h_{ps} x_p + \sqrt{P_s}\, h_{ss} x_s + n_{sh} \tag{5.6}$$

where n_{pv}, n_{ph}, n_{sv} and n_{sh} are additive white Gaussian noise with zero mean and variance σ_n^2. Now combining the signals y_{pv} and y_{ph} in proper way we can eliminate the interfering term involving x_s at the primary receiver.

$$y_p = y_{pv} - \sqrt{\alpha} y_{ph} = (1-\alpha)\sqrt{P_p}\, h_{pp} x_p + n_{pv} - \sqrt{\alpha}\, n_{ph} \tag{5.7}$$

Similarly the interfering term at the secondary receiver can be eliminated.

$$y_s = y_{sh} - \sqrt{\alpha} y_{sv} = (1-\alpha)\sqrt{P_s}\, h_{ss} x_s + n_{sh} - \sqrt{\alpha}\, n_{sv} \tag{5.8}$$

The signal-to-noise ratio at the primary receiver and secondary receiver are respectively given by $\gamma_p = \frac{(1-\alpha)^2 P_p |h_{pp}|^2}{(1+\alpha)\sigma_n^2}$ and $\gamma_s = \frac{(1-\alpha)^2 P_s |h_{ss}|^2}{(1+\alpha)\sigma_n^2}$. When M-PSK modulation is used the symbol error probability (SER) for the primary user is given by:

$$P_{se}(primary) \approx \alpha_M Q\left(\sqrt{\beta_M \gamma_p}\right) \tag{5.9}$$

where $\alpha_M = 2$ and $\beta_M = 2sin^2\left(\frac{\pi}{M}\right)$.

Average symbol error probability can be calculated as

$$\overline{P_{se}}(primary) \approx \int_0^\infty \alpha_M Q\left(\sqrt{\beta_M \gamma_p}\right) p(\gamma_p) d\gamma_p \tag{5.10}$$

For Rayleigh fading $p(\gamma_p) = \frac{1}{\overline{\gamma_p}} e^{-\gamma_p/\overline{\gamma_p}}$ where $\overline{\gamma_p}$ is the mean value of γ_p which is given by $\overline{\gamma_p} = E(\gamma_p) = \frac{(1-\alpha)^2 P_p E\left[|h_{pp}|^2\right]}{(1+\alpha)\sigma_n^2}$. Since $|h_{pp}|$ follows Rayleigh distribution with parameter σ_h, $|h_{pp}|^2$ will follow exponential distribution with mean $2\sigma_h^2$. Therefore $E\left[|h_{pp}|^2\right] = 2\sigma_h^2$. Hence $\overline{\gamma_p} = E(\gamma_p) = \frac{2(1-\alpha)^2 P_p \sigma_h^2}{(1+\alpha)\sigma_n^2}$. From (5.10) it can be derived [11]:

$$\begin{aligned}\overline{P_{se}}(primary) &= \frac{\alpha_M}{2}\left[1 - \sqrt{\frac{0.5\beta_M\overline{\gamma_p}}{1+0.5\beta_M\overline{\gamma_p}}}\right] \\ &= \frac{\alpha_M}{2}\left[1 - \sqrt{\frac{(1-\alpha)^2\beta_M P_p \sigma_h^2}{(1+\alpha)\sigma_n^2 + (1-\alpha)^2\beta_M P_p \sigma_h^2}}\right]\end{aligned} \tag{5.11}$$

The outage probability P_{out} is defined as the probability that the received SNR falls below a given threshold corresponding to the maximum allowable symbol error probability. The outage probability of the primary user relative to the threshold γ_{p0} is given by

$$P_{out}(primary) = P(\gamma_p \leq \gamma_{p0}) = \int_0^{\gamma_{p0}} \frac{1}{\gamma_p} e^{-\gamma_p/\overline{\gamma_p}} d\gamma_p = 1 - e^{-\gamma_{p0}/\overline{\gamma_p}} = 1 - e^{\frac{2(1-\alpha)^2 P_p \sigma_h^2}{(1+\alpha)\sigma_n^2 \gamma_{p0}}} \tag{5.12}$$

In the similar way the average symbol error probability and the outage probability relative to the threshold γ_{s0} of the secondary user can be calculated.

$$\overline{P_{se}}(secondary) = \frac{\alpha_M}{2}\left[1 - \sqrt{\frac{(1-\alpha)^2\beta_M P_s \sigma_h^2}{(1+\alpha)\sigma_n^2 + (1-\alpha)^2\beta_M P_s \sigma_h^2}}\right] \tag{5.13}$$

$$P_{out}(secondary) = 1 - e^{\frac{2(1-\alpha)^2 P_s \sigma_h^2}{(1+\alpha)\sigma_n^2 \gamma_{p0}}} \tag{5.14}$$

5.4 Interference Mitigation with Dual Polarized Antenna at the Secondary Transmitter and Using Relay Cooperation

The system model is shown in the Fig. 5.3. In this case the secondary transmitter is equipped with a dual polarized antenna one horizontal and one vertical. It is assumed that both the receivers know all the channel state information (CSI). The secondary user serves as a decode-and-forward (DF) relay of the primary signal in this scheme. The transmission takes place in two phases. In the first phase the primary transmitter sends the signal x_p with power P_p and the secondary transmitter transmits the signal x_s with power P_s through its horizontally polarized antenna. Simultaneously the vertically polarized antenna of the secondary transmitter also receives the primary transmission and decodes it. And in the second

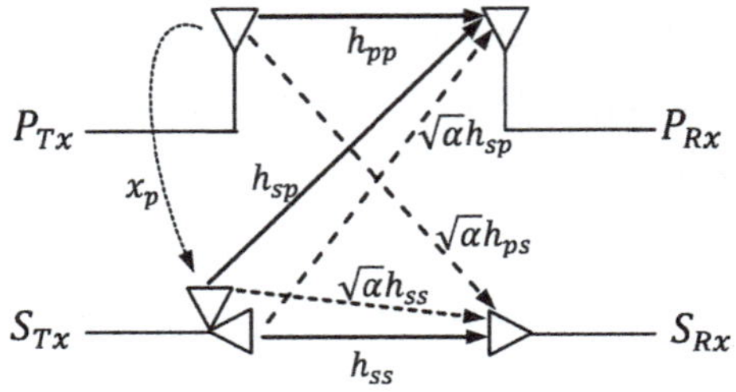

Fig. 5.3 A 2 × 2 cognitive radio model using relay cooperation and space–time coding

phase, the primary transmitter remains silent but the secondary transmitter transmits the signal x_p^* with power P_p' through its vertically polarized antenna and the signal $-x_s^*$ with power P_s through its horizontally polarized antenna. In the first transmission phase the signal received by the primary and secondary receiver are respectively

$$y_{p1} = \sqrt{P_p}\, h_{pp} x_p + \sqrt{\alpha P_s}\, h_{sp} x_s + n_{p1} \tag{5.15}$$

$$y_{s1} = \sqrt{P_s}\, h_{ss} x_s + \sqrt{\alpha P_p}\, h_{ps} x_p + n_{s1} \tag{5.16}$$

and in the second transmission phase the signal received by the primary and secondary receivers are respectively:

$$y_{p2} = \sqrt{P_p'}\, h_{sp} x_p^* - \sqrt{\alpha P_s}\, h_{sp} x_s^* + n_{p2} \tag{5.17}$$

$$y_{s2} = -\sqrt{P_s}\, h_{ss} x_s^* + \sqrt{\alpha P_p'}\, h_{ss} x_p^* + n_{s2} \tag{5.18}$$

where n_{p1}, n_{p1}, n_{p2} and n_{p2} are additive white Gaussian noise with zero mean and variance σ_n^2. Similar to Alamouti's scheme we can now properly combine the received signals in the two phase transmission to minimize the interference. From (5.15), (5.16), (5.17) and (5.18) we can write:

$$h_{sp}^* y_{p1} + h_{sp} y_{p2}^* = \left(\sqrt{P_p} h_{pp} h_{sp}^* + \sqrt{P_p'} |h_{pp}|^2\right) x_p + h_{sp}^* n_{p1} + h_{sp} n_{p2}^* \tag{5.19}$$

$$\begin{aligned} h_{ss}^* y_{s1} - h_{ps} y_{s2}^* = {} & \left(\sqrt{P_s}|h_{ss}|^2 + \sqrt{P_s} h_{ss}^* h_{ps}\right) x_s + \sqrt{\alpha} h_{ps} h_{ss}^* \left(\sqrt{P_p} - \sqrt{P_p'}\right) x_p \\ & + h_{ss}^* n_{s1} + h_{ps} n_{s2}^* \end{aligned} \tag{5.20}$$

From these expressions it is evident that the interference at the primary receiver due to secondary transmission is completely removed but there is some interference at the secondary receiver from the primary transmission because of the slight difference in estimated power P_p' and P_p. If the secondary transmitter can estimate the primary transmitted power exactly then there will be no interference at the secondary receiver also. Assuming $P_p' = P_p$, the signal-to-noise ratio at the primary receiver and secondary receiver are then respectively given by $\gamma_p = \frac{P_p \left| h_{pp} h_{sp}^* + \left| h_{sp} \right|^2 \right|^2}{2\, h_{sp}|^2 \sigma_n^2}$ and $\gamma_s = \frac{P_s \left| |h_{ss}|^2 + h_{ss}^* h_{ps} \right|^2}{(|h_{ss}|^2 + |h_{ps}|^2)\sigma_n^2}$. Using Rayleigh distribution of the channels it can be derived that $\overline{\gamma_p} = \frac{3P_p\sigma_h^2}{\sigma_n^2}$. Similarly it can be shown $\overline{\gamma_s} = \frac{3P_s\sigma_h^2}{\sigma_n^2}$. When M-PSK modulation is used the symbol error probability (SER) for the primary and the secondary users can be calculated in the similar way as before

$$\overline{P_{se}}(primary) = \frac{\alpha_M}{2}\left[1 - \sqrt{\frac{1.5\beta_M P_p \sigma_h^2}{\sigma_n^2 + 1.5\beta_M P_p \sigma_h^2}}\right] \tag{5.21}$$

$$\overline{P_{se}}(secondary) = \frac{\alpha_M}{2}\left[1 - \sqrt{\frac{1.5\beta_M P_s \sigma_h^2}{\sigma_n^2 + 1.5\beta_M P_s \sigma_h^2}}\right] \tag{5.22}$$

The outage probability of the primary and the secondary users relative to the corresponding threshold γ_{p0} and γ_{s0} are given by

$$P_{out}(primary) = 1 - e^{-\gamma_{p0}/\overline{\gamma_p}} = 1 - e^{\frac{3P_p\sigma_h^2}{\gamma_{p0}\sigma_n^2}} \tag{5.23}$$

$$P_{out}(secondary) = 1 - e^{-\gamma_{s0}/\overline{\gamma_s}} = 1 - e^{\frac{3P_s\sigma_h^2}{\gamma_{s0}\sigma_n^2}} \tag{5.24}$$

5.5 Simulations and Discussion

The schemes mentioned above have been simulated in MATLAB. We set the values of the following simulation parameters such as the parameter of the Rayleigh faded channels $\sigma_h^2 = 1$ and the threshold for outage $\gamma_{p0} = \gamma_{s0} = 2\,\text{dB}$. In Fig. 5.4 the average symbol error probability or symbol-error-rate (SER) performance of the primary or the secondary user (since the primary and the secondary users have similar performance they can be interpreted interchangeably) is plotted with respect to the signal-to-noise ratio (P_p/N_0 or P_s/N_0) for different values of α for the first scheme with dual polarized receivers. Here we use 8-PSK modulation for simulation. Figure 5.5 shows the plot outage probability relative to threshold $\gamma_{p0} = \gamma_{s0} = 2\,\text{dB}$ with respect to the signal-to-noise ratio for different values of α. From the plots we see that for a particular signal-to-noise ratio both the error rate and the outage probability increase as α is increased.

Figure 5.6 shows the same plot of SER performance for the second scheme which uses relay cooperation. It has been assumed that the secondary transmitter can estimate the primary transmitted power exactly. The outage probability versus signal-to-noise ratio is plotted in Fig. 5.7. The error performance and the outage performance of the primary and secondary users are quite independent of α.

Fig. 5.4 Error performance of the primary or the secondary user for the first scheme with α as parameter ($\gamma_{p0} = \gamma_{s0} = 2\,\text{dB}$)

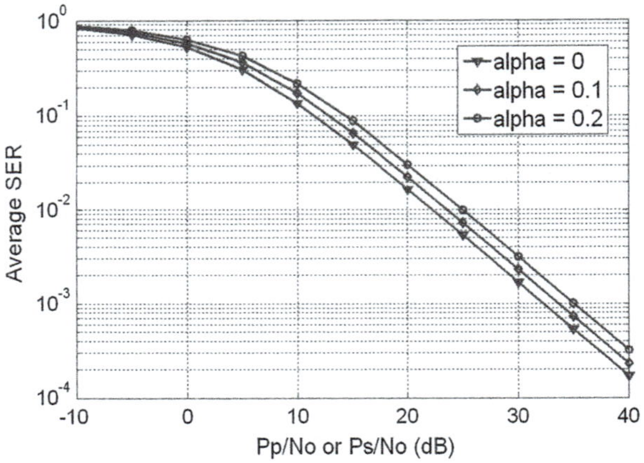

Fig. 5.5 Outage performance of the primary or the secondary user for the first scheme with α as parameter ($\gamma_{p0} = \gamma_{s0} = 2\,\text{dB}$)

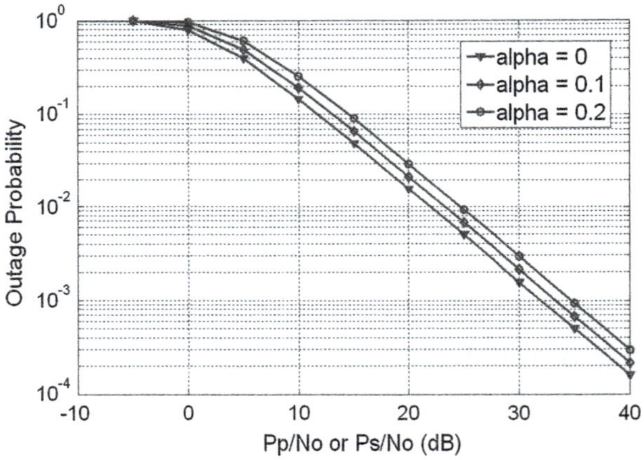

Fig. 5.6 Error performance of the primary or the secondary user for the second scheme with α as parameter ($\gamma_{p0} = \gamma_{s0} = 2\,\text{dB}$)

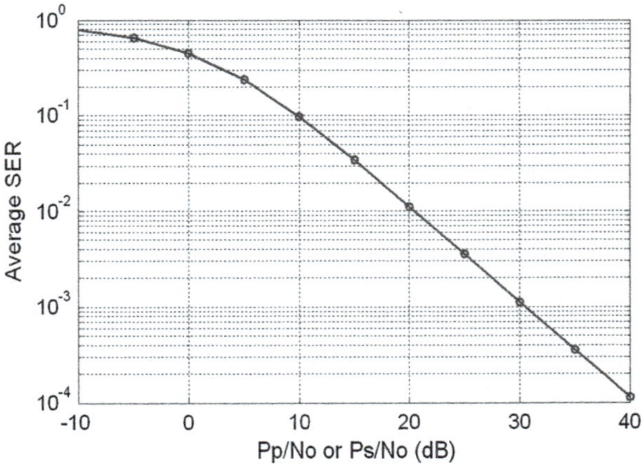

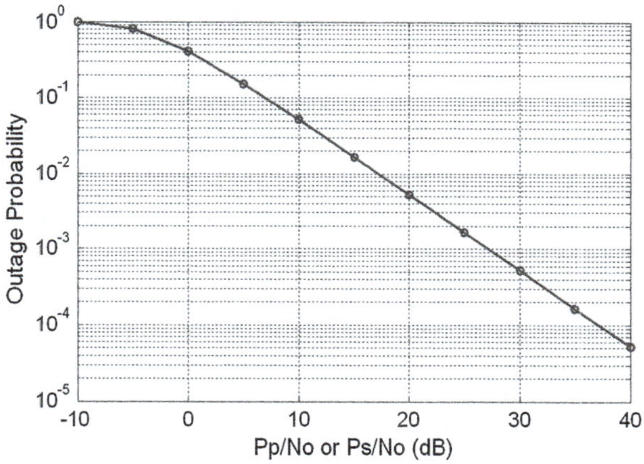

Fig. 5.7 Outage performance of the primary or the secondary user for the second scheme with α as parameter ($\gamma_{p0} = \gamma_{s0} = 2\,\text{dB}$)

Clearly from the plots it is evident that the second scheme shows a slightly better performance in the same channel conditions. To achieve an average symbol error probability of 10^{-3} the second scheme requires about 30 dB signal-to-noise ratio whereas the first scheme requires nearly 32 dB signal-to-noise ratio in an ideal case i.e. when $\alpha = 0$ and the required SNR increases as the value of α increases. Also at an outage probability of 10^{-3} required signal-to-noise ratio for the second scheme is nearly 27 dB compared to 32 dB at $\alpha = 0$ for the first scheme.

5.6 Conclusions

In this work, we have proposed two efficient schemes to mitigate the interference in a 2×2 overlay cognitive radio by using the polarization of the signal. In the first scheme diversity-combining approach has been applied at the receivers to cancel out the interference whereas in the second scheme a decode-and-forward relay cooperation approach has been proposed. The error performance and the outage performance for each of the cases have also been shown. From the results it is evident that the second scheme performs slightly better than the first scheme.

Acknowledgments The work is undertaken as part of Media Lab Asia project entitled "Mobile Broadband Service Support over Cognitive Radio Networks".

References

1. Srinivasa S, Jafar SA (2006) The throughput potential of cognitive radio: a theoretical perspective. In: Fortieth Asilomar conference on signals, systems and computers ACSSC' 06
2. Devroye N, Mitran P, Tarokh V (2006) Achievable rates in cognitive radio channels. IEEE Trans Inf Theory 52(5):1813–1827

3. Zhang L, Liang YC, Xin Y (2008) Joint beamforming and power control for multiple access channels in cognitive radio networks. IEEE J Sel Areas Commun 26(1):38–51
4. Yiu S, Vu M, Tarokh V (2008) Interference reduction by beamforming in cognitive networks. In: Proceedings of IEEE global communication conference (GLOBE-COM), pp 1–6
5. Vaughan R (1990) Polarization diversity in mobile communications. IEEE Trans Veh Technol 39(3)
6. Nabar RU, Bölcskei H, Erceg V, Gesbert D, Paulraj AJ (2002) Performance of multiantenna signaling techniques in the presence of polarization diversity. IEEE Trans Signal Process 50(10)
7. Erceg V, Soma P, Baum DS, Catreux S (2004) Multiple-input multiple-output fixed wireless radio channel measurements and modeling using dual-polarized antennas at 2.5 GHz. IEEE Trans Wireless Commun 3(6)
8. Oestges C, Clerckx B, Guillaud M, Debbah M (2008) Dual-polarized wireless communications: from propagation models to system performance evaluation. IEEE Trans Wireless Commun 7(10)
9. Coldrey M (2008) Modeling and capacity of polarized MIMO channels. In: IEEE vehicular technology conference VTC Spring 2008
10. Bohara VA, Ting SH, Han Y, Pandharipande A (2010) Interference-free overlay cognitive radio network based on cooperative space time coding In: Proceedings of the fifth international conference on cognitive radio oriented wireless networks and communications (CROWNCOM)
11. Goldsmith A (2005) Wireless communications. Cambridge University Press, Cambridge

Chapter 6
Experimental Study and Analysis of Security Threats in Compromised Networks

Usha Banerjee and K. V. Arya

Abstract Intrusion Detection Systems (IDSs) are an indispensable part of a network infrastructure where inordinate attacks such as Distributed Denial-of-Service (DDoS) and metasploits have posed a major problem to the public and private computer networks. IDS assist the network administrators to monitor activities like gaining unauthorized access, session hijacking etc. These unlawful activities can result in losses to an enterprise, both in terms of money and resources. In this paper we detect and prevent one of the commonly occurring server attacks and follow it up with a fatal attack that can fully immobilize and destroy a server. We study and analyze the responses of the intrusion detection server when the network is exploited and the security of the network is compromised. Several dissimilar exploits are made on various Linux distributions hence, assisting the network administrators relying on the IDS to take appropriate action.

Keywords Attacks · Denial of service · Operating system · Server attacks · SNORT · Vulnerability analysis

6.1 Introduction

Nowadays, large and medium scale corporations are facing the threat of increasing cyber attacks in both public and private networks. Hence, in today's era, the area of cyber security is gaining popularity. Primarily there are two methods in which a

U. Banerjee (✉)
Department of Computer Science, College of Engineering Roorkee, Roorkee, India
e-mail: ushaban@gmail.com

K. V. Arya
Department of ICT, ABV-IIITM Gwalior, Gwalior, India
e-mail: kvarya@iiitm.ac.in

S. Sengupta et al. (eds.), *Emerging Trends in Computing and Communication*, Lecture Notes in Electrical Engineering 298, DOI: 10.1007/978-81-322-1817-3_6,

malicious node can get unauthorized access to a network: from inside or outside the organization. Wireless networks can be described as a type of network in which the nodes are not connected through cables. On the other hand, wired network connects similar or dissimilar devices in the network using cables. In comparison to wired networks, besides the disadvantages of speed and reliability, wireless networks are more susceptible to security threats due to their dynamic nature whereas in case of wired networks it is possible to connect to network only through physical access. Thus, concatenating the security policies made by the security administrator with the technology-based security solutions [1] due to which the IDS can easily detect and identify attacks in real time. Signature based intrusion detection to detect intruders on the intranet are too elementary intrusions to detect [2]. Hence, this area of research is still wide and open to researchers and analysts.

The essence of research in this field focuses primarily on detecting attacks that are evolving with technological advancements. According to [3], IDSs are specifically designed for the purpose of reliably detecting range of attacks like DoS, U2R and data attacks against all the available Operating Systems such as Redhat Enterprise Linux, Fedora, and Windows XP etc. in such a manner that will produce low false rates.

6.1.1 Prior Work

Several attempts have been made by security experts and researchers to create a well defined rule base but none of have been able to implement it up to an exemplary level because of the increase in complexity of these attacks. Recently, several mechanisms have been devised for attack correlation and attack scenario analysis. In a survey [4] of numerous Collaborative Intrusion Detection network (CIDN), the stalwartness of these networks have been analyzed. The authors in [4] categorized the network intrusions, insider attacks and IDSs for CIDNs and have followed it up with the sophisticated technology used. Koch [5] proposed an overview of a modern system that enforces a demarcation line between different computer networks and probes their loopholes and breaches. The upcoming cyber-threats are analyzed and their impact on the enterprise's private network is evaluated. Kleinwaechter [2] provides an overall picture of present day's security systems and probes their flaws by analyzing the latest security related threats and methods of handling these threats. Kleinwaechter [2] presented the loopholes and challenges that the current IDSs have to face like onerous configuration and impact of large volume of traffic including false alerts as generated by the IDSs on the bandwidth. According to [2], the competence of the IDS largely depends on the rate of updation of the alerts and response-times of the answerable department of the association.

Various search algorithms have been successfully coordinated in the present day scenario with the IDSs so that the response time can be reduced, throughput can be increased and memory usage can be made as low as possible which is a prime concern in computer science [6]. In [7] the despicable problem of alert correlation with the careful evaluation of the attack scenario developed by the intruder. The authors of [7] use clustering method to transform low level alerts into high level alerts based on this information the relationships among the alerts are determined. In [8] approaches have been developed for detecting intrusions of encrypted information in public as well as private networks by recognizing the features of cryptographic protocols like Kerberos, transaction layer security and IPsec.

The authors in [9] synthesized a new architecture for network based brute force detection method in environments where the overall transactions are encrypted and hence the detection is quite necessary in such cases. Zhang et al. [10] explores the critical vulnerabilities in a wireless ad-hoc network due to which it can be exploited by the intruders and why there is a need of intrusion detection systems in the networks. Zhang et al. [10] elucidates the ineffectiveness of the current methods when applied directly. Zhang et al. [10] proposed a new intrusion detection and response mechanism involving every node to participate in intrusion detection so that it can detect intrusions locally and on their own. Todd et al. [11] inspects the numerous responses produced by the server when four different types of exploits are made from four different Linux distros. Todd et al. [11] shows new alert verification methods that can bypass attacks i.e. they are unable to breach the network even after they are successful in exploitation.

6.2 Attacks

An exhaustive list of attacks and threats can be found in [12]. In the context of this paper we first start by describing what Denial of Service (DoS) and Metasploit attacks are. We do this by setting up simple experimental setups using simple network topologies. Using simple experiments we prove that such attacks can hamper the regular functioning of the network: wired or wireless and hence need to be detected and if possible, prevented.

6.2.1 Distributed Denial of Service Attack

A Denial of Service (DoS) attack is simply a method to make a machine, a server or certain resources unavailable for users. The simplest method of this attack is saturating the sever machine with requests such that it is unavailable to respond to logical and legitimate requests. It is a deliberate attempt to make the potential network rescurces of an enterprise inaccessible to its legitimate users [11]. The consequences of this attack can pose great problem to the network because it can compromise the bandwidth of even a router between the Local Area Network (LAN) and the Wide Area Network (WAN).

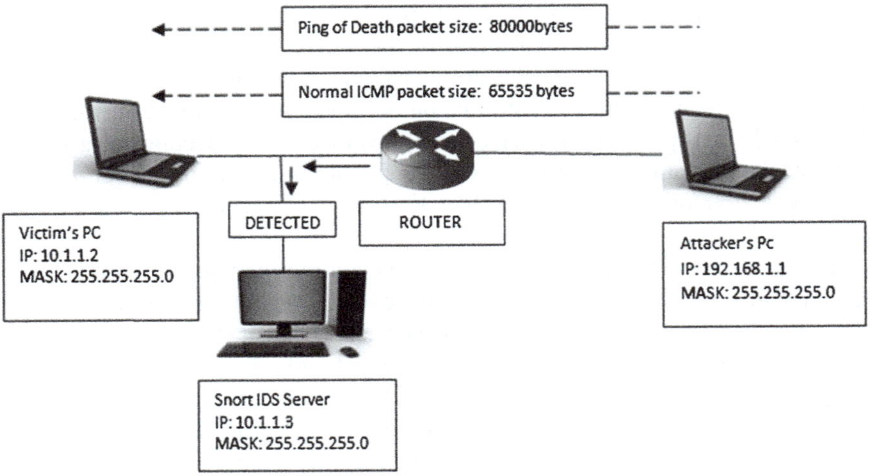

Fig. 6.1 Network Topology showing Ping of Death

6.2.2 Ping of Death

A ping of death attack is a category of DoS attack in which the attacker sends a malicious "ping" to a server. In normal circumstances a ping usually consists of 32 bytes of data. Ping of death implies sending huge sized ping data packets up to the tune of a few thousands of bytes. This results in a buffer overflow and ultimately leads to crashing the server.

As shown in Fig. 6.1, a simple experimental setup has been designed using three machines where one machine acts as the server on which SNORT is installed. A second machine is the victim and the third is the attacker. All the machines are connected via a router. A ping of death data packet of about 80,000 bytes is sent from the attacker machine to the victim via the router. The sever, with SNORT installed in it detects this attack and is able to send an alert to the network administrator providing details of the attacker's and victim's IP address and time of the attack.

In this paper, the purpose of the simulation of this attack is to show the capability of SNORT IDS to detect exploits. Figure 6.3 shows a snapshot for the detection of ping of death with the assistance of the SNORT IDS server.

Figure 6.2 is a snapshot that furnishes details to the network administrator that an attack is deployed by a malicious node whose host's IP address is 192.168.1.1 on a system whose IP address is 10.1.1.2 present in the private network of the organization. The snapshot is taken from the Snort IDS server with IP address 10.1.1.3. As soon as the network administrator receives the heavy alerts on the Snort IDS server, a rule can be created on the Server which is connected to the router directly which blocks the traffic from this particular IP address. The rule is as follows:

```
root@ redhat ~]#iptables –A INPUT –s 192.168.1.1 –j DROP
```

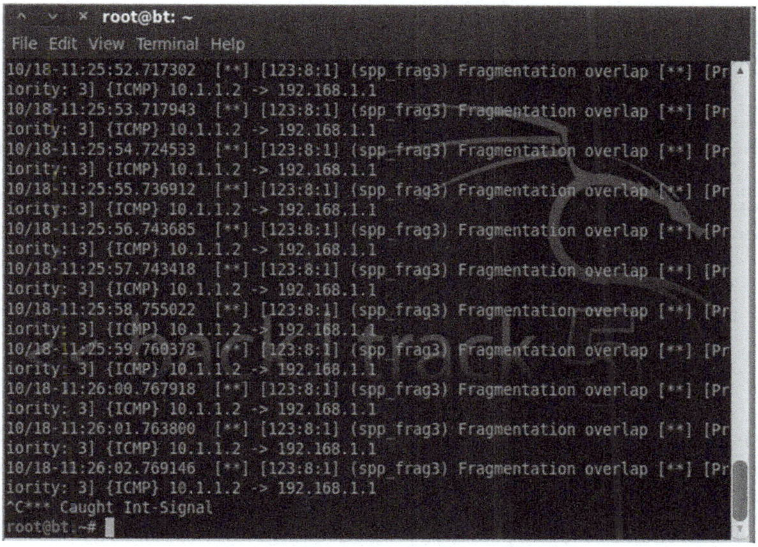

Fig. 6.2 Snapshot of an attack deployed to a malicious node whose IP address is 192.168.1.1

6.3 Experimental Design

A simple network topology was established as shown in Fig. 6.3. In this topology, one computer is the attacker's system and is running Backtrack 5 as its operating System [13]. The attacker's computer is configured to use the Metasploit [14]. Metasploit an open source framework that provides the exploits. Wireshark [15] is installed on the sever machine and is configured to monitor and capture the network traffic.

The IDS server is also running Backtrack 5 as its operating System. Snort IDS is installed and configured on it. The primary reason for choosing SNORT as a tool for intrusion detection is that SNORT logs the attacks in clear text files and also gives alerts on the terminal of the IDS server. The victim machines are not just the Workstations but the File Server also.

6.3.1 Vulnerability Analysis

We carry our research one step forward by doing vulnerability assessment of the target network to first identify the type of Operating Systems running on the victim machines in the private network of the organization. It vigilantly assesses the various risks associated with the ports that are found open after finding them in the process of scanning. According to [16], it is a practice of finding vulnerabilities in numerous areas including IT, banking and defense. In our research, vulnerability

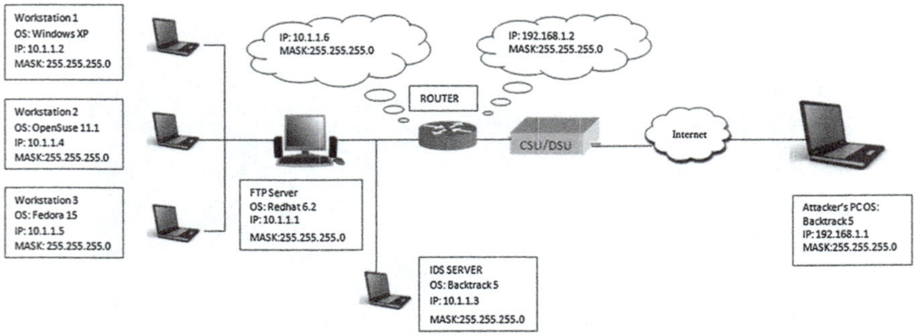

Fig. 6.3 Experimental setup

analysis is done by using Nessus [17]. Nessus is open source software capable of performing probabilistic analysis of the possible breaches. The reason why Nessus is used for vulnerability assessment is because of its capability in exploring the true vulnerabilities that are present in our network. It is a heavily used software program used by security professionals. The remote host is affected by several vulnerabilities in the server that may allow an attacker to execute arbitrary code or perform a denial of service against the remote host. These vulnerabilities depend on access to a shared drive, but do not necessarily require credentials.

6.3.2 Metasploit

It is a hacking framework written in Ruby [18]. Built with the desire to help make writing and running exploits as simple as possible. It is well known for its evasive and anti-forensic built-in tools. In this paper, the purpose of deploying this attack is to examine the reaction of Snort IDS towards it. The Snort IDS successfully alerts the network administrator and also furnishes the IP address of the Attacker's system as shown in Fig. 6.4.

Figure 6.4 exposes the malicious activity of the attacker with IP address 192.168.1.1 who is trying to gain unauthorized to the victim's system which can be any of the workstations or the servers in the server farm, but in this research as can be seen from the snapshot the victim is the workstation with IP address 10.1.1.2 running Windows XP as its OS.

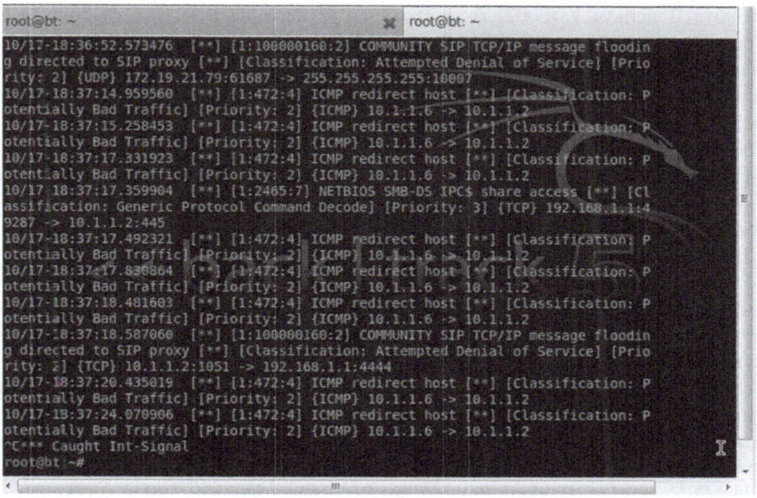

Fig. 6.4 Snapshot of a metasploit attacker

6.4 Results and Discussion

In this paper, a detailed study and analysis and detection of security threats ranging from simple (Distributed Denial of Service (DDoS) Ping of death) to advance (Metasploit) was presented. We have presented that both Windows and Linux OS are vulnerable to attacks in both wired and wireless networks and how this attack can be prevented if attacks are carefully analyzed by the concerned person as observation and evaluation of alerts is an indispensable part of intrusion detection. As with the evolution of new technologies, more and more catastrophic vulnerabilities can be found to breach the enterprise network. Hence, in future we will continue our research to find more efficient detection and prevention measures to curb such malicious activities in the network.

Acknowledgments The Usha Banerjee wishes to acknowledge the support of a WOS-A project (ref. no. : SR/WOS-A/ET-20/2008) funded by the Department of Science and Technology, Government of India.

References

1. Cuppens F (2001) Managing alerts in a multi-intrusion detection environment. In: Proceedings of ACSAC
2. Kleinwaechter J (1998) The limitations of intrusion detection on high speed networks. In: First international workshop on the recent advances in intrusion detection (RAID'98), Louvain-La-Neuve, Belgium

3. Marinova Boncheva V (2007) A short survey of intrusion detection systems. Institute of Information Technologies, 1113 Sofia, pp 23–30
4. Fung C (2011) Collaborative intrusion detection networks and insider attacks. J Wirel Mob Netw Ubiquit Comput Dependable Appl 2(1):63–74
5. Koch R (2011) Towards next-generation intrusion detection. Institut für Technische Informatik (ITI), Universität der Bundeswehr, Munich
6. Singaraju S, Parsi K (2012) A precise survey on intrusion detection systems. Int J Adav Res Comput Sci Softw Eng 2(9):243–247
7. Xinzhou Q, Lee W (2003) Statistical causality analysis of infosec alert data. In: Proceedings of the 6th international symposium on recent advances in intrusion detection (RAID 2003), pp 73–93
8. Yasinsac A, Goregaoker S (2002) An intrusion detection system for security protocol traffic. Technical report, Department of computer science, Florida State University, Tallahassee, Florida 32306-4530
9. Koch R (2012) Fast network-based brute-force detection. In: 8th advanced international conference on telecommunications (AICT)
10. Zhang Y, Lee W, Huang Y (2003) Intrusion detection techniques for mobile wireless networks. ACM Wirel Netw J 9(5):545–556
11. Todd AD, Raines RA, Baldwin RO, Mullins BE, Rogers SK (2007) Alert verification evasion through server response forging. In: Proceedings of the 10th international conference on recent advances in intrusion detection (RAID'07), pp 256–275
12. Banerjee U, Swaminathan A (2011) A taxonomy of attacks and attackers in MANETs. Int J Res Rev Comput Sci 2:437–441 (Academy Publishers)
13. BackTrack Linux (2011) http://www.backtracklinux.org/
14. Metsaploit framework, http://en.wikipedia.org/wiki/Metasploit_Project
15. Wireshark available at http://www.wireshark.org/
16. Vulnerability assessment, http://en.wikipedia.org/wiki/Vulnerability_assessment
17. Nessus available at http://www.nessus.swri.org/
18. Ruby—an open source programming language, http://www.ruby-lang.org/en/

Part II
Image Processing and OCR

Chapter 7
A Weighted Counter Propagation Neural Network for Abnormal Retinal Image Classification

J. Anitha and D. Jude Hemanth

Abstract Artificial Neural Networks (ANN) is one of the primarily used computing techniques for medical image classification applications. However, ANN like Counter Propagation Neural Network (CPN) is less accurate which limits the usage for medical image analysis. In this work, this problem of low accuracy of CPN is tackled by performing suitable modifications in the conventional approach. The concept of weight assignment is used in the distance calculation procedure of Kohonen layer of CPN to enhance the performance of the overall system. The weight assignment procedure is based on the textural features estimated from the input images. This approach is called as Weighted Counter Propagation Neural network (WCPN) and the applicability of this approach is explored in the context of abnormal retinal image classification. Retinal images from four abnormal categories are used in this work. Suitable textural features are extracted from the input images and supplied as input for the ANN. The experiments are performed with the conventional CPN and the proposed WCPN. The performance measures used in this work are classification accuracy and convergence time. Experimental analysis shows promising results for the proposed approach.

Keywords Counter propagation neural network · Image classification · Retinal images · Accuracy and convergence time

7.1 Introduction

Medical image classification is one of the applications in which the role of ANN is highly significant. Being a medical application, the results obtained from ANN must be highly accurate. Literature survey reveals the availability of ANN for

J. Anitha · D. Jude Hemanth (✉)
Department of ECE, Karunya University, Coimbatore, India
e-mail: jude_hemanth@rediffmail.com

S. Sengupta et al. (eds.), *Emerging Trends in Computing and Communication*, Lecture Notes in Electrical Engineering 298, DOI: 10.1007/978-81-322-1817-3_7,

various medical imaging applications. Artificial neural network based keratoconous detection in abnormal retinal images is implemented in [1]. The combined FCM and neural techniques for exudates detection in diabetic retinal images is developed by [2]. This methodology is implemented on color retinal images. A hybrid approach of fuzzy theory and ART neural network is used by [3]. Hard exudates are detected from the abnormal DR images using this approach. Supervised algorithm for differentiating different retinal diseases is designed by [4]. RBF classifier based disease identification in abnormal retinal images is proposed by [5].

The application of multi layer perceptron for ophthalmologic disease classification is explored by [6]. Initially, features are extracted and then Principal Component Analysis is used for feature reduction. But the success rate of this methodology is based on input feature set. The differentiation of different stages of Diabetic Retinopathy (DR) retinal images is performed by [7]. Back Propagation Neural Network (BPN) is used as the classifier in this work. The application of multi-layer perceptron for hard exudates detection in retinal images is explored by [8]. Auto associative neural network based retinal disease identification is implemented by [9]. A suitable feature set is extracted from the pre-processed images and supplied to the neural classifier for classification. Linear Discriminant Analysis based abnormal retinal image classification is performed by [10]. Different varieties of abnormal images are used to check the applicability of this approach. A detailed description of various ANN and their applications are given in [11].

However, the conventional ANN does not guarantee high accuracy which opens the necessity for modified ANN. In this work, a modified ANN is proposed for retinal image classification. This modified ANN is framed with an objective to improve the accuracy without compromising the convergence rate.

7.2 Proposed Methodology

The framework of the proposed automated system is shown in Fig. 7.1.

Real world retinal images collected from the ophthalmologists are used in this work for image classification. 220 images from four abnormal categories such as Choroidal Neo-Vascular Membrane (CNVM), Central Serous Retinopathy (CSR), Central Retinal Vein Occlusion (CRVO) and Non-Proliferative Diabetic Retinopathy (NPDR) are employed in the ANN based experiments. Five textural features such as Mean, Standard Deviation (SD), Energy, Entropy and Skewness are extracted from these retinal images. A weight value is assigned to features based on uniqueness of features for each category.

These weighted features are supplied as input to the CPN network for the training process. After training, the proposed approach is tested with the unknown images and the performance measures are calculated. The conventional CPN is also trained and tested with the features without any weights. A comparative analysis is performed between these two techniques to highlight the optimal approach for retinal image classification.

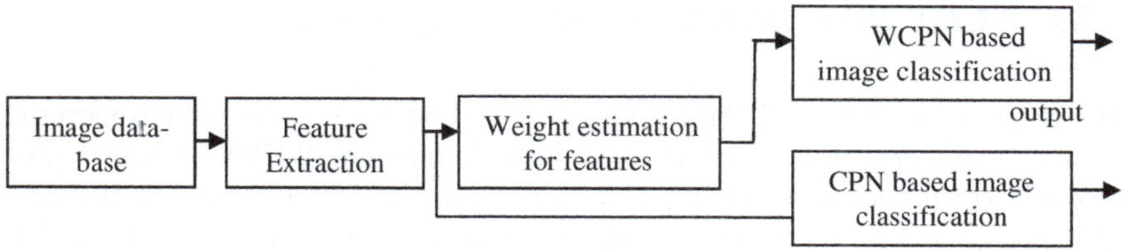

Fig. 7.1 Flow diagram of the automated system

7.3 Feature Extraction

Feature extraction is the significant part of any automated classification system. The purpose of feature extraction is twofold: (a) Representing each abnormal category in a unique way and (b) dimensionality reduction of the input images. These objectives enhance the accuracy of the system besides reducing the computational complexity of the system. These features are unique for each category which aids the classification process of the artificial neural networks. The textural features are extracted using mathematical equations. A detailed explanation is given in [12].

7.4 Weight Estimation for Different Features

All the extracted features do not guarantee high accuracy in any classification system. Hence, it is highly necessary that the optimal features must be given higher importance in the classification process and the less significant features must be given lower importance in the process. Thus, these features are assigned a weight value between 0 and 1 to show the measure of importance of each feature. This measure of importance can be determined from the experimental results of feature extraction. The features which are largely different between categories but similar within the same category images are assigned the highest value of 1. The rest of the features are given the weight values of 0.8, 0.6, 0.4, 0.2 respectively based on this procedure. Thus, the weight values impact the accuracy of the overall classification system.

7.5 ANN Based Retinal Image Classification

In this work, two ANN such as WCPN and conventional CPN are implemented for retinal image classification. Initially, the WCPN is analyzed in detail followed by a brief illustration on conventional CPN.

7.5.1 WCPN Based Retinal Image Classification

The proposed WCPN is a hybrid ANN in which both supervised and unsupervised training methodologies are used for weight estimation. WCPN is a two layer network which consists of the input Kohonen layer which uses the "winner take-all" strategy and the output Grossberg layer which uses the error signal for weight adjustment. The architectural details and the training algorithm are described below. The topology of the WCPN consists of two set of weight matrices. Hence, it is essential to find the optimal set of weight matrices using the training algorithm of WCPN.

7.5.1.1 Training Algorithm

The training algorithm of the WCPN is explained with the following procedural steps.

Step 1: The Euclidean distance d_j between the weight vector $\overline{U}$and the input vector $\overline{X}$ is calculated using the following formula

$$d_j = \sum \left(u_{ij} - x_i\right)^2 \times m_r \quad j = 1, 2, \ldots, 12 \tag{7.1}$$

In the above equation, m corresponds to predefined weights and r varies from 1 to 5.

Step 2: Each neuron in the competition layer competes with the other neurons and the node with the shortest Euclidean distance wins. The output of the winning node is set to 1 and the rest to 0. Thus, the output of the jth node in the competition layer is

$$\begin{aligned} Z_j &= 1.0 \quad \text{if } d_j \text{ is minimum} \\ Z_j &= 0.0 \quad \text{Otherwise} \end{aligned} \tag{7.2}$$

Step 3: The weight adjustment between input layer and the competition layer is given by

$$U_{ij}^{t+1} = U_{ij}^{t} + \alpha(x - U^t)Z_j \tag{7.3}$$

where 't' is the iteration number and 'α' is the learning coefficient.

Step 4: After the weight matrix $\overline{U}$ have stabilized, the output layer begins the learning procedure. The weight adjustments for the output layer is given by

$$V_{jk}^{t+1} = V_{jk}^{t} + \alpha(T - V^t)Z_j \tag{7.4}$$

where T is the target vector.

Step 5: This process is repeated for the specified number of iterations and the stabilized weight matrices are observed.

Thus, WCPN involves the concept of weight assignment in Eq. (7.1) which ensures the success of the subsequent steps of the training algorithm. After the training process, the stabilized network is tested with the unknown images. The same experiments are also repeated with the conventional CPN. The difference between the WCPN and CPN is the presence of importance measure for each feature in WCPN which is not available in CPN. A detailed explanation on conventional CPN is available in [12].

7.6 Experimental Results and Discussions

The experiments are conducted on 220 real world retinal images collected from four categories. The specifications of the processor used for the experiments are 1 GB RAM and 2 GHz clock frequency. The software used for the implementation is MATLAB. Initially, an analysis is performed on extracted features for weight assignment to the features. Table 7.1 shows the experimental results of feature extraction and the details of weights assigned to the features.

The average "within class" values must be low and "between class" values must be high for the best feature. From Table 7.1, it is evident that Entropy is better than all other features and Mean is the least preferred feature. Hence a weight value of 1 is assigned to entropy feature and a value of 0.2 is assigned to the mean feature. Using this procedure, SD, energy and skewness are allotted the values of 0.4, 0.6 and 0.8 respectively. These weighted features are used for training and testing the proposed approach. The conventional CPN is trained and tested with the non-weighted features. Table 7.2 shows the confusion matrix of WCPN and CPN.

The above analysis is performed only on the testing images. 25 images from each category are used for training and the remaining images are used for testing. From Table 7.1, it is clearly visible that the level of misclassification rate in WCPN is lesser than the CPN. The Classification Accuracy (CA) is calculated from the confusion matrix which is shown in Table 7.3. The number of iterations and time requirement for convergence is also shown in Table 7.3.

The performance measures of the proposed approach are significantly better than the conventional approach. Thus, this research suggests suitable alternate for the conventional ANN for medical applications.

Table 7.1 Feature extraction analysis

	Mean	SD	Energy	Entropy	Skewness
Average "within class" difference value	5.05	4.3	3.2	0.8	2.8
Average "between class" difference value	10.12	12.6	13.8	25.5	14.5

Table 7.2 Confusion matrix of WCPN and CPN

	WCPN				CPN			
	Class 1	Class 2	Class 3	Class 4	Class 1	Class 2	Class 3	Class 4
CNVM	27	1	0	1	23	2	2	3
CSR	2	26	1	1	3	22	3	2
CRVO	0	1	27	1	2	3	24	1
NPDR	0	1	1	28	0	3	5	22

Table 7.3 Performance analysis of the classifiers

	Average CA (%)	No. of iterations required	CPU time (seconds)
WCPN	90	135	925
CPN	73	262	1,100

7.7 Conclusion

A suitable alternate approach for conventional CPN is proposed in this work for retinal image classification. The concept of weight assignment is used in the proposed approach to enhance the performance of the overall system. Experimental results have highlighted the various benefits of the proposed technique in terms of the performance measures.

Acknowledgments The authors thank Dr. A. Indumathy, Lotus Eye Care Hospital, Coimbatore, India for her help regarding database validation. The authors also wish to thank Council of Scientific and Industrial Research (CSIR), New Delhi, India for the financial assistance towards this research (Scheme No: 22(0592)/12/EMR-II).

References

1. Agostino AP, Stefano P (2003) Neural network based system for early keratoconus detection from corneal topography. J Biomed Inform 35:151–159
2. Alireza O et al (2003) Automated identification of diabetic retinal exudates in digital colour images. Br J Ophthalmol 87:1220–1223
3. Jayakumari C, Santhanam T (2007) Detection of hard exudates for diabetic retinopathy using contextual clustering and fuzzy ART neural network. Asian J Inf Technol 6(8):842–846

4. Meindert N et al (2007) Automated detection and differentiation of drusen exudates and cotton wool spots in digital color fundus photographs for diabetic retinopathy diagnosis. Invest Ophthalmol Vis Sci 48(5):2260–2267
5. Acharya R et al (2007) Automatic identification of anterior segment eye abnormality. Proc ITBM-RBM 28(1):35–41
6. Povilas T, Vydunas S (2007) Neural network as an ophthalmologic disease classifier. Inf Technol Control 36(4):365–371
7. Wong LY et al (2008) Identification of different stages of diabetic retinopathy using retinal optical images. Inf Sci 178:106–121
8. Maria G et al (2009) Neural network based detection of hard exudates in retinal images. Comput Methods Programs Biomed 93:9–19
9. Jayanthi D, Devi N, Swarna PS (2010) Automatic diagnosis of retinal diseases from color retinal images. Int J Comput Sci Inf Secur 7(1):234–238
10. Clara IS et al (2008) A novel automatic image processing algorithm for detection of hard exudates based on retinal image analysis. Med Eng Phys 30:350–357
11. Freeman JA, Skapura DM (2004) Neural networks, algorithms, applications and programming techniques. Pearson Education, New Jersey
12. Fausett L (2006) Fundamentals of neural networks; architectures, algorithms and applications. Pearson Education India, New Delhi

Chapter 8
A Language Independent Hybrid Approach for Text Summarization

Vishal Gupta

Abstract This paper discusses an algorithm for language independent hybrid text summarization. There are seven features used in this text summarizer: (1) words form similarity with title line (2) n-gram similarity with title (3) Normalized NTF-PSF feature (4) Position feature (5) Relative length feature (6) Extraction of number data (7) User specified domain specific keywords feature. All the features are statistical based (i.e. language independent features) and no language specific feature is taken except stop words list and stemmer for that language. In this language independent summarizer we consider every feature as equally important so no weights are assigned to different features. Top scored 20 % sentences are extracted and rearranged according to their appearance in the input for maintaining sentence coherence.

Keywords Language independent summarizer · Hybrid text summarizer · Extractive summarization · Information extraction · Text mining

8.1 Introduction

Automatic text summarization [1–3] is the technique of compressing the original text into smaller form but maintaining the original information and meaning of a document or multiple documents. Extractive text summarization [4] involves identifying certain relevant sentences from the source text based on mixture of statistical and language dependent text features. On the other hand, Abstractive summarization techniques involves natural language understanding based techniques for making the summary in same manner as human beings usually make the

V. Gupta (✉)
University Institute of Engineering and Technology, Panjab University, Chandigarh, India
e-mail: vishal@pu.ac.in; vishal_gupta100@yahoo.co.in

S. Sengupta et al. (eds.), *Emerging Trends in Computing and Communication*,
Lecture Notes in Electrical Engineering 298, DOI: 10.1007/978-81-322-1817-3_8,

summary. These methods are difficult than text extraction because these involve understanding of text rather than just extracting relevant sentences. Moreover abstractive summary may contain words and sentences which might not be present in the original text.

This paper discusses an algorithm for language independent hybrid text summarization. For developing it, we have taken language independent features from four papers proposed by Krishna et al. [5], Fattah and Ren [6], Bun and Ishizuka [7] and Lee and Kim [8]. There are seven features used in this text summarizer: (1) words form similarity with title line (2) n-gram similarity with title (3) Normalized NTF-PSF feature (4) Position feature (5) Relative length feature (6) Extraction of number data (7) User specified domain specific keywords feature. All the features are statistical based (i.e. language independent features) and no language specific feature is taken except stop words list and stemmer for that language. For each sentence, scores of each feature is calculated and final scores of sentences are determined from equation:

$$\text{Value}(\text{feature}_1) + \text{Value}(\text{feature}_2) + \text{Value}(\text{feature}_3) + \cdots \text{Value}(\text{feature}_7)$$

In this language independent summarizer we consider every feature as equally important so no weights are assigned to different features. Top scored 20 % sentences are extracted and rearranged according to their appearance in the input for maintaining sentence coherence.

8.2 Language Independent Hybrid Text Summarization

There are seven features used in language independent hybrid text summarization: (1) words form similarity with title line (2) n-gram similarity with title (3) Normalized NTF-PSF feature (4) Position feature (5) Relative length feature (6) Extraction of number data (7) User specified domain specific keywords feature. Before calculating these features input text is pre processed and boundary is identified for sentences and words. Then stop words are removed from sentences. It is the only language specific component that is added along with stemmer for that language. Stop words are un-important words with very high frequency. We have applied Porter's stemmer [9] for English text to convert all the words in input into their stem words. Duplicate sentences are deleted from input.

8.2.1 Words Form Similarity of Sentences with Title Line

We have applied the word form similarity measure as described by Krishna et al. [5]. Word form similarity is meant for calculating similarity between a given sentence and title line, is measured by the number of same words in two sentences. The sentences are preprocessed to filter the words from the overall words in the sentence. Consider there are two sentences s1 and s2, the word form similarity is determined as follows:

$$\text{Word form similarity}(s1, s2) = (2 * \text{Number of Same words } (s1, s2))/(\text{Length } (s1) + \text{Length } (s2))$$

The value of word form similarity of a sentence with title lies between 0 to 1.

Example: The following example shows how the word form similarity is calculated between a sample Title and a sentence in a document

Title T1: Electronic design has integrated circuits.
Sentence S1: Integrated circuits are parts of electronic.
Title is partitioned into tokens: Electronic |design| has | integrated | circuits.
After applying stop list: Electronic |design| integrated | circuits.
Applying Stemming: Electronic |design| integrate | circuit.
Sentence is partitioned into tokens: |Integrated| circuits| are| parts| of| electronic.
After applying stop list: Integrated| circuits |electronics
Applying stemming algorithm: Integrate| circuit| electronic.
Calculation of Word Form Similarity for above two sentences:

$$\text{Word form similarity } (T1,\ S1) = (2\ *\ \text{Number of Same words } (T1,\ S1))/(\text{Len } (T1)\ +\ \text{Len } (S1))$$

Similarity $(T1,\ S1) = ((2\ *\ 3)/(4+3))\ *\ 100$
Similarity $(T1,\ S1) = 6/7 = 0.8571\ *\ 100\ =\ 85.71\ \%$
Similarity $(T1,\ S1) =\ 85.71\ \%$

Therefore we can say that the above two sentences T1 and s1 are almost similar.

8.2.2 N-Gram Based Similarity of Sentences with Title Line

We have applied the word form similarity measure as described by Krishna et al. [5]. The n-gram based similarity can be calculated after removing the stop words from the two sentences by using the formulae.

$$\text{N-gram similarity } (s1,\ s2) = (2\ *\ (\text{no of common N-grams in } s1\ \&\ s2))/(\text{Total no of N-grams in } s1\ \&\ s2)$$

The value of N-grams similarity of a sentence with title lies between 0 to 1.

Example: The following example shows how the N-gram based similarity is calculated between a title line and a sentence in a document

Title T1: Electronic design has integrated circuits.
Sentence S1: Integrated circuits are parts of electronic.
Title is partitioned into tokens: Electronic |design| has | integrated | circuits.

After applying stop list: Electronic |design| integrated | circuits.
Split the Title T1 into bi-grams
Sentence is partitioned into tokens: |Integrated| circuits| are| parts| of| electronic.
After applying stop list: Integrated| circuits |electronics
Split the sentence S1 into bi-grams

N-gram similarity (T1, S1) =
(2 * (no of common bi-grams in s1 & s2))/(Total no of bi-grams in s1 & s2) * 100

Therefore we can say that the above two sentences are almost similar.

8.2.3 Normalized NTF-PSF Feature for Keywords Extraction

For extracting the keywords from a text document we are using the NTF-PSF measure which is modified improved form of conventional TF-ISF [10, 11], measure. NTF is normalized Term Frequency as defined by Lee and Kim [8]. PSF [7] is Proportional Sentence Frequency. TF is count of number of occurrence of any term in each sentence. Normalized Term frequency is obtained by taking ratio of frequency of a given word in a document to the total frequency of all terms in every line of that document. PSF (proportional sentence frequency) is calculated as defined by Bun and Ishizuka [7]. PSF is calculated by taking exponential of frequency of lines possessing the given word to the total frequency lines in the document. Words which will appear in many lines will be more relevant than the words which appear in less number of lines. Terms with higher score of NTF-PSF are treated as key-terms. NTF-PSF value is determined for every term of each line. Then NTF-PSF values of all terms in each line are summed up for obtaining final NTF-PSF value of each line. To normalize the NTF-PSF score for any sentence it is divided by maximum NTF-PSF score of a sentence.

$$\text{Final Normalized NTF-PSF Score of a sentence} = \frac{\text{NTF-PSF Score of that sentence}}{\text{Max (NTF-PSF) score of any sentence in document}}$$

8.2.4 Position Feature for Sentences

Starting first three sentences of any paragraph are always more relevant. The paragraph lines are ranked according to their position in that Paragraph [6]. So we have considered maximum three positions. We have given the score 3/3 to Ist line in a paragraph. To the second line we have given the score of 2/3 and to the 3rd line of any paragraph the score given is 1/3. For rest of lines in any paragraph the are given score of zero for this feature because those lines are not of much relevance according to this feature.

8.2.5 Relative Length Feature of Sentences

According to us short lines possess little information than lengthy ones normally. So we prefer the long sentences for including in our summary because they can give more information [10]. The relative length of any sentence is determined as:

$$\text{Sentence-Relative-Length} = \text{Words frequency in a line/Frequency of words of longest line.}$$

8.2.6 Numeric Data Identification

Number data is always more relevant. So the lines containing number data are also very relevant in point of view of summary [6]. We have given importance to digits and Roman-numbers This value of this feature is determined by dividing the frequency of number data in any line by the length of that sentence.

8.2.7 User Specified Domain Specific Keywords Extraction

For any language in the world user can give certain domain specific keywords [10] and sentences containing more number of these domain specific keywords are important. This feature is calculated as follows:

Score of User Specific Keywords Feature = Frequency of unique domain specific keywords present in any line/Sentence-Length.

8.2.8 Final Calculation of Scores of Sentences

Finally scores of sentences are determined from line feature equation: $\text{Value}(\text{feature}_1) + \text{Value}(\text{feature}_2) + \text{Value}(\text{feature}_3) + \cdots \text{Value}(\text{feature}_7)$

In this language independent summarizer we consider every feature as equally important so no weights are assigned to different features. Top scored 20 % sentences are extracted and rearranged according to their appearance in the input for maintaining sentence coherence.

Table 8.1 Intrinsic and extrinsic summary evaluation

Compression ratio (in %)	Intrinsic and extrinsic summary evaluation			
	Avg. F-score	Avg. Cosine similarity	Avg. Jaccard coeff.	Question answering task
20 %	87.56	0.844	0.857	86.34

8.3 Algorithm for Hybrid Language Independent Summarizer

This language independent hybrid algorithm starts by splitting the text into words and sentences.

Step 0: Initially for all sentences set their scores as zero.
Step 1: Redundant lines are eliminated in the input text.
Step 2: All Stop words are eliminated from input text.
Step 3: Apply Porter's Stemmer [8] for converting words into their stem words.
Step 4: Calculate words form similarity of sentences with title line
Step 5: Calculate N-gram based similarity of sentences with title line
Step 6: Calculate NTF-PSF Feature for all the sentences.
Step 7: Calculate position feature score for all sentences.
Step 8: For all lines determine the relative length score.
Step 9: Calculate number feature score for all sentences.
Step 10: Finally scores of sentences are determined from line feature equation:

$$\text{Value}(\text{feature}_1) + \text{Value}(\text{feature}_2) + \text{Value}(\text{feature}_3) + \ldots \text{Value}(\text{feature}_7)$$

Step 11: Top scored lines are extracted at 20 % compression ratio.
Step 12: Rearrange these top ranked 20 % sentences according to their position in the input for maintaining sentence coherence. This is final summary.

8.4 Results and Discussions

The language independent hybrid summarizer has been tested over 50 English documents covering news articles, general articles and stories. Data set contains 6,314 sentences and 73,458 words. These 50 test documents were collected from popular punjabi websites: www.wikipedia.org, and websites of English news papers The Tribune, Hindustan Times and Indian Express. The results of intrinsic and extrinsic summary evaluation [12] for our summarizer are shown in Table 8.1 and Fig. 8.1.

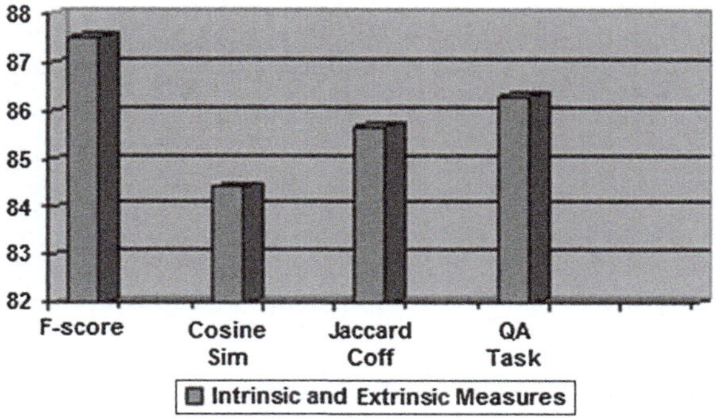

Fig. 8.1 Intrinsic and extrinsic measures for language independent summarizer

References

1. Kyoomarsi F, Khosravi H, Eslami E, Dehkordy PK (2008) Optimizing text summarization based on fuzzy logic. In: IEEE international conference on computer and information science, University of Shahid Kerman, UK, pp 347–352
2. Gupta V, Lehal GS (2010) A survey of text summarization extractive techniques. In: Proceedings of JETWI, vol 2. Academy Publisher, Chicago, pp 258–268
3. Lin J (2009) Summarization. In: Liu L, Özsu MT (eds) Encyclopedia of database systems. Springer, Heidelberg
4. Gupta V, Lehal GS (2012) Automatic Punjabi text extractive summarization system. In: International conference COLING-2012, IIT Bombay, India, pp. 191–198
5. Krishna RVVM, Kumar SYP, Reddy CS (2013) A hybrid method for query based automatic summarization system. Int J Comput Appl Comput Appl 68:39–43
6. Fattah MA, Ren F (2008) automatic text summarization. World Acad Sci Eng Technol 27:192–195
7. Bun KK, Ishizuka M (2002) Topic extraction from news archive using TF-PDF algorithm. In: Proceedings of 3rd international conference on web information system engineering WISE 02, pp 73–82
8. Lee S, Kim HJ (2008) News keyword extraction for topic tracking. In: Proceedings of 4th international conference on networked computing and advanced information management, pp 554–559
9. http://tartarus.org/martin/PorterStemmer/
10. Kaikhah K (2004) Automatic text summarization with neural networks. In: IEEE international conference on intelligent systems, Texas, USA, pp. 40–44
11. Neto JL, Santos AD, Kaestner CAA, Freitas AA (2000) Document clustering and text summarization. In: International conference on practical application of knowledge discovery & data Mining, London, pp 41–55
12. Hassel M (2004) Evaluation of automatic text summarization. Licentiate Thesis. Stockholm, Sweden, pp 1–75

Chapter 9
Facial Expression Recognition Using PCA and Various Distance Classifiers

Debasmita Chakrabarti and Debtanu Dutta

Abstract Information Technology is playing a very big role in today's world. Our interaction with IT is mainly through advanced human computer interfaces. In this regard we seek to enhance the interface in a way such that it can take into account the human facial expression and respond according to a person's feelings in a broader sense. We propose here a simple yet efficient way of facial expression recognition using Eigenspaces and dimensionality reduction techniques and multiple classifiers. It is a modified approach to the original Eigenface method by Turk and Pentland (J Cogn Neurosci, 1991) [1] for face recognition, where using the standard JAFFE database we classify each test image as belonging to one of the six basic expression classes—anger, disgust, fear, happiness, sadness or surprise. In the process we put to test four different classifiers—Euclidean distance, Manhattan distance and Cosine distance and Mahalanobis distance classifiers. In this paper we present with experimental evidence the accuracy of our method and the comparative results yielded by each of the classifiers.

Keywords Facial expression recognition · Eigenspace · PCA · Classifier

D. Chakrabarti (✉)
Department of Computer Science, Sir Gurudas Mahavidyalaya, Kolkata, India
e-mail: noko.chakrabarti@gmail.com

D. Dutta
IBM India Pvt. Ltd., Kolkata, India
e-mail: dutta.debtanu@gmail.com

S. Sengupta et al. (eds.), *Emerging Trends in Computing and Communication*, Lecture Notes in Electrical Engineering 298, DOI: 10.1007/978-81-322-1817-3_9,

9.1 Introduction

A lot of things are involved as part of human communication, like voice and speech, tone, gesture, expression and body language. Also prior knowledge about the person comes into play when we receive such information. A conventional human computer interface [2] can record the speech and some other inputs but normally does not take into account all the other information. If the interface is enhanced in such a way that it can capture the facial expression of an individual and if possible the body language, gesture, etc. then a lot more information will be available to work with resulting in better overall experience. This can be utilized by intelligent education systems [3], AI games, etc. which can adjust their pace according to the expression feedback received from the interface based on the user's mood reflected by his acceptance level. This forms the main motivation behind our study.

There are six expressions—anger, disgust, fear, happiness, sadness, surprise and a neutral expression which have been identified by psychologists to be basic and universal across all cultures. Our study is based on the classification of images based on these classes. In this paper we use Eigenspaces for the recognition of facial expressions. Eigen-space method has been previously used in face detection and sometimes even been extended to expression recognition. In our proposed approach we have modified this method to perform this classification in a simple, fast and accurate way using PCA on a large standard dataset.

9.2 Why Eigenvectors

Even relatively small grey-scale images of dimension 256 × 256 when represented as matrices can lead to very large matrix operations which are quite costly. To optimize this situation, we need to extract only the relevant information out of a face image, encode it efficiently to reduce the operational complexity and compare such an encoded face image with a training set of similarly encoded face images. An image can be considered as a vector in a high dimensional space. We need to find the eigenvectors of the covariance matrix of the set of those images. These eigenvectors—also called feature vectors, represent a set of features which characterize the variation between the images. If we display these eigenvectors as images, they show up sometimes as blurry faces which are called eigenfaces. Input face images can be fully represented as a linear combination of eigenfaces. The higher valued eigenvectors represent greater variation than the lower ones. So we can ignore some lower valued eigenvectors to decrease the dimensionality of the problem. If we consider a set H of highest valued eigenfaces then we obtain an H-dimensional face space of all possible images.

A 256 × 256 image can be considered as a matrix of dimension N^2 ($N = 256$), that is of dimension 65,536. It is thus a point in a 65,536 dimensional space.

The training set of images can be considered as points in this huge space. It will be very resource consuming to carry out mathematical operations on such huge dimensions—the covariance matrix formed out of this will have dimension $N^2 \times N^2$. As the face images will be similar in nature they will not be scattered randomly along this huge space. This gives us the opportunity to represent the actual face space in much smaller dimensions. Considering only the largest H eigenfaces, our problem then reduces to only H points in this N^2 dimensional space and there will be only H—1 meaningful eigenvectors, rest of the eigenvectors having eigenvalue zero and hence ignored. In this way we can reduce our problem to H × H dimensions which is reasonably lighter to work with.

9.3 Distance Classifiers

Once the dimensionality reduction is taken care of, we need to find the distance between the projection of each test image and projections of all the training images. We test our program using four different distance classifiers:

Euclidean distance which is given by Eq. (9.1)

$$(x, y) = \|\mathrm{x} - \mathrm{y}\|^2 = \sum_{i=1}^{k} (x_i - y_i)^2 \tag{9.1}$$

Cosine distance which is given by Eq. (9.2)

$$x \cdot y / |x| \, |y| \tag{9.2}$$

Manhattan distance which is given by Eq. (9.3)

$$(x, y) = |\mathrm{x} - \mathrm{y}| = \sum_{i=1}^{k} |x_i - y_i| \tag{9.3}$$

Mahalanobis distance which is given by Eq. (9.4)

$$\mathrm{d}(x, y) = \sqrt{\sum\nolimits_{(n=1)}^{k} (m_i - n_i)} \tag{9.4}$$

where $\mathrm{m_i} = \mathrm{x_i}/\sigma_\mathrm{i}$ $\mathrm{n_i} = \mathrm{y_i}/\sigma_\mathrm{i}$ σ being the standard deviation.

9.4 Proposed Method

Our method is a modified approach to the well-known method of face detection using Eigenspaces [4] but here we have used it for expression recognition instead of identification of the person. To represent the expression recognition problem

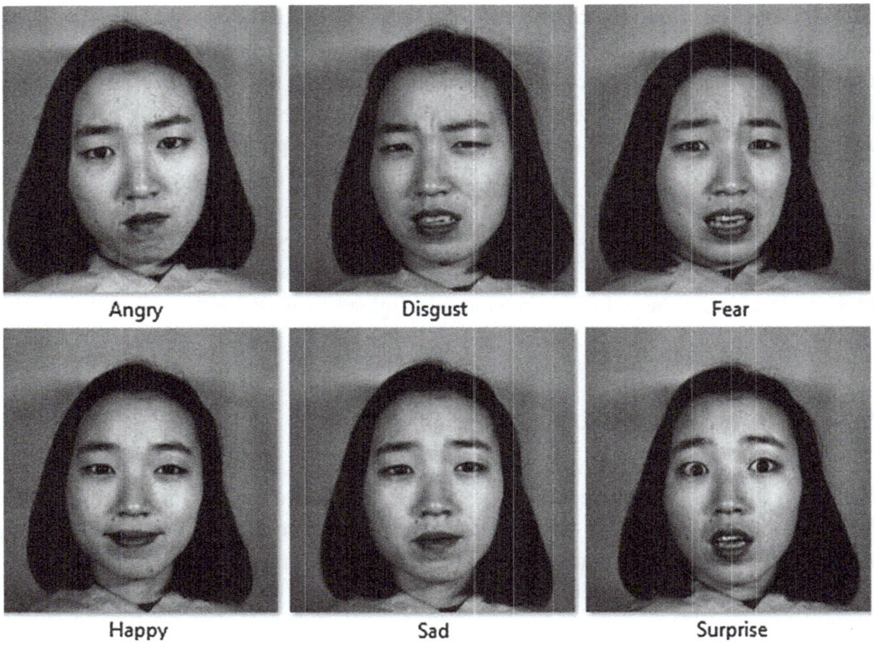

Fig. 9.1 The six universal expressions represented by one of the models from JAFFE database

using Eigenspaces, we have used the PCA reconstruction [5] for dimensionality reduction along with the snapsort method. The images taken from a standard image database were segregated into six different sets each representing one of the six universal classes. Then the Eigenspaces of each class have been computed. Figure 9.1 shows the six universal expressions. The test image is classified as belonging to the particular class to which it has the most similarity. Before processing the images, they are slightly cropped to extract only the centre face and leave out the fringe parts like ears, hair, etc. which do not take part in conveying expression.

9.5 Working with Eigenspaces

In the Eigenspaces approach for expression recognition, one possibility is to calculate the feature vectors of each facial expression from a labeled database of different persons—which is how the face detection method works. Project a test image onto the Eigenspace of the training set and select the closest matching projection—the class of the corresponding image being the class of the test image. However the difference here is that the person whose facial expression needs to be classified is unknown. The same expression may vary across different persons, but we should be able to still classify a smile as a smile.

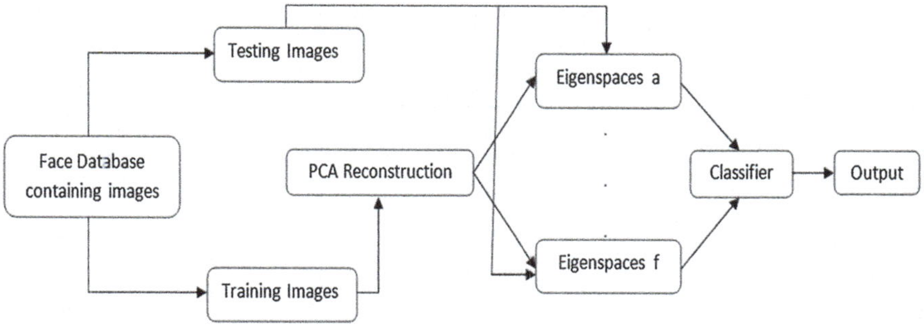

Fig. 9.2 General process flow

In order to deal with this problem, we have modified the original Eigenspace method. A separate subspace is now formed for each expression class instead of having a common overall Eigenspace. Now instead of projecting each image into the common Eigenspace, we project it onto each of the six subspaces corresponding to the six universal expressions. The distance between the projection of the test image and that of each of the images of the six classes is calculated—the minimum distance determining the closest match. A major difference between this method and the original approach is that in the original method the distance within the same subspace is considered whereas here it is the distance between the test image vector and the vector subspace of each class of expression. The general process flow is shown in Fig. 9.2.

9.6 Result Analysis

Our results have been tested using the standard JAFFE database [6] of facial expressions. This database contains 213 images seven facial expressions posed by ten Japanese female models.

We have used one image for each expression of each model for our training set which thus comprises of 60 images. For testing we have used 120 images (20 test images per class) other than those used for training. Table 9.1 shows the result in terms of percentage of images correctly recognized by each of the three classifiers [7] for each of the six expressions as well as the overall match.

The performances of the distance classifiers are very close to each other with Manhattan distance giving the best result. Euclidean and Cosine metrics have given the same overall result although there are subtle variations within the different expression classes.

Analysis of the result suggests that Anger is the best recognized class of expression followed by surprise and fear. Sad and happy are the least recognized expressions which may be because sad is difficult to pose for and for happiness it

Table 9.1 Accuracy for each class using our proposed method

Expr. class	Accuracy in % for each classifier			
	Euclidean	Cosine	Manhattan	Mahalanobis
Anger	95	95	95	95
Disgust	80	80	80	80
Fear	90	85	90	90
Happy	80	75	80	80
Sad	75	80	80	75
Surprise	85	90	90	85
Overall	84.16	84.16	85.83	84.16

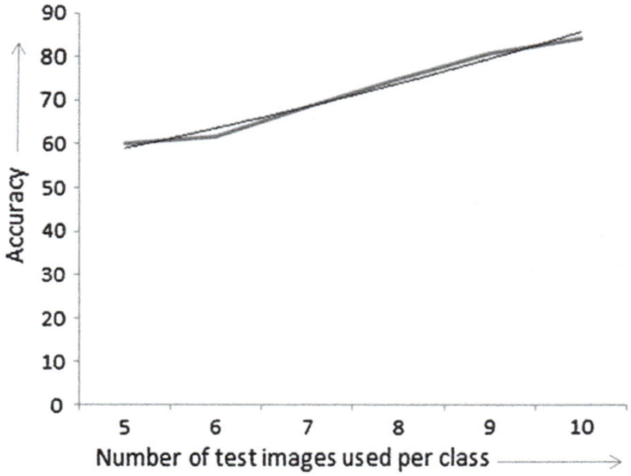

Fig. 9.3 Graph of number of training images used versus accuracy in %

involves largely varying degrees of smile. Disgust fear and sad are sometimes confused with each other. Also happy and surprise are confused amongst each other mainly due to open mouth in both cases.

While performing our analysis we observed that the accuracy is directly proportional to the total number of samples used, that is the greater the number of training images used per class greater is the accuracy. This behavior is graphically represented by Fig. 9.3 where the blue line is the direct plot obtained and the black line shows the trend.

An explicit analysis has also been made on the representation of the reconstructed data of the Eigenspaces and its variation over the number of non-zero eigenvalues taken into account. We varied the number of non-zero eigenvalues used and observed that the best results are obtained when we take into account all available values. But as we gradually decrease the lower valued eigenfaces the performance remains very close to maximum for a very large range. The entire behavior is depicted by the graph in Fig. 9.4. This observation gives us the opportunity to further optimize the process by ignoring about two thirds of the eigenvectors from our processing yet obtaining almost similar results.

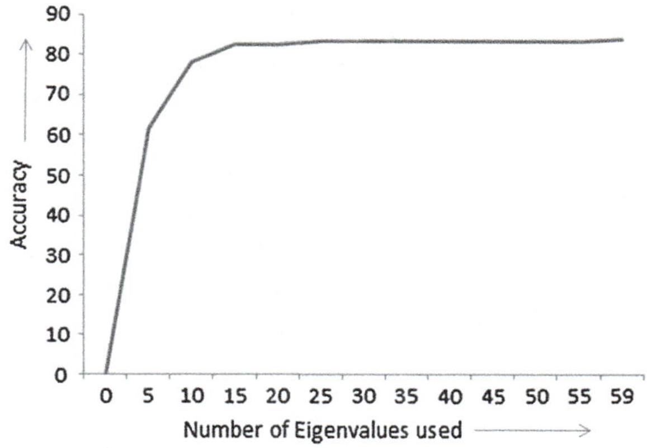

Fig. 9.4 Graph of number of eigenvalues used versus accuracy in %

9.7 Conclusion

In normal human interaction between two individuals it is not only the visual image of the faces but also various other factors like gesture, body language, tone of speech, previous knowledge about the person, etc. which come into play. To get close to this, the modern human computer interfaces will have to be able to capture and analyze all this available information. This is quite a challenge and thus also a scope for future development and progress in this field. This paper focuses on the capturing of facial expression information from still images using a modified recognition method providing good accuracy as evident from the experimental results. Observations have been performed extensively on the result and a way to further reduce the dimensionality of the problem has been discussed.

References

1. Turk M, Pentland A (1991) Eigenfaces for recognition. J Cogn Neurosci
2. Kumbhar M, Jadhav A, Patil M (2012) Facial expression recognition based on image feature. Int J Comput Commun Eng 1(2)
3. D'Mello SK, Picard RW, Graesser AC (2007) Towards an affect-sensitive autotutor. IEEE Intell Syst, Spec Issue Intell Educ Syst 22(4)
4. Murthy GRS, Jadon RS (2007) Recognizing facial expressions using eigenspaces. In: Proceedings of IEEE international conference on computational intelligence and multimedia applications, Dec 2007, Sivakasi, Tamilnadu, India
5. Kaur M, Vashisht R, Neeru N (2010) Recognition of facial expressions with principal component analysis and singular value decomposition. Int J Comput Appl (0975–8887) 9(12):36–40
6. Lyons M, Kamachi M, Gyoba J (1997) Japanese female facial expressions (JAFFE). Database of digital images
7. Shih FY, Chuang CF, Wang PSP (2008) Performance comparisons of facial expression recognition in JAFFE database. Int J Pattern Recognit Artif Intell 22(3):445–459

Chapter 10
A Face Recognition System Based on Back Propagation Neural Network Using Haar Wavelet Transform and Morphology

Krishna Gautam, Nadira Quadri, Abhinav Pareek and Surendra Singh Choudhary

Abstract Today it is necessary to design an efficient security system which can protect unauthorized access on any system using extremely secure and excellent system for face recognition. In this paper, a robust face recognition system approach is proposed for image decomposition using Haar wavelet transform, feature detection using Successive Mean Quantization Transform (SMQT) and splipt up sparse network of window (SNoW) classifier and after detected face is sent for feature extraction using gray-scale morphology then extracted feature is sent for recognition using Backpropagation neural network which provide verification of face images. Average recall rate of up to 98.5 % for the database of 200 images. The efficiency of the proposed system obtained as 98.5 %. In This paper we use MATLAB to detect and recognize the respective face.

Keywords Back propagation neural network (BPNN) · Haar wavelet transform · Face recognition · SMQT features · Snow classifier · Gray-scale morphology

K. Gautam (✉) · N. Quadri (✉) · A. Pareek (✉) · S. S. Choudhary (✉)
Department of Computer Engineering, Engineering College Bikaner, Bikaner, Rajasthan, India
e-mail: krishnagautam3@gmail.com

N. Quadri
e-mail: nadira.quadri88@gmail.com

A. Pareek
e-mail: ab199002@gmail.com

S. S. Choudhary
e-mail: surendra2060@gmail.com

S. Sengupta et al. (eds.), *Emerging Trends in Computing and Communication*, Lecture Notes in Electrical Engineering 298, DOI: 10.1007/978-81-322-1817-3_10,

10.1 Introduction

Face image is a biometrics physical feature that is used to verify the identity of people. The important components involved in the face image space include mouth, nose, and eyes. The goal of this work is to develop an efficient, real-time face recognition system that would be able to recognize a person as soon as he/she will be in front of camera.

In this paper, firstly captured image from camera are decomposed into regions by haar wavelet transform then we used a framework for face detection that is proposed using illumination insensitive features gained from the local Successive Mean Quantization Transform (SMQT) features and rapid detection is achieved by the splipt up sparse network of window (SNoW) classifier. Then we use morphological operation which is used as a tool for extracting image components. It is about adding and removing pixels from a binary image according to certain rules depending on neighborhood patterns. Dilation, Erosion, Closing, and Opening are the more common morphological operations. After that we use median filter to remove the paper or salt noise. During back-propagation neural network training process, Back Propagation Neural Network (BPNN) learning algorithm adjusts the weights (knowledge base) and bias of each of the neurons. The layers contain identical computing neurons associated such that the input of every neuron in the next layer receives the signal from the output neuron in the previous layer. We use Olivetti Research Laboratory (ORL) database in this approach and we compare ORL database recognition rate with respect to local database recognition rate. The recognition rate was approximately 98.5 and 94 % respectively.

10.2 Proposed Approaches

1. Image Acquisition from camera.
2. Face are detected using Local SMQT Features and Split up SNoW Classifier.
3. Face Recognition:
   (a) Facial features are extracted using Gray-Scale Morphology
   (b) Classification process using BPNN.

Since the output of each step is the input to the next, the functional parts must execute in sequence Fig. (10.1).

10.2.1 Image Decomposition Using 2-Level Haar Wavelet Transforms

The Discrete Wavelet Transform includes a technique Haar wavelet transform which is used to decompose a grayscale image into four regions. These regions are

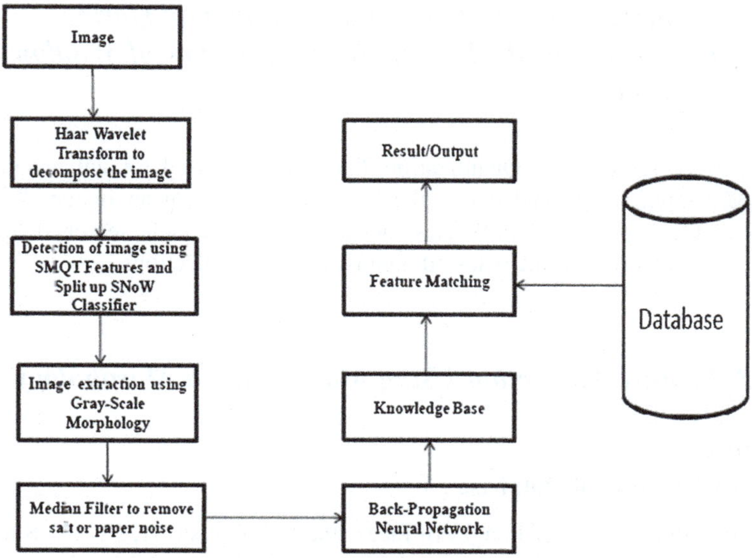

Fig. 10.1 Flow chart of the whole system

horizontal high frequencies region, vertical high frequencies region, dialog high frequencies region, and approximate region that contains minimum frequencies feature. Now at next decomposition level, image's upper left corner is further festered into four smaller regions.

Algorithm

1. Firstly we choose an input image.
2. Then we define the number of decompositions and calculate values of the the Horizontal detail coefficient storage, the Vertical detail coefficient storage, the Diagonal details coefficient storage and Approximation coefficient storage.
3. After that 2-level decompositions are applied here.
4. Now we rescale an input matrix to a particular range for display.
5. Then we convert to the outcome of above step into unsigned 8-bit integer to display.

The 2×2 Haar matrix that is associated with the Haar wavelet is

$$H = \begin{bmatrix} 1 & 1 \\ 1 & -1 \end{bmatrix}$$

10.2.2 Local Successive Mean Quantization Transform Features and SPLIPT up Sparse Network of Window Classifier

SMQT (Successive Mean Quantization Transform) is used for enhancement of gray-scale input image and detect the facial features from an image. With the SNoW and the splipt up SNoW classifier, a pertained a table searched for face. Every each detection result is tested against all other detections.

10.2.3 Feature Extraction Using Mathematical Morphology

Algorithm

Step 1 Creating of Database

1. We coordinate a set of images of faces (the training set S1, S2, …, Sm).
2. Now we reshape the 2D images of the training database into 1D column vectors. After that, we put these 1D column vectors in a row to construct 2D matrix "S".
3. Then we allocate numbers to all the images in the training database.
4. Now we construct a 2D matrix from 1D image vectors.
5. After that we choose the image name in database as a corresponding number and reshape the 2D image into 1D image vectors and the 2D matrix "S" grows after each turn.

Step 2 Calculate feature vector

1. We use mathematical morphology to determine the important discriminating features between face images.
2. In its first step, we perform hit-miss transform that means a combination of gray-scale erosion and dilation.
3. The gray-scale dilation of f by structuring element b, denoted f $\oplus$ b, is denoted as

$$(f \oplus b)(x,y) = \max\{f(x-x',\, y-y') + b(x',y')|(x',y') \in D\}$$

where D is the domain of b and f(x, y) is assumed to equal $-\infty$ outside the domain of f. In practice gray-scale dilation usually is performed using flat structuring elements. In this situation the value (height) of b is 0, and the simplified gray-scale dilation is

$$(f \oplus b)(x,y) = \max\{f(x-x',\, y-y')|(x',y') \in D\}$$

4. The gray-scale erosion of f by structuring element, denoted $(f \ominus b)$ is defined as

$$(f \ominus b)(x,y) = \min\{f(x+x',\, y+y')|(x',y') \in D\}$$

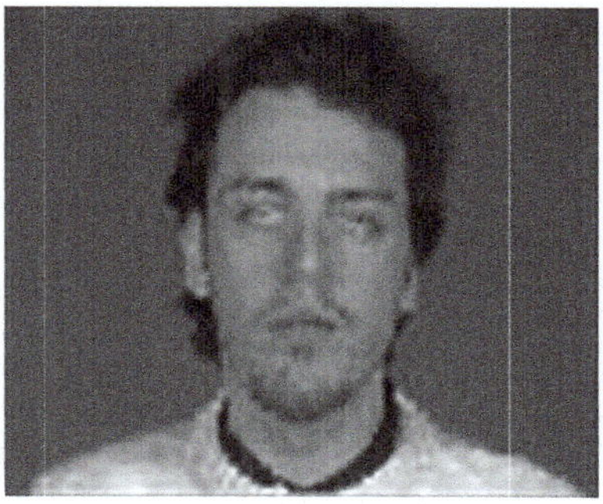

Fig. 10.2 **a** Original image

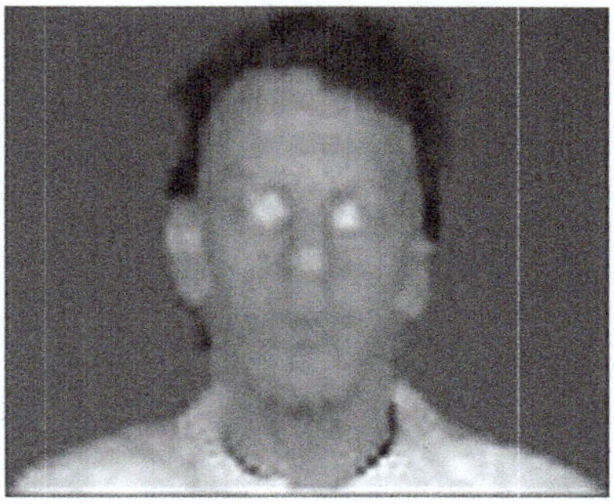

Fig. 10.2 **b** Dilated image

where D is the domain of b and f(x, y) is assumed to equal $+\infty$ outside the domain of f. In practice gray-scale dilation usually is performed using flat structuring elements. In this situation the value (height) of b is 0, and the simplified gray-scale dilation is

$$(f \ominus b)(x, y) = \min\{f(x + x', y + y') | (x', y') \in D\}$$

Dilation and erosion can be combined to achieve a result. For instance, subtracting an eroded image from its dilated version produces a "morphological gradient" which is a measurement of local gray-scale level variation in the image:

$$\text{Morphological gradient} = (f \oplus b) - (f \ominus b)$$

Thus we find extracted feature result in 1D. Now it is preceded for filtering Figs. (10.2a, b and 10.3a, b).

Fig. 10.3 **a** Original image

Fig. 10.3 **b** Eroded image

10.2.4 Median Filtering

Now we apply median filtering process which is a nonlinear process useful in reducing impulsive or salt-and-paper noise. In a median filter, a window slides along the image being preceded. All type noising errors are removed from above outcome.

10.2.5 Face Recognition Using Back Propagation Neural Networks

BPNN is a multilayered feed forward Artificial Neural Network. BPNN learning algorithm manages the weights (knowledge base) and bias of each of the neurons during its training process.

First of all according to random value range from −1.0 to 1.0 we set the value of all weights.

1. Then we set an input binary values pattern into the neurons of the input layer of network.

Table 10.1 The results of face recognition

Techniques	Test cases	Recognized face images	Unsuccessful face images
Local database	200	188(94 %)	12(6 %)
ORL database	200	197(98.5 %)	3(1.5 %)

2. At this step, we adjust each and every neuron of the following layer in active mode:
   (a) We multiply this neuron with the output values of the preceding neurons to weight (knowledge base) values of the connections leading and we summarize these values.
   (b) Now this result is passed to an activation function, which calculates the output value of the neuron.
3. After that we repeat all this procedure until the final output layer is achieved in this methodology.
4. Now we compare the desired objected pattern to the computed output approach pattern and calculate value of square error.
5. Next we change the values of all weight of each weight using the formula

Learning Rate * Output Error * Output (Neuron j) * Output (Neuron j + 1) * (1 − Output) + Weight (old)

(Neuron j + 1) where Weight (old) is previous layer weight, Learning Rate where database is being trained, Output Error is the error which occurs in previous steps.

6. After that go to the step 1.
7. Finally if these all output patterns are matched with their desired outcomes then the algorithm ends.

10.3 Result and Discussion

At first to check the reliability of the system we use ORL database for testing the system. As well as our locally created database,where the recognition rate was approximately 98.5 and 94 % respectively. For the online purpose we strictly try to reduce recognition rather than unrecognizing. Due to the use of very high threshold (0.98) on the test output the chance of appearing false recognizing result is reduced but the rate of showing unrecognizing result was going high Table (10.1).

10.4 Conclusion

In this paper, we have proposed a technique for designing fast, secure and robust face recognition system. This technique reduces the time required to recognize an image to decompose it into 2-level sub images bands. Then we apply morphology for extracting feature vector. And finally BPNN is used for image classification, training and recognition. So this combined approach develops a more accurate approach compared to the existing techniques. It gives the better performance for recognition and makes a more secure, reliable and robust face recognition system.

References

1. Agarwal P, Prakash N (2013) An efficient back propagation neural network based face recognition system using haar wavelet transform and PCA
2. Podder PK, Sarker DK, Kundu D (2012) Real-time face recognition system based on morphological gradient features and ANN. Glob J Res Eng 12(2-A)
3. Sivanandam SN, Paulraj M (2003) Introduction to artificial neural networks
4. Nilsson M, Nordberg J, Claesson I (2007) Face detection using local SMQT features and Split up SNoW classifier. In: IEEE international conference on acoustics, speech and signal processing (ICASSP)
5. Rowley H, Baluja S, Kanade T (1996) Neural network-based face detection. In: Proceedings of Computer Vision and Pattern Recognition, June 1996, pp 203–208
6. Roth D, Yang M, Ahuja N (2000) A snow-based face detector. In: Advances in neural information processing system 12 (NIPS 12). MIT Press, Cambridge, pp 855–861
7. Saudagare PV, Chaudhari DS (2012) Efficient face recognition system using artificial neural network. Int J Comput Appl 41(21):12–15
8. Adebayo Daramola S, Sandra Odeghe O (2012) Facial expression recognition using neural network—an overview. Int J Soft Comput Eng (IJSCE) 2(1):224–227
9. Benzaoui A, Bourouba H, Boukrouche A (2012) System for automatic faces detection. In: Proceedings of the image processing theory, tools and applications, IEEE
10. Bhatt HS, Bharadwaj S, Singh R, Vatsa M (2013) Recognizing surgically altered face image using multi-objective evolutionary algorithm. IEEE Trans Inf Forensics Secur 8(1):89–100
11. Yu H, Yang J (2001) A direct LDA algorithm for high-dimensional data with application to face recognition. Pattern Recogn 34:2067–2070
12. Barrlett MS, Movellan JR, Sejnowski TJ (2002) Face recognition by independent component analysis. IEEE Trans Neural Netw 13(6):1450–1464

Chapter 11
Video Watermarking Scheme Resistant to Rotation and Collusion Attacks

Amlan Karmakar, Amit Phadikar and Arindam Mukherjee

Abstract In this paper, a blind watermarking algorithm is proposed for Moving Picture Experts Group 4 (MPEG-4) videos, which is perceptually invisible and robust against rotation and collusion attacks. The goal is achieved by (1) embedding the watermark in discrete cosines transform domain (DCT) of luminance channel. The rotation invariance property of Complex Zernike moments is exploited to predict the rotation angle of the video at the time of extraction of watermark bits, (2) design of the scheme is done in such a way that the embedding blocks will vary for the successive frames of the video. A pseudo random number (PRN) generator and permutation vector are used to achieve the goal. This makes the scheme robust to collusion attack. Simulation results have shown the validity of the above claims.

Keywords Video watermarking · Complex zernike moments · Rotation attack · Collusion attack

11.1 Introduction

In recent times, MPEG-4 video standard has been found widespread applications in internet streaming, digital High-definition (HD) handy-cams as well as in mobile phones. So many researchers are working in the field of digital MPEG-4

A. Karmakar · A. Phadikar (✉)
Department of Information Technology, MCKV Institute of Engineering Liluah, Howrah 711204, India
e-mail: amitphadikar@rediffmail.com

A. Karmakar
e-mail: amlan.karma@gmail.com

A. Mukherjee
Institute of Computer Engineers, Techno India Group, Kolkata 700071, India
e-mail: arindam26509580@gmail.com

S. Sengupta et al. (eds.), *Emerging Trends in Computing and Communication*, Lecture Notes in Electrical Engineering 298, DOI: 10.1007/978-81-322-1817-3_11,

video watermarking [1]. At the time of developing a video-watermarking algorithm, the researchers should concentrate on the two most important things, i.e. imperceptibility and robustness. In the most of the video-watermarking schemes it is seen that the concentration of robustness is given mainly onto temporal attacks like frame dropping, frame inserting, and frame rate changes etc. [2, 3]. While not too many work has been done yet to achieve the robustness against geometrical attacks [1, 4] especially rotation of video frames in any random angle. Zernike moments have a rotation invariance property. One can find out the angle of rotation from the phase information of the Zernike moments. So it is widely used in the field of image as well as video watermarking.

Besides temporal and geometrical attacks, another type of attack for watermarked video is collusion attack. In some situations, it is possible for an attacker to obtain multiple watermarked data. The attacker can often exploit this situation to remove watermarks without knowing the watermarking algorithm. This kind of attack is known as collusion attack. There are two types of collusion attack i.e. Type 1 and Type 2 [5, 6]. The collusion attack is a different kind of attack and very less video watermarking scheme can resists against collusion attack [5, 6].

In this paper, a blind MPEG-4 video watermarking algorithm is proposed in DCT domain. The Complex Zernike moment is used to make the scheme robust against rotation attack. At the same time, the design of the watermark embedding algorithm is made in such a way that robustness is also achieved against collusion attacks. The superiority of the proposed scheme is verified by simulation results and compared with selected other methods.

11.2 Proposed Scheme

The scheme is divided into two major parts i.e. watermark embedding and watermark extraction.

11.2.1 Watermark Embedding

Step 1: Extract frames from the original video one by one.

Step 2: Change the color model of the extract frame from RGB (red, green, and blue) to YC_bC_r. In RGB color space the perceived color quality of a video frame is dependent on all components. Thus, embedding watermark bits into one component independently of the other RGB components is not the best choice. On the other hand, the YC_bC_r permits to extract uncorrelated components and it favor the separation of the achromatic part from the chromatic parts of the color image.

Step 3: A square block of size (P × P) is chosen in each luminance component's center which is considered as the target embedding area. The change in the intensity of chrominance components is the most sensitive to human eyes whereas for luminance components are least sensitive. Thus, the proposed scheme uses the luminance component for embedding the watermark. Moreover, data embedding into the central area of the frame makes the scheme resistant against cropping. The block (P × P) is divided into non-overlapping sub-blocks of size (8 × 8).

Step 4: Then Q number of distinct (8 × 8) blocks are selected pseudo-randomly, where the watermark information is to be embedded.

Step 5: Apply 2D DCT on each selected blocks (8 × 8). This is due to the fact that MPEG-4 uses DCT transformation. To make the proposed scheme, compliant with MPEG-4 coded we have used DCT domain for data embedding.

Step 6: Select n no. of AC components pseudo randomly for each DCT blocks (8 × 8). The modification of AC components is done according to the following rule:

```
if (W(k)==1)
   if (mod(C(i,j),δ)≤α)
          Cw(i,j)  = C(i,j)-mod(C(i,j),δ)-α
   else
          Cw(i,j)= C(i,j)-mod(C(i,j),δ)+γ
   endif
  elseif (W(k)== -1)
   if (mod(C(i,j),δ)≥γ)
          Cw(i,j)=C(i,j)-mod(C(i,j),δ)+ε
   else
          Cw(i,j)= C(i,j)-mod(C(i,j),δ)+α
   endif
endif
```

where, W(k) is the watermark bit to be embedded, C(i, j) is the original AC component, $C_w(i, j)$ is the watermarked AC components. Here, α, β, γ and δ are the embedding strength and considered as global constants. The relations between those global constants are $\beta = 2\alpha$, $\gamma = 3\alpha$, $\delta = 4\alpha$ and $\varepsilon = 5\alpha$. The values are taken based on the large number of experimentation.

Step 7: Compute the Zernike moments ($Z_{m,\ n}$) for some specific values of 'm' and 'n' and save those values as keys [7].

Step 8: Repeat Step 1–6 until all the video frames are considered.

Step 9: Convert YC_bC_r to RGB and merge all frames to construct the watermarked video.

11.2.2 Watermark Extraction

Step 1: Frames are extracted from the watermarked video one by one.
Step 2: Change the color model of the extract frame from RGB to YC_bC_r.
Step 3: Compute the Zernike moments $\left(Z^R_{m,n}\right)$ of the rotated watermarked video frame (Y channel) for the same values of 'm' and 'n' that was used during embedding. Then angle of rotation is detected and frames are reverse rotated for geometrical restoration [7].
Step 4: A square block of size (P × P) is selected from the luminance component's center which was used as the embedding area. Then the block is divided into non-overlapping sub-blocks of size (8 × 8).
Step 5: Same pseudo random number (PRN) generator is used to select the same Q number of (8 × 8) blocks, where the watermark information was embedded.
Step 6: Apply 2D DCT on each selected blocks (8 × 8).
Step 7: Select the same n no. of AC components pseudo randomly which was modified at the time of embedding and the watermark bit is detected according to the following rule:

if (mod $(C_{w(i,j),\delta)>\beta))}$

```
W(k)=1
else
W(k)=-1
endif
```

where, $W(k)$ is the extracted watermark bit and $C_w(i, j)$ is the watermarked AC components.
Step 8: Repeat Step 2–7 until all watermark bits are extracted.

11.3 Performance Evaluation

Two standard MPEG-4 videos viz. "akiyo.mp4", and "foreman.mp4" are used for the experiment. The size of binary logo is (50 × 25). During simulation, we have used 150 frames of each video. The size of the square luminance block, where the watermark is embedded is (176 × 176). This study uses the peak–signal-to-noise-ratio (PSNR) and the mean-structure-similarity-index-measure (MSSIM) [8] as distortion measures for the watermarked video frame, whereas the relative entropy distance (Kullback–Leibler distance) [9] is used as measure of security (ε). Figure 11.1a, b show one of the original and watermarked video frame along with the PSNR, MSSIM and ε values. Figure 11.1c shows the original watermark image, while Fig. 11.1d shows the extracted watermark from all the watermarked video.

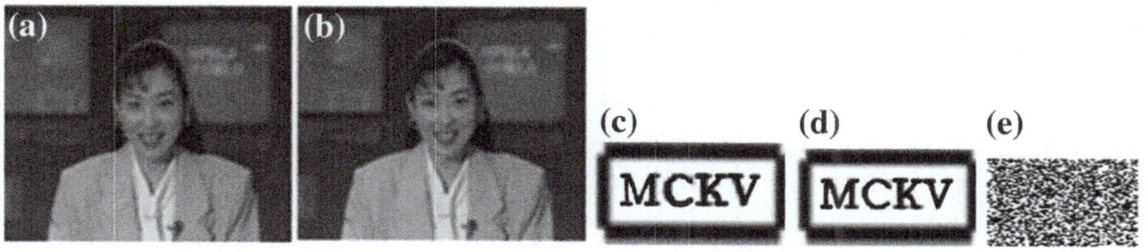

Fig. 11.1 **a** Original frame of Akiyo. **b** Watermarked frame of Akiyo. **c** Original watermark image (50 × 25). **d** Extracted watermark (50 × 25) from all watermarked video with NCC = 1. **e** Extracted watermark using fake key. (P, M, ε) above each image represents the PSNR (in dB), MSSIM and security values of the watermarked video frame. (P : 43.99, M : 0.98, $\varepsilon =$ 0.0011)

Table 11.1 Robustness against rotation attack

Rotation angle	Akiyo	Foreman
10°	0.992	0.978
20°	0.976	0.989
45°	0.996	0.985
60°	0.987	0.974

Table 11.2 Robustness against collusion attack of Type-1

No. of averaged frame		2	5	7	10
Akiyo	NCC	0.996	0.956	0.885	0.810
Foreman		0.994	0.954	0.884	0.806

Table 11.3 Robustness against collusion attack of Type-2

Videos		Extracted using PV1	Extracted using PV2
Akiyo	NCC	1.000	0.5536
Foreman		0.5464	1.000

PV1 permutation vector 1 (used in the video—Akiyo). *PV2* permutation vector 2 (used in the video—Foreman)

Robustness of the proposed scheme is shown in Tables 11.1, 11.2, and 11.3. The high normalized cross correlation (NCC) [7] values in Tables 11.1, 11.2, and 11.3 depict that the scheme is robust to frame rotation and collusion attacks.

The robustness performance of the proposed method is also compared with previously reported works [1, 4, 5] to demonstrate the performance comparison. It is observed from the results of Fig. 11.2a, b that the proposed method offers better gain in term of NCC over other methods.

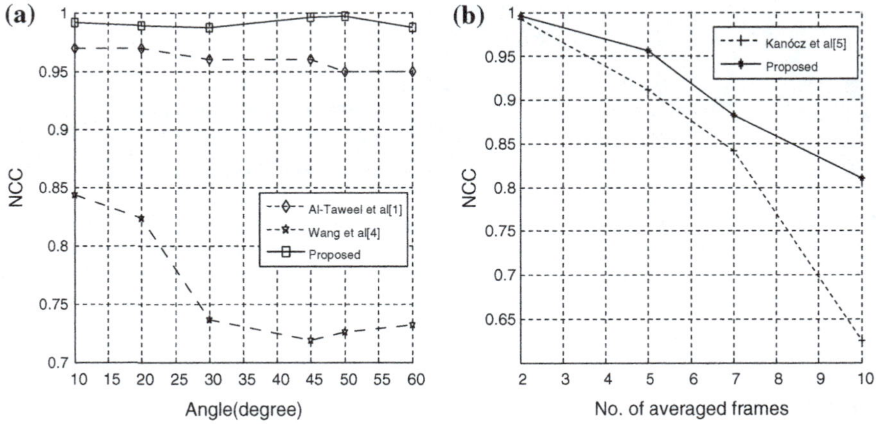

Fig. 11.2 Comparative performance: **a** rotation attack, **b** collusion attack

11.4 Conclusion

In this paper, a DCT based rotation and collusion attack resistant blind MPEG-4 video watermarking technique is proposed. The experimental results show that the scheme provides robustness against rotation of video in any angle, collusion attacks of Type-1 and Type-2. However, the robustness against other geometrical attacks viz. scaling, translation etc. is yet to be proved and kept as future plan.

References

1. Al-Taweel SAM, Sumari P, Alomari SAK (2010) Robust video watermarking algorithm using spatial domain against geometric attacks. Int J Comput Sci Inf Secur 8(2):51–58
2. Chao C, Tie-gang G, Li-zong L (2008) A compressed video watermarking scheme with temporal synchronization. In: IEEE congress on image and signal processing, China. pp 605–612
3. Al-Taweel SAM, Sumari P, Alomari SA, Hussain AJA (2009) Digital video watermarking in discrete cosine transform domain. J Comput Sci 5(8):536–543
4. Wang Z, Ye X, Xiao N (2008) Robust watermarking based on norm quantization singular value decomposition and zernike moments. In: IEEE Pacific-Asia workshop on computational intelligence and industrial application. pp 1005–1008
5. Kanócz T, Tokár T, Levický D (2009) Robust frame by frame video watermarking resistant against collusion attacks. In: IEEE international conference on radioelektronika, Bratislava. pp 99–102
6. Saxena V, Gupta JP (2007) Collusion attack resistant watermarking scheme for colored images using DCT. Int J Comput Sci (IAENG) 34(2):1–7
7. Guo-juan X, Rang-ding W (2009) A blind video watermarking algorithm resisting to rotation attack. In: IEEE international conference on computer and communications security, Hong Kong. pp 111–114

8. Wang Z, Bovik AC, Sheikh HR, Simoncelli EP (2004) Image quality assessment: from error measurement to structural similarity. IEEE Trans Image Process 13:1–14
9. Maity SP, Nandy P, Das TS, Kundu MK (2004) Robust image watermarking using multiresolution analysis. In: India annual conference, 2004. Proceedings of the IEEE INDICON 2004, pp 174–179

Chapter 12
Development of Crossover and Encryption Based Text Steganography (CEBTS) Technique

Tarun Kumar, Abhinav Pareek, Jyoti Kirori and Maninder Singh Nehra

Abstract This paper presents new text steganography approach named as Crossover Encryption Based Text Steganography (CEBTS) approach. This approach is a combination of random character sequence and feature coding method. In this approach in order to hide the secrete message, we first encrypt the message and then mix it in the cover text. Cover text is generated by random character sequence. In this approach the key is generated from the very first three character of the cover text by using crossover technique between these three characters. This generated key will be used to encrypt the plain text and this key will also used to decide that how the encrypted message will hide with cover text. For this approach we have developed two algrithms, one is for key generation using crossover technique and then encrypt the message using this key and hide the encrypted message into cover text using this key and second is used to retrieve the secrete message. We are also presenting comparison of our proposed approach with some of the previous popular text steganographic approaches with execution time and also the length of cover text which will be required to encipher using n bit cover text. At the last in our conclusion section we are showing that how our approaches are performing best in the existing approaches.

Keywords Crossover · Encryption · Feature encoding · Steganography · Genetic algorithms

T. Kumar (✉) · A. Pareek (✉) · J. Kirori (✉) · M. S. Nehra
Department of Computer Engineering, Government Engineering College Bikaner, Bikaner, India
e-mail: ertarunkumar@yahoo.co.in

A. Pareek
e-mail: ab199002@gmail.com

J. Kirori
e-mail: jyotikirori8@gmail.com

M. S. Nehra
e-mail: maninder4unehra@yahoo.com

S. Sengupta et al. (eds.), *Emerging Trends in Computing and Communication*, Lecture Notes in Electrical Engineering 298, DOI: 10.1007/978-81-322-1817-3_12,

12.1 Introduction

Steganography is the art and science of writing hidden messages in such a way that no one, apart from the sender and intended recipient, suspects the existence of the secret message. Steganography algorithms uses cover media such as image, text, audio and video etc. to hide the secret data. User relies on change in the structure of these mediums and features of the target medium in such a manner as is not identifiable by human. However, using the text as the target medium is relatively difficult as compared to the other target media. This difficulty is generally observed because of the lack of redundant information in a text file, as compared to an image or a sound clip which contains a lot of redundancy that is exploited by the steganography algorithms. Cryptography and Steganography are two ways to secure data transfer over the Internet Whereas Cryptography technique is used to protect the contents of message using encryption which will send it to the intended recipient and Steganography can be used to hide the existence of a message using cover media. Cryptography techniques have some limitation as the third party is always aware of the communication because of the unintelligible nature of the text. So any third party can fetch the original content from encrypted text by using some techniques as it is aware of the communication. Whereas steganography overcomes this limitation by hiding message in an innocent looking object called cover media which can be text, image, video etc. In the steganography techniques plain text should be hide into cover media in such manner that structure and feature changes would not be identified just by looking through human eyes. However, text medium is relatively difficult as compared to other target media because of lack of available redundant information in text data. In this paper, we present an overview of our proposed text steganography method and various existing text-based steganography methods. Our main target for any steganography approach should be that the cover media should be minimum to hide any data and timing overhead also should be very less as it would not take much time for encryption or decryption rather in transfer data. In the next section we are showing our proposed approaches algorithms for key generation using crossover techniques, message hiding, message retrieving. After that we showed and example for better understanding of our approaches. Finally at the last we have showed the experimental result of our technique and also conclude that how our proposed approaches is far better than all existing approaches using random cover media.

This paper presents new text steganography approach named as Crossover Encryption Based Text Steganography (CEBTS) approach. This approach is a combination of random character sequence and feature coding method. In this approach to hide the secrete message, we first encrypt the message and then mix it in the cover text. Cover text is generated by random character sequence. In this approach the key is generated from the very first three character of the cover text by using crossover technique between these three characters. The generated key will be used to encrypt the plain text and this key will also used to decide that how the message will be hide with cover text. For this approach we have developed three algorithms one for key generation using crossover technique, second is for hiding the message in cover media using generated key and third is used to retrieve the secret message. We are showing the comparison of our approaches with existing approaches with graph in which we can conclude that how our approach giving best result in form of cover text length and timing overhead.

12.2 Related Work

In Steganography, when our aim is used to hide some significant data in a document to protect it, it is necessary to use some other redundant data as a cover media for the existing valid data. The probable media that can be used as a cover can be text, image or a movie clip or text file. Out of these different media files, a text file normally uses much more as compare to other cover media as text file contains lesser storage so we can hide much more data using text media. In addition, this process needs less cost for printing, too as it takes less space to hide more information. On the other hand, images or video file is not much use as cover media because the structure of these cover media is different than what we actually see, because more attributes about the data is required to be stored. This makes, handling such files more complicated as compared to handling a text file. Because of this simplicity and fast processing text data is always preferred for being used in Steganography, such an approach is known as Text Steganography. Steganography sometimes is used when encryption is not allowed. Or, more commonly, steganography is used to supplement encryption. An encrypted file may still hide information using steganography, so even if the encrypted file is deciphered, the hidden message is not seen have original text then it is very easy to disclose the method of steganography just by comparing these two document files Text steganography can be broadly classified into three types: Format based Random, Statistical generation, and Linguistic methods.

In Text steganography there are many methods are available. Some methods change the format of text while some other methods change the actual word to hide secret data as per cover text. In format based methods we alter physically the format of the text to conceal the information. In an Open Space method and white space is used to hide our secret data. Some other methods available which uses white spaces for hiding secret message are Inter-Sentence space methodand End-of-line space method and Inter-word space method. These methods are not used much as data would be lost if incase format of file is changed. Then other method is Semantic method which is used to hide secret message by changing actual words. In this method Synonyms are used to hide secret data such as 0 and 1.

Where as Syntactic method uses punctuations to hide secret data like commas(,) and full stop(.). This method is better than previous one but this would change meaning of our original text when we have more punctuation in our text file. In an acronyms and semantic method, meaning of information can be changed because these methods uses actual word replacement or punctuation method to hide secret data. After this technique a new terminology comes which uses characteristics of that particular language to hide data such as in Persian/Arabic Text steganography and Hindi Text steganography.

Another method for concealing information is the one which uses random looking sequence of characters. In another method, the statistical properties of word length and letter frequencies are used in order to create words which will appear to have same statistical properties as actual words in the given language.

Linguistic steganography is another method of steganography that specifically considers the linguistic properties of generated and modified text, and in many cases, uses linguistic structure as the space in which messages are hidden.

In semantic method, secret information is hidden by changing actual words by their synonyms. In this method, synonyms of words are use to hide secret data. Actual word can be used to hide "1" bit while synonyms of that word can be used to hide "0" bit. This method is better than other method because data can survive in case of retyping text or re-formatting text. Problem also occurs in this method because of word changing

So we have proposed a new approach which is far better than other existing approaches. In the next section we will describe our approach with implementation details that how our method works. We will also show that our proposed approaches take very less time overhead and memory overhead as compared to existing approaches. Also we can hide more number of bytes using proposed approach. Required cover text size is also very small in proposed approach which will be equal to our original text.

12.3 Proposed Approach

In Crossover Encryption Based Text Steganography (CEBTS) approach we are using a key to encrypt our secret message and the same key will be used to choose hiding method into the cover text. Our first step is to generate a key from random cover text. We are generating our key using crossover technique between the first three characters of our cover text. Firstly we will use 3×5 crossover techniques between first and second characters of our random cover text which will generate a key let us assume it's key1 then next we will apply 5×3 crossover techniques between key1 and third character of random cover text. This new generated key will be used to encrypt our secret message using simple $\times$-Or Operation and number of

1's in the generated key will decide that how we have to mix up our encipher data between cover text. If we have x no. of one's in our generated key then we will hide the encrypted data after every x characters of our random cover text.

This method is describe in below implementation section. In which we have described two algorithms one for encryption and another for decryption. We have also showing an example for better understanding of our algorithms.

12.4 Implementation

For implementation of our proposed approaches we have developed three algorithms for generating key, hiding plain text into cover media and unhide our original text from received message. We have implemented this techniques in Java and tested time overhead and space overhead using java profiler.

12.4.1 Pseduo Code for Key_Generation

Algorithm for Hiding Secrete Message

```
INPUT: A secrete message msg and cover text cvt.
OUTPUT: final hidden message finalmsg.
1. First read three characters from cvt as ch1,ch2,ch3.
2. Compute 3X5 bit wise crossover between ch1 and ch2 as
cross1.
3. Compute 5X3 bit wise crossover between cross1 and ch3
as key.
4. Encrypt the msg with key using EX-OR operation as
encyrptmsg.
5. Compute the stegno key from key by counting on-bits as
pos.
6. For each character in encryptmsg do
J=pos;
Read one character from encryptmsg;
Place this character in cvt after jth position;
J=j+pos;
7. Finalmsg=cvt;
8. Output finalmsg.
```

Algorithm for Retrieving Secrete Message From Hidden Message

```
INPUT: A hidden message as hiddenmsg.
OUTPUT: secrete message msg.
1. First read three characters from hiddenmsg as ch1,ch2,ch3.
2. Compute 3X5 bit wise crossover between ch1 and ch2 as cross1.
3. Compute 5X3 bit wise crossover between cross1 and ch3 as key.
4. Compute the stegno key from key by counting on-bits as pos.
5. For each character in hiddenmsg do
J=pos;
Read jth character from hiddenmsg as ch.
Encypmsg=encypmsg+ch;
J=j+pos;
6. decrypt the encypmsg with key using EX-OR operation as msg.
7. Output msg.
```

12.4.2 Example Description for Encryption

1. Let us assume that our plain text is "I am Indian" which is shown in Fig. 12.1.
2. Generated random cover text which is shown in Fig. 12.2
3. Take first three character of our cover text for generating the key Kt which is shown in Fig. 12.3.
4. Apply 3 × 5 crossover between first and second character of key Kt which is shown in Fig. 12.4.
5. Apply 5 × 3 crossover between key1 and third character of key Kt as shown in Fig. 12.5.
6. Encrypt the message with the key using Ex-or operation and mix with Cover text (Fig. 12.6).

12.5 Experimental Results

We have implemented this method in java and tested it using java profiler for time overhead and space consumption. In this section we are showing our proposed approaches result and compare it with existing text steganography approaches. We are showing the comparison of these approaches in form of time overhead and how

Fig. 12.1 Plain Text

Fig. 12.2 Generated Random text

Fig. 12.3 First 3 characters generating key Kt

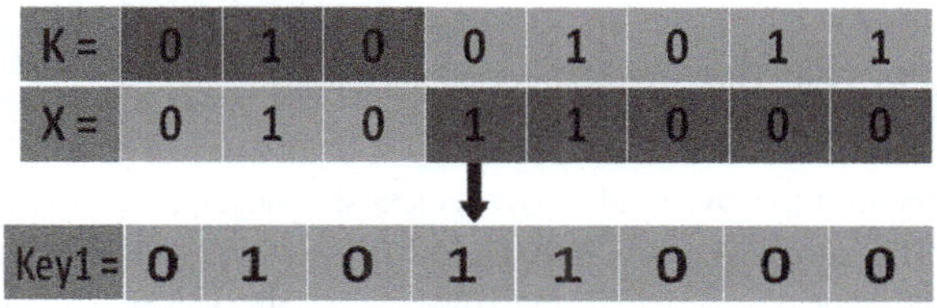

Fig. 12.4 Application of 3x5 crossover

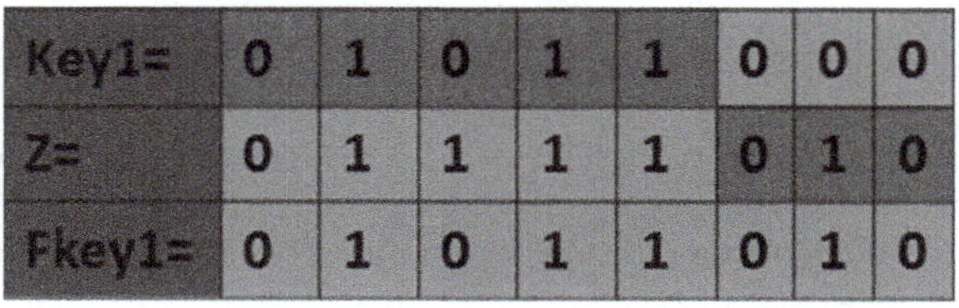

Fig. 12.5 Application of 5x3 crossover

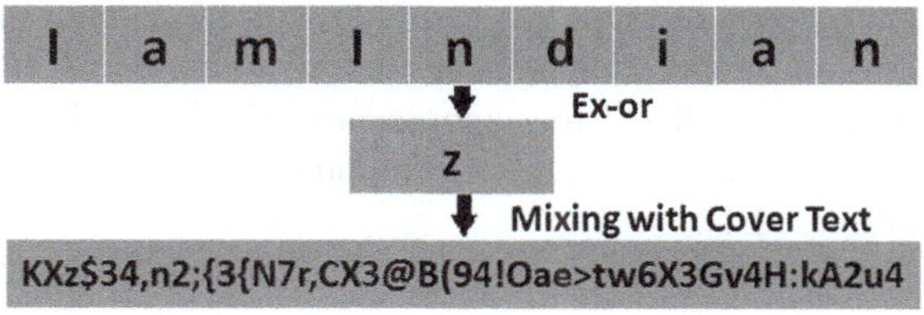

Fig. 12.6 Text Encryption

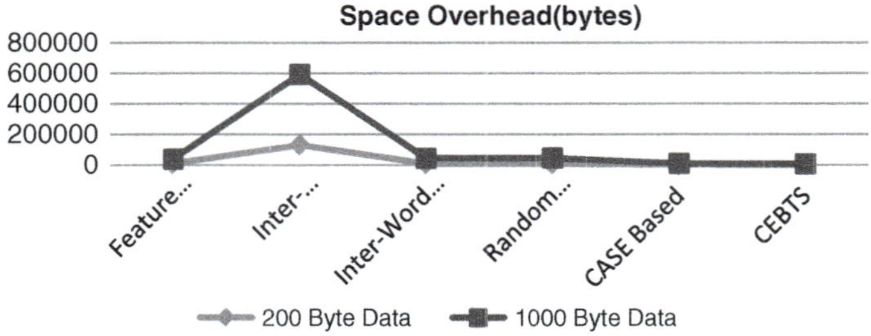

Fig. 12.7 Maximum cover text required to hide 200 and 10,000 bytes

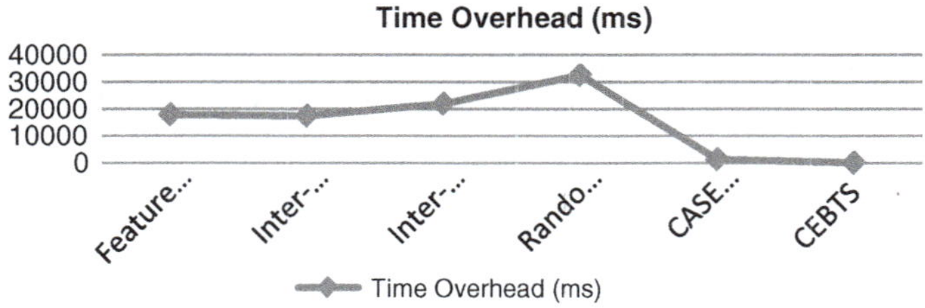

Fig. 12.8 Time overhead for all approaches

Table 12.1 Overhead in various text steganography approaches

Text steganography techniques	Message text size(bytes)	Cover text size(bytes)	No. of bytes hidden (bytes)	Time overhead (ms)
Feature coding	200	660	13	18,158
Inter sentence space	200	660	1	19,276
Inter word space	200	660	14	20,906
Random character sequence	200	660	14	28,100
CASE based Text stenography	200	660	66	1,672
CEBTS (Average case)	200	660	161	31.3

many bytes of plain text can hide with fix amount of cover text. We can see that how our proposed approaches hiding more than 70 % of our plain text which is far better than all existing approaches. The best thing in this technique is reducing time overhead as this technique takes very less time (Figs. 12.7, 12.8), (Table 12.1).

12.6 Conclusion

In this paper, we have proposed a new text steganography approach which would be basically used for English characters. In this approach, we are generating a key from first three characters of cover text using crossover techniques and then next encrypt the plain text using this key and the same key will be used for choose the method that how encrypted data will mix up with cover text. Based on our survey of the existing Text Steganography approaches, we have shown that our proposed approach can hide more number of bytes and it has also very small cover text and required very less time overhead as compared to other techniques. In addition, our proposed approach is also immune to retyping and reformatting of text. This Approach generating an non-sense message but taking very less time for encryption and decryption of the plain text and it is also hard to fetch secret message from received message. Because of fast processing this approaches used in cloud computing for store backup data in large amount so there encryption and decryption techniques will not take too much time and data transfer rate will be high.

References

1. Chaudhary S, Mathur P, Kumar T (2013) A capital shape alphabet encoding (CASE) based text steganography. In: Sharma R (ed) Conference on advances in communication and control systems 2013 (CAC2S 2013), India
2. Agarwal M (2013) An efficient dual text steganographic approach: hiding data in a list of words. In: Computer networks and communications (NetCom), vol 131. Lecture notes in electrical engineering Springer, New York
3. Shirali-Shahreza M (2008) Text steganography by changing words spelling. In: 10th IEEE international conference on advanced communication technology, Korea
4. Shirali-Shahreza M, Shirali-Shahreza MH (2007) Text steganography in SMS. In: IEEE international conference on convergence information technology
5. Khan F (2009) Enhanced text steganography in SMS. In: 2nd IEEE international conference on computer, control and communication
6. Shirali-Shahreza M, Shirali-Shahreza MH (2006) A new approach to Persian/Arabic text steganography. In: 5th IEEE/ACIS international conference on computer and information science (ICIS COMSAk'06), pp 310–315
7. Bender W, Gruhl D, Morimoto N, Lu A (1996) Techniques for data hiding. IBM Syst J 35(3 & 4):313–336
8. Alla K, Shivramprasad R (2009) An evolution of Hindi text steganography. In: 6th IEEE international conference on information technology
9. Provos N, Honeyman P (2003) Hide and seek: an introduction to steganography. IEEE Security & Privacy, pp. 32-44, May/June 2003

Chapter 13
Automatic Color Image Segmentation Using Spatial Constraint Based Clustering

Abu Shama and Santanu Phadikar

Abstract Color image segmentation is a much talked about topic in image processing, where there is plenty of scope for improvement. A cluster validation index based novel method for automatic color image segmentation is proposed here. To identify the number of segments automatically cluster validity indices (Partition Coefficient, Partition Entropy, Xie-Beni index, Kwon's index and Fuzzy hyper-volume index) have been used. Image has been segmented into the number of segments identified by cluster validation indices using modified Fuzzy C-means (FCM) algorithm, which not only uses the color values, but also the spatial relation of the pixels to identify the segment. The performance of the proposed segmentation algorithm has been evaluated using the benchmark data from Berkeley image segmentation dataset and also been compared with existing Otsu's method, K-means algorithm and FCM algorithms based segmentation method using Jaccard Index (JI). Experimental results show that the proposed method gives better segmentation results both subjective and in terms of JI values.

Keywords Color image segmentation · Fuzzy C-means clustering · Cluster validation index · Automatic segmentation

13.1 Introduction

Image segmentation implies the procedure of grouping the similar pixels of an image into distinctive sets or sections, has a variety of applications in a variety of domains [1, 2]. Though different segmentation algorithms are available in the

A. Shama · S. Phadikar (✉)
Department of CSE, West Bengal University of Technology, BF-142,
Sector-I, Salt Lake, Kolkata 700064, India
e-mail: sphadikar@yahoo.com

A. Shama
e-mail: abushama.malda@gmail.com

S. Sengupta et al. (eds.), *Emerging Trends in Computing and Communication*,
Lecture Notes in Electrical Engineering 298, DOI: 10.1007/978-81-322-1817-3_13,

literatures [1], still image segmentation is a challenging task because, segmentation objective for different application is different. Not only that, presence of different textures in the image and improper illumination makes the segmentation more difficult. Discontinuity and similarity/homogeneity are two basic properties of the image pixels used in many segmentation methods.

Cluster based segmentation is significant approaches among the different approaches of segmentation available in literature. FCM, a clustering algorithm is widely used for image segmentation [3, 4]. Xia [5] defined the process of image segmentation as clustering of spatial patterns, as image can be modeled as set spatial patterns on rectangular lattice. Zhang and Jiang [6] proposed a Gaussian kernel function based weighted FCM algorithm which considered the effect of neighboring pixels to make the algorithm robust towards noisy images. Shasidhar et al. [7] proposed a modified FCM algorithm to produce better results on normal as well as noisy images.

In [8] Yannis and Stavros proposed an adaptive fuzzy clustering algorithm for image segmentation with neighborhood information. Cai et al. [9] proposed a fast and robust modification of standard FCM which incorporated a spatial-grey similarity measure to obtain robustness to noise and improved segmentation results.

Though a various variation of FCM is available in literature all of them require the number of clusters to be specified manually. Also the algorithms do not consider the spatial features of the pixels as a strong parameter in segmentation, which is a major drawback as different images have different area of interest. Hence a method which can segment color images in a very efficient way and also doesn't require any type of previous knowledge (e.g. number of segments) about the image had to be devised. So in this paper a novel method has been proposed to automatically segment images which incorporate the spatial feature of the image pixels.

Section 13.2 describes the proposed algorithm and the method of segmentation. Experimental results have been discussed in Sect. 13.3 and finally Sect. 13.4 concludes the paper.

13.2 The Proposed Method

Determining the number of clusters is a crucial task for segmenting an image correctly. Though some unsupervised learning based [10] and fuzzy method based algorithms [3, 4] are available for segmenting images, no such methods are available which first determine the number of segments automatically and then cluster the image. To identify the number of segments, cluster validation indices based method is used.

13.2.1 FCM Algorithm

The FCM algorithm is an unsupervised clustering method; this algorithm makes soft partitions where a datum can belong to different clusters with a different membership degree to each cluster. This clustering method is an iterative algorithm which uses the necessary condition to achieve the minimization of the objective function J_m for fuzzyfier [11] 'm' as given in Eq. (13.1).

$$J_m = \sum_{i=1}^{c} \sum_{j=1}^{n} U_{ij}^{m} \left\| x_j - v_i \right\|^2 \qquad (13.1)$$

Where $x_1, x_2, \ldots, x_n$ are the values of the image pixels, $v_1, v_2, \ldots, v_c$ are the centers of the clusters, $\boldsymbol{U_{ij}}$ is the degree of membership of pixel x_j in the cluster v_i, 'c' is the number of clusters to be produced and 'n' is the number of data points available as input. The parameter 'm' is called the fuzzyfier which controls how fuzzy the clusters are going to be; here $m = 2$ has been used. The above objective function is minimized when datum close to the center points are assigned higher values in the U matrix. The elements of U are fractions between 0 and 1 which are the probability of how much a given datum belongs to a specific cluster. At the first iteration the membership matrix is initialized and the centroids are chosen as random data points. The cluster centroids $V = [v_1, v_2, \ldots, v_c]$ *for* $1 \le i \le c$ is updated using Eq. (13.2).

$$v_i = \frac{\sum_{j=1}^{n} (U_{ij})^m x_j}{\sum_{j=1}^{n} (U_{ij})^m} \qquad (13.2)$$

And the membership matrix U for each of $v_i (1 \le i \le c)$ is updated using Eq. (13.3).

$$U_{ij} = \frac{1}{\sum_{k=1}^{c} \left(\left\| x_j - v_i \right\| / \left\| x_j - v_k \right\| \right)^{\frac{-2}{(m-1)}}} \qquad (13.3)$$

The distance function $\|x_j - v_i\|$ used in the above equations is the Euclidian distance between the two data points. The above process is repeated until the algorithm converges to a locally minimum solution for v_i. Convergence can be tested by the difference in the membership matrix in each iteration; $\max_{ij}\left\{ \left| U_{ij}^{k+1} - U_{ij}^{k} \right| \right\} < \varepsilon$, where ε is a termination criterion, $0 < \varepsilon < 1$ and k is the iteration step. In this way the algorithm iteratively converges on the most optimum cluster centers.

13.2.2 Determining the Number of Clusters

Cluster validity index provides an objective measurement of a clustering result and its optimal value is often used to indicate the best possible choice for the number of clusters [12]. Thus cluster validation indices are used here for identifying the optimal number of segments in an image. Since the use of single validation index may lead to biasness, five widely used cluster validity indices Partition coefficient (PC) [12], Partition entropy (PE) [11, 13], Xie-Beni index (XB) [14], Kwon's index (K) [15], Fuzzy hyper volume (FH) [16] are used here.

To find out the number of segments present in the image we first cluster the image using the standard FCM assuming the cluster values 2 through 10. Five different cluster validation indices are computed for each segmented image, obtained by segmenting the image using standard FCM taking cluster values 2, 3, … , 10 respectively. Then for each of the cluster validation indices optimal number of cluster is determined which gives the best value (minimum value for PE, XB, K, FH and maximum value for PC) of that index. The optimal number of clusters is computed by the majority of five indices consider here. If tie occur then optimal number of cluster is considered as the number of cluster determined by XB.

13.2.3 Modified Spatial Constraint Based FCM

Once the number of segment is identified, next task is to segment the image into the appropriate number of segment. To do this modified FCM, which includes the spatial information along with the color level information is proposed here.

The basic method for clustering the data in the proposed algorithm differs from the standard FCM in the way it measures the distance between two pixels. As FCM is not optimized for color image segmentation and can only take grey values. If all three R, G, B values are used in clustering inspite of using only one grey value, it will produce better clusters. Also the standard FCM neglects the spatial feature of the pixels. Here in the proposed algorithm distance function is incorporated in a manner so that it utilize the spatial feature of the pixels; here $\|p - v\|$ is the Euclidian distance between two given pixels (on individual properties), which is calculated using Eq. (13.4).

$$\|p - v\| = \left(a \times \sqrt{(p_X - v_X)^2 + (p_Y - v_Y)^2}\right) + \left(b \times \sqrt{(p_R - v_R)^2 + (p_G - v_G)^2 + (p_B - v_B)^2}\right) \tag{13.4}$$

Here x and y suffixes are used to denote the X and Y coordinate values of the pixel respectively and suffixes R, G, and B signify the Red, Green and Blue color component values of the pixel respectively. The weighting parameters '**a**' and '**b**' are used to control the influence of the spatial and the color components of the image in the clustering process determined as follow.

13.2.4 Determining the Values for 'a' and 'b'

The values of the parameters 'a' and 'b' were determined by evaluating large number of standard images from Berkeley segmentation dataset [17]. The segmentation result obtained by using different values of the parameters 'a' and 'b' were evaluated using Jaccard index [18]. The result shows that the number of clusters suggested by cluster validity indices varied with the number of picks present in the histogram of the image. Thus experimentally, for an image that has only one prominent peak in the histogram the parameter values are chosen as a = 0.4, b = 0.6. For images having two or three prominent peaks in their histogram a = 0.3, b = 0.7 is used. And for images having more than 3 peaks in their histograms a = 0.2, b = 0.8 is used.

13.2.5 The Proposed Segmentation Method

The procedure of segmenting an image is described step by step below.

Step 1: Cluster the given image using standard FCM.
Step 2: Estimate the values of different validation indices for this cluster.
Step 3: Repeat step 1 and 2 assuming number of clusters to be 2, 3... 10.
Step 4: For each of the cluster validation indices optimal number of cluster is determined which gives the best value (minimum value for PE, XB, K, FH and maximum value for PC) of that index.
Step 5: *Optimal number of clusters* is computed by the majority of five indices consider here. If tie occur then optimal number of cluster is consider as the number of cluster determined by XB.
Step 6: Determine the number of peaks in histogram and assign values to parameters 'a' and 'b'.
Step 7: Use *MSCFCM* to segment the image taking *optimum number of clusters* determined in step 5.

The procedure can best be understood by looking at the flow-chart in Fig. 13.1.

13.3 Experimental Data Analysis and Results

The proposed Modified spatial constraint based FCM (MSCFCM) algorithm has been applied on 600 images available in the Berkeley image segmentation dataset [17]. For explaining the procedure the aircraft image labeled as '3096.jpg' in the Berkeley image Segmentation dataset is used and shown in Fig. 13.2a. Taking the cluster value from 1 to 10 and segmenting it using standard FCM, five different

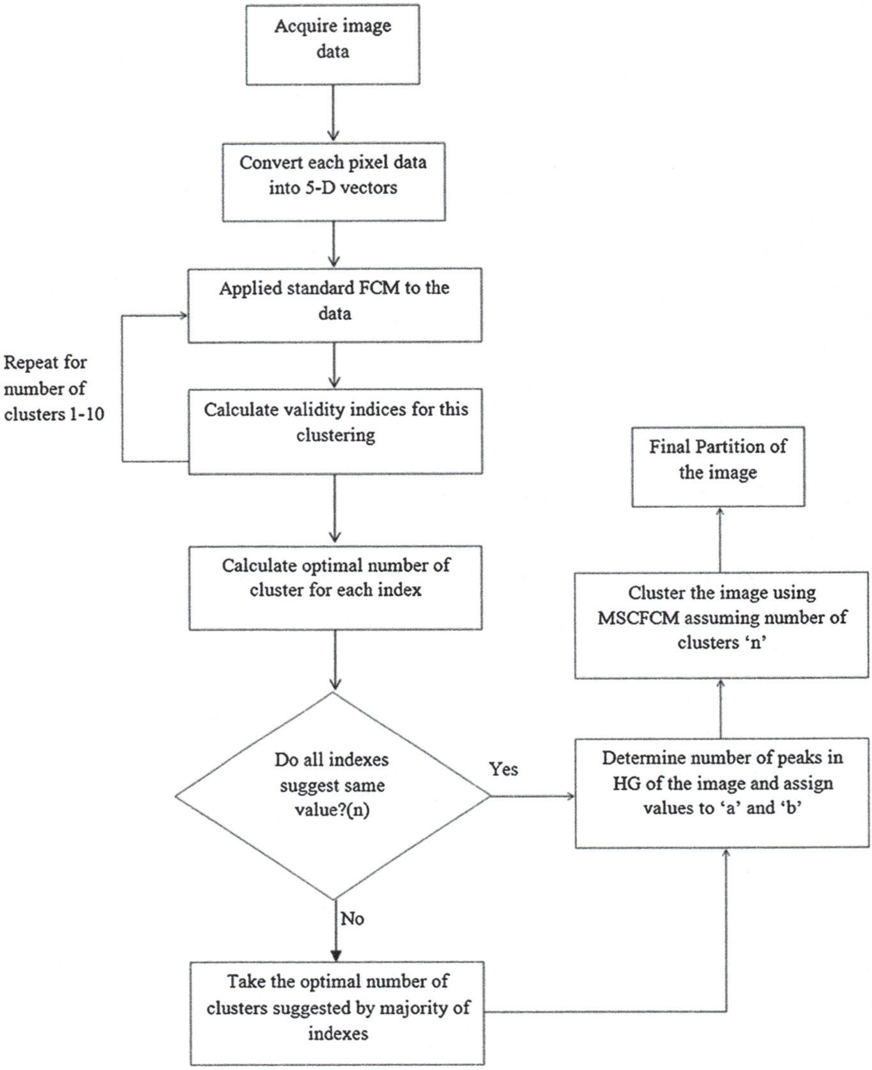

Fig. 13.1 Flow chart describing the proposed method

cluster validation indices (PC, PE, XB, K and FH) are computed and results are given in Table 13.1.

From the column two of Table 13.1, it is found that optimum value (maximum value) of PC is 0.9455 which corresponds to number of cluster = 2. Similarly optimum number of cluster for PE, FH, XB and K are 2, 3, 2, and 2 respectively, obtained by considering the minimum values of the columns 3, 4, 5, and 6 respectively as smaller values of these indices represents better cluster. The indices PC, PE, XB and K gives the vote for number of cluster = 2, which clearly shows

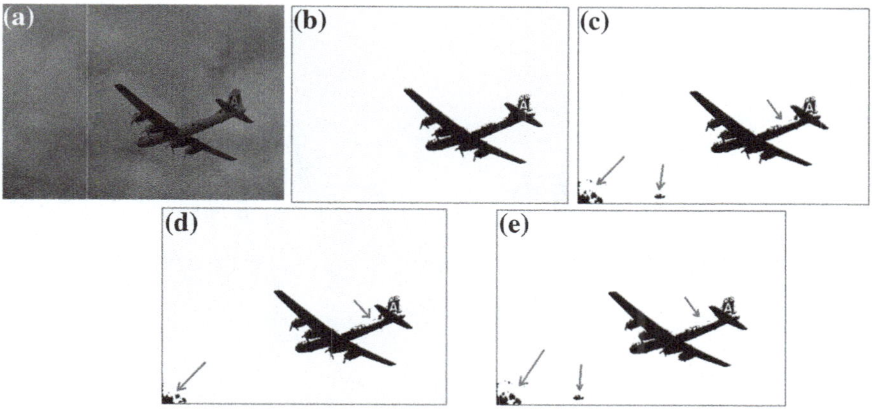

Fig. 13.2 Performance of the algorithms for '3096.jpg': **a** '3096.jpg' original image. **b** O/P of the MSCFCM. **c** O/P of Otsu's method. **d** O/P of FCM. **e** O/P of K-means

Table 13.1 Validation index values for '3096.jpg'

Number of clusters	PC	PE	FH	K	XB
2	**0.9455**	**0.1048**	71.86	**10,220**	**0.0220**
3	0.8115	0.3275	**68.763**	55,727	0.1202
4	0.7653	0.4365	106.28	70,839	0.1529
5	0.7401	0.5037	105.59	59,766	0.1290
6	0.7165	0.5681	116.30	81,604	0.1761
7	0.7035	0.6134	129.67	10,019	0.2162
8	0.7011	0.6229	168.74	77,343	0.1669
9	0.6645	0.7083	159.76	11,166	0.2409
10	0.6598	0.7208	206.99	10,172	0.2194

the majority for cluster 2, and is treated as the optimal cluster value. Then the MSCFCM is applied on the image taking number of cluster = 2 and the result is shown in Fig. 13.2b.

To show the efficiency of the proposed method, the image in Fig. 13.2a is segmented using the proposed method, the existing Otsu's method [19], the standard fuzzy C-Means algorithm [11] and the K-means algorithm [20] and the results are shown in Fig. 13.2b–e respectively. The noises (unwanted pixels) produce by different methods are shown by arrows in the Fig. 13.2.

13.3.1 Performance Comparison

In this section we present an example image segmentation samples using all the four algorithms compared above. The objective evaluation of the proposed method is done using the Jaccard Index [18] computed suing the Eq. (13.5).

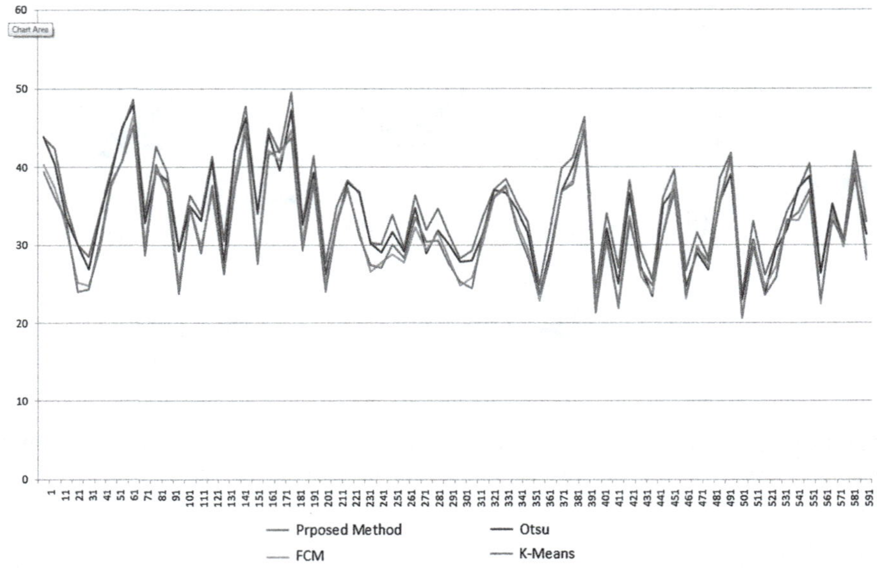

Fig. 13.3 Chart showing performance comparison of the MSCFCM method with others

$$J(A, B) = \frac{A \cap B}{A \cup B} \tag{13.5}$$

Where A is the segmented ground truth images (the 'correct' segmentation) given in the Berkeley image segmentation dataset of the image and B is the segmented image generated by the segmentation algorithm. The JI values computed for the MCSFCM, Otsu's, K-means and standard FCM segmentation algorithm using Berkeley image segmentation dataset is shown in Fig. 13.3.

Clearly the graph shows the superiority of the proposed MSCFCM over the existing Otsu's, FCM and K-means algorithms.

13.4 Conclusion and Future Work

The paper presents a method for automatic color image segmentation using cluster validity indices. Some well-known validation indices have been used to accurately calculate the optimal number of clusters present in an image. It is observed that the proposed method is able to automatically segment given images into an optimal number of segments. Performance of the proposed MSCFCM was compared with well-known segmentation algorithms such as Otsu's, standard FCM, and K-means using standard data set, and the results are satisfactorily better than those algorithms. proposed method is evaluated using the standard image set, applicability of the method for a specific field is also open. The improvements may also be done by using the other validation indices and selecting the value of parameter 'a' and 'b' in the Eq. (13.4) based on the application field.

References

1. Pal NR, Pal SK (1993) A review on image segmentation techniques. Pattern Recogn 26(9):1277–1294
2. Haralick RM, Shapiro LG (1985) Image segmentation techniques. Comput Vis Graph Image Proc 29(1):100–132
3. Cinque L, Foresti G, Lombardi L (2004) A clustering fuzzy approach for image segmentation. Pattern Recogn 37(9):1797–1807
4. Ng HP, Ong SH, Foong KWC, Goh PS, Nowinski WL (2006) Medical image segmentation using K-means clustering and improved watershed algorithm. In: IEEE Southwest symposium on image analysis and interpretation, IEEE, Mar 2006, pp 61–65
5. Xia Y, Wang T, Zhao R, Zhang Y (2007) Image segmentation by clustering of spatial patterns. Pattern Recogn Lett 28(12):1548–1555
6. Zhang XB, Jiang L (2009) An image segmentation algorithm based on fuzzy c-means clustering. In: International conference on digital image processing, Mar 2009, IEEE, pp 22–26
7. Shasidhar M, Raja VS, Kumar BV (2011) MRI brain image segmentation using modified fuzzy c-means clustering algorithm. In International conference on communication systems and network technologies (CSNT), June 2011, pp 473–478
8. Tolias YA, Panas SM (1998) Image segmentation by a fuzzy clustering algorithm using adaptive spatially constrained membership functions. IEEE Trans Syst Man Cybern Part A Syst Hum 28(3):359–369
9. Cai W, Chen S, Zhang D (2007) Fast and robust fuzzy c-means clustering algorithms incorporating local information for image segmentation. Pattern Recogn 40(3):825–838
10. Dong G, Xie M (2005) Color clustering and learning for image segmentation based on neural networks. IEEE Trans Neural Networks 16(4):925–936
11. Bezdek JC (1981) Pattern recognition with fuzzy objective function algorithms. Kluwer Academic Publishers, Dordrecht
12. Bezdek JC (1973) Cluster validity with fuzzy sets
13. Bezdek JC (1974) Numerical taxonomy with fuzzy sets. J Math Biol 1(1):57–71
14. Xie XL, Beni G (1991) A validity measure for fuzzy clustering. IEEE Trans Pattern Anal Mach Intell 13(8):841–847
15. Kwon SH (1998) Cluster validity index for fuzzy clustering. Electron Lett 34(22):2176–2177
16. Gath I, Geva AB (1989) Unsupervised optimal fuzzy clustering. IEEE Trans Pattern Anal Mach Intell 11(7):773–780
17. Martin D, Fowlkes C, Tal D, Malik J (2001) A database of human segmented natural images and its application to evaluating segmentation algorithms and measuring ecological statistics. In: Eighth IEEE international conference on computer vision, ICCV 2001. Proceedings, IEEE, vol 2, pp 416–423
18. Jaccard P (1901) Etude comparative de la distribution floraledansune portion desAlpeset du Jura. Impr. Corbaz
19. Otsu N (1975) A threshold selection method from gray-level histograms. Automatica 11(285–296):23–27
20. MacQueen J (1967) Some methods for classification and analysis of multivariate observations. In: Proceedings of the fifth Berkeley symposium on mathematical statistics and probability, vol 1, no 281–297, June 1967, p 14
21. Thilagamani S (2011) A survey on image segmentation through clustering. Int J Res Rev Inf Sci (IJRRIS), 1(1):16–19
22. Seize TK (1977) Student's t-test. South Med J 70(11):1299

Chapter 14
Historical Handwritten Document Image Segmentation Using Morphology

Bishakha Roy and Rohit Kamal Chatterjee

Abstract Automatic recovery of text from historical documents is a difficult task due to their degradation because of different types of noise. Applying a global threshold or a chosen threshold based on visual intuition misses the finer handwritten text with low intensity values. These low intensity text are actually considered as a part of background when applying global threshold and are neglected. A single threshold is unable to segment the whole image clearly as various levels of intensities are present in text because of degradation. For restoration of missing texts we propose a thresholding algorithm based on mathematical morphology, which generates very fine adaptive threshold. After applying global threshold, left out background image consists of some mixed image background and handwritten text intensities on which we apply mathematical morphology (opening and closing), which produces a smooth contour and gives an adaptive threshold. The resultant thresholded image have clear uniform background and foreground with enhanced character appearance.

Keywords Historical text segmentation · Adaptive thresholding · Mathematical morphology · Opening and closing

14.1 Introduction

The need for text segmentation from degraded old paper becomes immensely important to restore the historical documents in digitized format for archival purpose. Due to degradation of the old handwritten documents caused by ink splitting, blotting or added dirt, digitization clean paper media is difficult to attain.

B. Roy (✉) · R. K. Chatterjee
BIT, Mesra (Kolkata Campus), Ranchi, Jharkhand, India
e-mail: redsunshinerimp@yahoo.co.in

R. K. Chatterjee
e-mail: rkchatterjee@bitmesra.ac.in

S. Sengupta et al. (eds.), *Emerging Trends in Computing and Communication*, Lecture Notes in Electrical Engineering 298, DOI: 10.1007/978-81-322-1817-3_14,

Purpose of a good thresholding algorithm is to retain almost all the pixel intensities of text while cleaning the background. Various global (bi-level) thresholding methods have been proposed (e.g. Otsu's method [1], Pun's method [2, 3]) in literature. But a bi-level thresholding doesn't give the desired result, because global thresholding misses the finer gray-values in the handwritten documents.

Global thresholding methods like Otsu's, Kapur's entropy and Solihin & Leedham's Quadratic Integral Ratio (QIR) are compared and they are used also for multistage thresholding in [4]. A graph based approach for image background elimination and segmentation is discussed in [5]. In [6] it is shown how the documents are first decomposed into their features such as paper texture, colours, typewritten parts, pictures, etc. and reassembled generating an image visually close to the original document. Mello et al. proposes this new method to remove the shade of document images captured with digital cameras followed by a binarization algorithm. In [7] Leedham et al. discusses with some existing thresholding technique (Niblack's Method, Quadratic Integral ratio, Yanowitz and Bruckstein's method) and compares them with proposed Mean-Gradient technique and background subtraction method. Yan and Leedham [8] proposed a new thresholding technique called the decompose-threshold approach, where image is divided recursively into sub images and based on the features of each local region a threshold value is decided. But because of the shape of images being always square shaped and also due to various sizes of sub images the image after thresholding will not be clear and distinction of square sub images might also be visible. In [9] planes are fitted for each local neighbourhood and pixel values are taken on the approximated plane. These values become the adaptive threshold value for each pixel. We have proposed an adaptive threshold.

Adaptive threshold means getting a threshold for each block or for each pixel. Our proposed method approximates the background using opening and closing. Pixel values on this approximated curve are calculated and used to threshold the image. Opening can be considered as if a ball is rolled from inside the profile the farthest point that it reaches is determined as the boundary and it smoothes the narrow peaks. Similarly when ball is rolled outside the deep valleys are fused. In Fig. 14.1 it is shown how in opening and closing a structuring element is rolled outside and inside the profile. In Fig. 14.2 the curve after opening and closing is shown which is our adaptive threshold. Our proposed adaptive threshold gives a curve like profile and almost fits the profile of the original document. In Sect. 14.2 we discuss our algorithm, in Sect. 14.3 we have experimental results and discussion, conclusion and future work in Sect. 14.4.

14.2 Thresholding Algorithm

Degraded text documents contain information as well as immense noise. Noise is hard to remove by a single global threshold. A threshold for every pixel is proposed in this algorithm i.e. an adaptive threshold. Mathematical morphological operations like opening and closing is used.

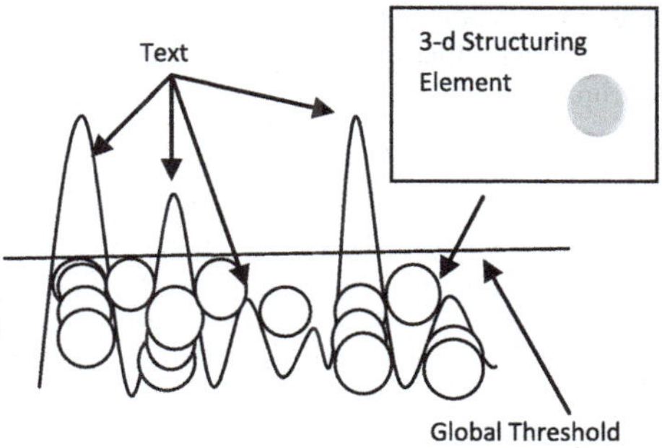

Fig. 14.1 Profile view along a line of a historical document

We consider each pixel of an image as 3-d object defined by P(x, y, z) where x and y identify a distinct pair of coordinate in image, z identifies intensity value. From Fig. 14.1 we can see a profile of a scan line of an image where texts are brighter than background. We binarize the image using global threshold and extract the background thereafter. Clipped off text pixel intensities are foreground which contains rich information are filled with white pixels. Left out background image contains mixed text and background intensities due to noise. It seems as if a plane of threshold has cut the intensities in two halves upper and lower. The lower portion contains mixed background and text intensities which are not distinguishable easily. The upper portion contained pure text or darkest intensities. We apply morphological operations opening and closing on background.

Equation of opening and Closing are given in (14.1) and (14.2) respectively,

$$f \circ b = (f \Theta b) \oplus b \tag{14.1}$$

$$f \bullet b = (f \oplus b) \Theta b \tag{14.2}$$

In Eqs. (14.1) and (14.2) opening and closing is defined. Opening and Closing are symbolized by $\ominus$ and $\bullet$ respectively. Erosion and Dilation are symbolized by Θ and $\oplus$ respectively. Opening of an image f(x, y) by b(structuring element) is defined by erosion of f by b followed by dilation of the same. Closing of an image f(x, y) by b(structuring element) is dilation of f by b followed by erosion. In (14.3) and (14.4) Erosion and Dilation of gray scale by a non flat structuring is defined where symbols have its usual meaning. Structuring element can be flat (e.g. square, rectangle, pair) and non flat.We have used a non flat ball shaped structuring element (e.g. ball).This element has radius and height.

Gray level erosion and dilation by non flat structuring element is given by,

$$[f \Theta b_N](x, y) = \min_{(s,t) \in b_N} \{f(x + s, y + t) - b_N(s, t)\} \tag{14.3}$$

$$[f \oplus b_N](x, y) = \max_{(s,t) \in b_N} \{f(x - s, y - t) + b_N(s, t)\} \tag{14.4}$$

In Fig. 14.2 opening of background is shown in the area marked 1, closing is shown in area marked 2. While opening by the 3-d object (e.g. ball) is rolled inside as it tries to fit the structuring element in the background profile.The intensity

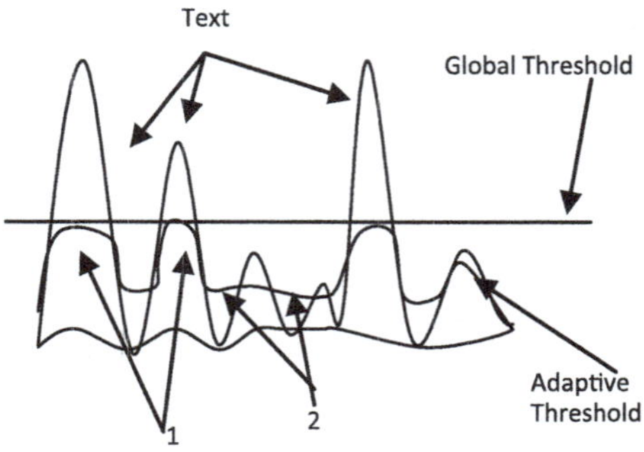

Fig. 14.2 *Adaptive threshold curve* along a line of the document

values it cannot reach is cut off and the farthest point it could reach is identified as boundary and is marked as adaptive threshold. Closing is the dual of opening. As opposite to opening the ball is rolled but outside the background curve, it tries to fit the ball in the curve from outside. Since the ball is rolling outside it cannot reach the darker intensities or the lowest point of the valleys in the curve. Instead of clipping above threshold intensities as in opening it just fits to the point it can reach which is much above the text pixel values and the lowest points where it could reach are identified as adaptive threshold. This threshold changes dynamically all over the image and fits the profile. After opening and closing intensity values above this resultant threshold are made brighter and thereby the text intensities get brighter. The intensities which were mixed with background, not the darkest and not brightest ones but some intensities which where somewhere close to background intensities but they were text gets segmented. The process described above is summarized by an algorithm given below.

```
Input: A gray scale image, Structuring Element
Output: Thresholded image
```

Algorithm

1. `Take an input image X (grayscale).`
2. `Apply global threshold [e.g. Otsu].`
3. `Extract the background which has some mixed low intensity text.`
4. `Apply opening then closing by a 3-d structuring element (e.g. ball).The resultant image is the adaptive threshold.`
5. `Binarize the image using adaptive threshold.`

14.3 Experimental Results and Discussion

A test image is arbitrarily taken for experimenting and validation purpose. Some measures like SSIM and RMS are used to compare thresholded images with reference image. The test image is considered as reference image as shown in

Fig. 14.3a Reference image

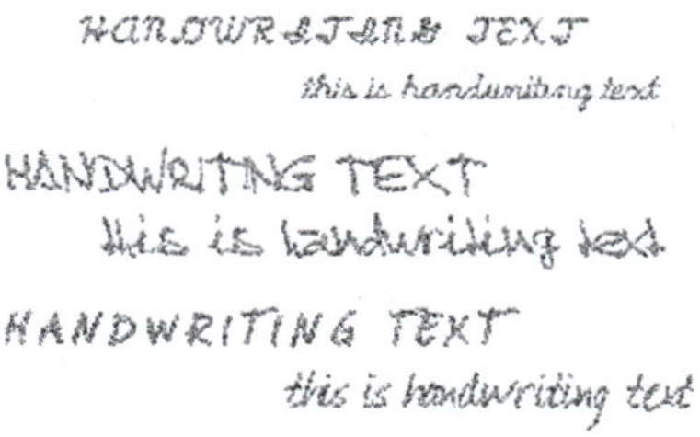

Fig. 14.3b An image with added Gaussian noise to reference image

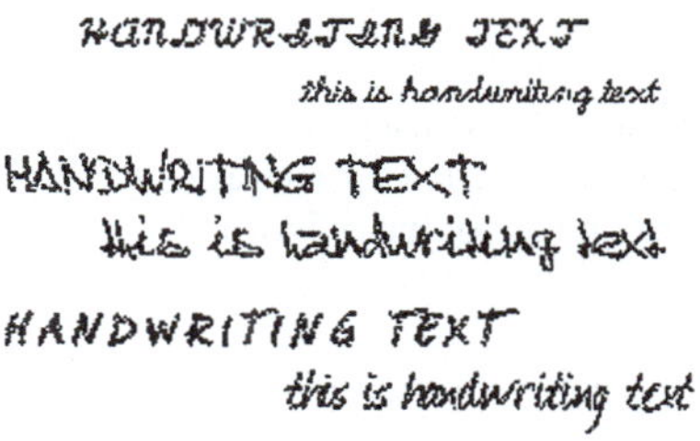

Fig. 14.3c Thresholded image using Otsu's

Fig. 14.3a. Adding some Gaussian noise to the reference image we get a noisy image in Fig. 14.3b. Applying thresholding techniques like Otsu's, Best-fit plane method and our proposed method on the noisy image, results in thresholded images given in Fig. 14.3c, d, e respectively.

The Structural Similarity (SSIM) index is a method for measuring the similarity between two images proposed by Wang et al. [10]. SSIM is expressed as,

$$SSIM(X,Y) = \frac{(2\mu_X\mu_Y + C_1)(2\sigma_{XY} + C_2)}{(\mu_X^2 + \mu_Y^2 + C_1)(\sigma_X^2 + \sigma_Y^2 + C_2)} \tag{14.5}$$

where μ_X and μ_Y symbolizes mean intensities of two images X and Y resp., σ_{XY} is the covariance of X and Y.σ_X and σ_Y are the standard deviation of image X,Y. C_1 and C_2 are constants. Here $C_1 = C_2 = 0$.This measure when equal to 1 means the images are completely similar and when equal to 0 are completely dissimilar. Root Mean Square (RMS) is method for measuring dissimilarity between two matrices. RMS is expressed as,

$$RMS(X,Y) = \sqrt{\frac{\sum_{i=1}^{N}(x_i - y_i)^2}{N}} \tag{14.6}$$

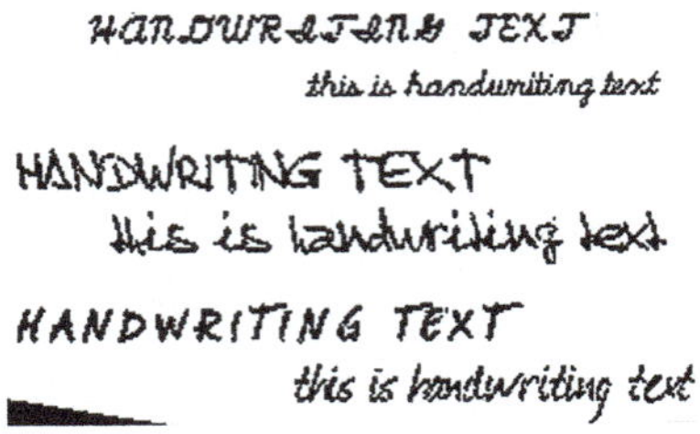

Fig. 14.3d Thresholded image using Best-fit plane method

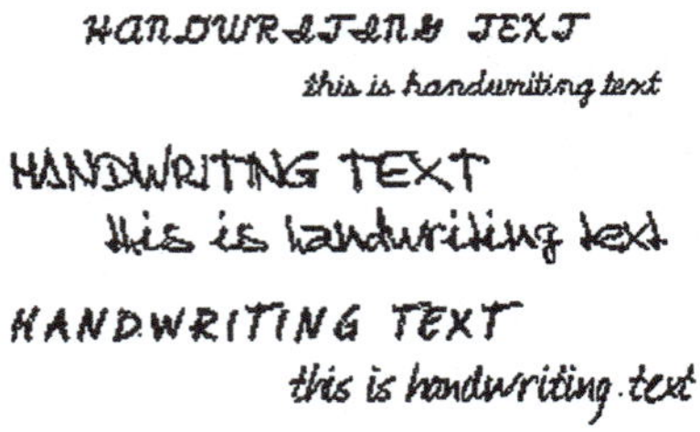

Fig. 14.3e Thresholded image using our proposed method

Table 14.1 Comparison between thresholded images with reference image based on two measures

Thresholding Technique ↓ Measure →	SSIM	RMS
Otsu	0.9477	0.0954
Best-fit plane	0.8189	0.1956
Proposed method	0.9919	0.0380

where, x_i and y_i are elements of matrices, N is the total number of pixels. This measure when equal to zero means two matrices have maximum similarity and least dissimilarity and as the value increases dissimilarity increases.

In Fig. 14.3c we can see the noisy part of the texts are missing whereas our proposed method gives a very clean output with enhanced text intensities. In Fig. 14.3d though plane fitting method enhances text but some unwanted noises appear.

In Table 14.1 a quantitative study of thresholded images with reference image are shown using SSIM and RMS. Comparing by SSIM our proposed method gives the best result as it is closest to 1. Comparing by RMS our proposed method gives better result as it is closer to 0 which signifies thresholded image is very similar to the reference image.

We have taken some more test images to visually compare thresholded image using our proposed method and the existing ones.

In Fig. 14.4a we have taken a sample document image full of noise. In Fig. 14.4b there is thresholded image using Otsu's global threshold. There is

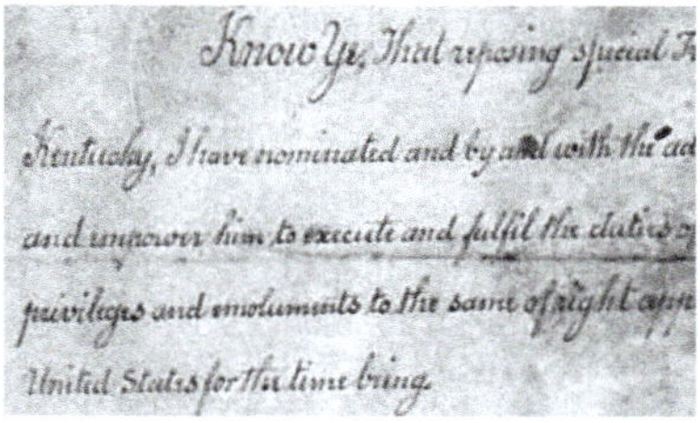

Fig. 14.4a A historical document image

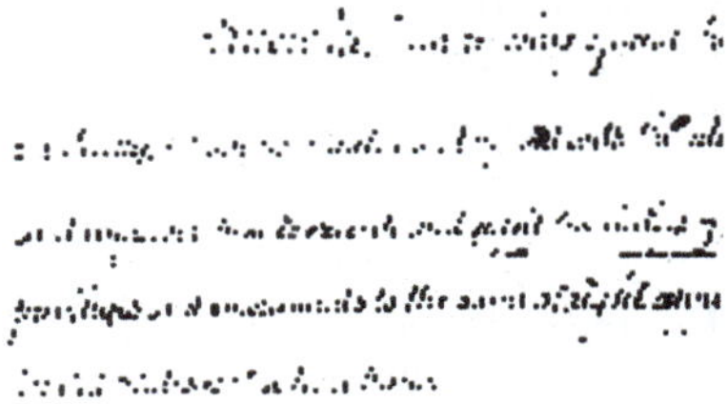

Fig. 14.4b A thresholded image using Otsu's method

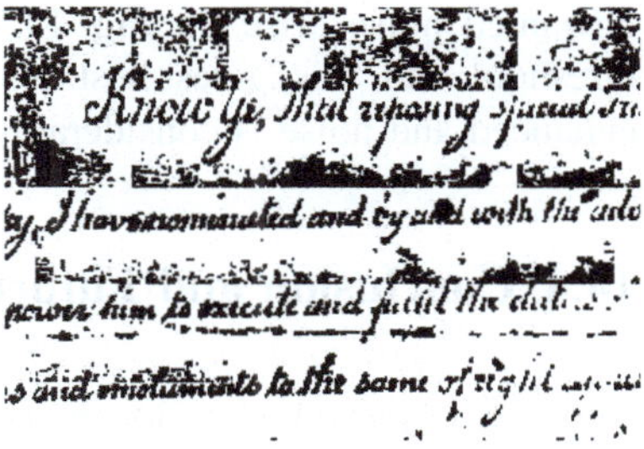

Fig. 14.4c Thresholded image using Best-Fit plane method

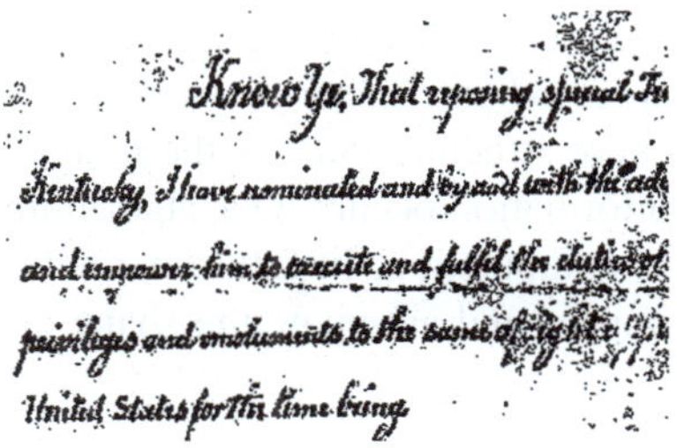

Fig. 14.4d Thresholded image using our proposed threshold

sufficient loss of data. Figure 14.4c is the thresholded image using best plane fit algorithm. Figure 14.4d is the thresholded image using our adaptive thresholding algorithm which recovers almost all the text information.

In Fig. 14.5a handwritten old image is taken for testing our algorithm. In Fig. 14.5b Otsu's global threshold is used to threshold the image which looses data. In Fig. 14.5c thresholded image using plane fitting is degraded where

Fig. 14.5a A handwritten text image

Fig. 14.5b Thresholded image using Otsu's method

Fig. 14.5c Thresholded image using Best-fit plane method

Fig. 14.5d Thresholded image using our proposed threshold

unwanted noise arise. In Fig. 14.5d the image is binarized using our adaptive threshold and this gives best result for the text image. Text characters gets enhanced and noise is considerably removed.

14.4 Conclusion and Future Work

Age old degraded text documents contains rich information but reading and interpreting them manually is a hard task. A single global threshold is not effective in all kinds of documents. Different algorithms are being proposed by many researchers. We have proposed an algorithm to get multiple thresholds for different areas which seems to work successfully. Our algorithm produces threshold of curve structure that fits the image almost and after thresholding very little loss of information occurs. Our algorithm not only works successfully it takes little time in computation. In future we will work to apply this algorithm in more complex images and automate the shape of the structural element size.

References

1. Otsu N (1978) A threshold selection method from grey level histogram. IEEE Trans Syst Man Cybern SMC8:62–66
2. Pun T (1989) A new method for gray-level picture threshoding using entropy of the histogram. Signal Process 2:223–237
3. Pun T (1981) Entropy thresholding: a new approach. Comput Vis Graphics Image Process 16:210–239

4. Leedham G, Varma S, Patankar A, Govindaraju V (2002) Separating text and background in de-graded document images—a comparison of global thresholding techniques for multi-stage thresholding. In: Proceedings of eighth international workshop on frontiers of handwriting recognition, Sept 2002, pp 244–249
5. Mallikarjunaswamy BP, Karunakara K (2011) Graph based approach for background elimination and segmentation of the image. Res J Comput Syst Eng 02(02)
6. Mello CAB, Lins RD (2002) Generation of images of historical documents by composition. In: ACM symposium on document engineering, McLean, VA, USA, p 127–133
7. Leedham G, Yan C, Takru K, Tan JHN, Mian L (2003) Comparison of some thresholding algorithims for text/background segmentation in difficult document images. In: Proceedings of the seventh international conference on document analysis and recognition(ICDAR 2003), IEEE
8. Yan C, Leedham G (2004) Decompose-threshold approach to handwriting extraction in degraded historical document images. In: Proceedings of the 9th international workshop on frontiers in handwriting recognition (IWFHR-9 2004), IEEE
9. Shi Z, Govindaraju V (2004) Historical document image enhancement using background light intensity normalization, ICPR 2004. In: 17th international conference on pattern recognition, Cambridge, United Kingdon, 23–26 Aug 2004
10. Wang Z, Bovik AC, Sheikh HR (2004) Image quality assessment: from error visibility to structural similarity. IEEE Trans Image Process 13(4):600–612

Chapter 15
Comparative Analysis of Offline Character Recognition Using Neural Network Approaches

Pramit Brata Chanda, Santanu Datta, Soham Mukherjee, Subhamoy Goswami and Sritama Bisi

Abstract Optical Character Recognition is a type of document image analysis where scanned digital image that contains either machine printed or handwritten script given as input into the software and translating it into an editable machine readable digital text format. Today, Optical character recognition (OCR) becomes an important area in pattern recognition and image processing. Today Neural Networks are mostly used for Pattern Recognition task. The paper describes the behaviours of different Models of Neural Network used in OCR. OCR is widespread use of Neural Network. We have considered parameters like number of Hidden Layer, size of Hidden Layer and epochs. We have used Multilayer Feed Forward network with Back propagation. Here we have done comparative study on character recognition using different neural network algorithm. The results showed that the MLP networks trained by the error back propagation algorithm are more superior in calculating recognition accuracy and also memory usage. The result indicates that the back propagation network provides better recognition accuracy of more than 80 % of for these English characters.

P. B. Chanda
Computer Science and Engineering, Kalyani Government Engineering College, Kalyani, West Bengal, India
e-mail: pramitcse@gmail.com

S. Datta (✉) · S. Mukherjee · S. Goswami · S. Bisi
Computer Science and Engineering, Academy of Technology, Adisaptagram, West Bengal, India
e-mail: santanudatta@yahoo.com; santanudattaaot@gmail.com

S. Mukherjee
e-mail: soham040288@gmail.com

S. Goswami
e-mail: subhamoygoswami2@gmail.com

S. Bisi
e-mail: sritama21061988@gmail.com

S. Sengupta et al. (eds.), *Emerging Trends in Computing and Communication*, Lecture Notes in Electrical Engineering 298, DOI: 10.1007/978-81-322-1817-3_15,

Keywords Character recognition · Hidden layer · Neural network · ANN · Neuron · Gradient descent back propagation · Epochs · Character matrixes · SSE

15.1 Introduction

Today character recognition is among the subjects which receive special attention to everyone. Character Recognition is a zone of pattern recognition that has become the familiar topic for the researcher during recent period of time. Neural network is playing an important role in character recognition because of different types of applications in various fields, many of which are related with our everyday life. Its applications include recognition of postal envelopes, bank checks, examination forms, Visa application forms and medical prescriptions etc. The necessity of character recognition software has enhanced much more rather than previous time. Optical Character Recognition (OCR) is a very well-studied problem in the huge area of pattern recognition. The character recognition software breaks the image into different sub-images, each containing a single character. The sub-images are then translated from an image format into a binary format, where each 0 and 1 represents an individual pixel of the sub image. The binary data is then fed into a neural network that has been trained to make the association between the character image data and a numeric value that corresponds to the character. Therefore, it is very important to develop a proper and automatic character recognition system for English language. In this paper, efforts have been given to make automatic character recognition system for English language with high recognition accuracy and minimum training and classification time.

15.2 Our Used Recognition Approach

Here we design a network and trained to recognize the 26 letters of the alphabet and 10 numbers. An imaging system that basically digitizes each letter centred in the system's field of vision is available. The result is that each letter is represented as a 5×8 grid of Boolean values. However, the imaging system is not perfect and the letters may suffer from noise. [Note Without noise, we only need 26 (long) if statements to justify the input vector into one of the 26 alphabets.] Perfect classification of ideal input vectors is very much required, and reasonably accurate classification of noisy vectors. The Offline Character Recognition System must first be created through a few simple steps in order to prepare it for presentation into MATLAB. The matrixes of each letter of the alphabet must be created along with the network structure. These are consisting of different steps which are given in Fig. 15.1.

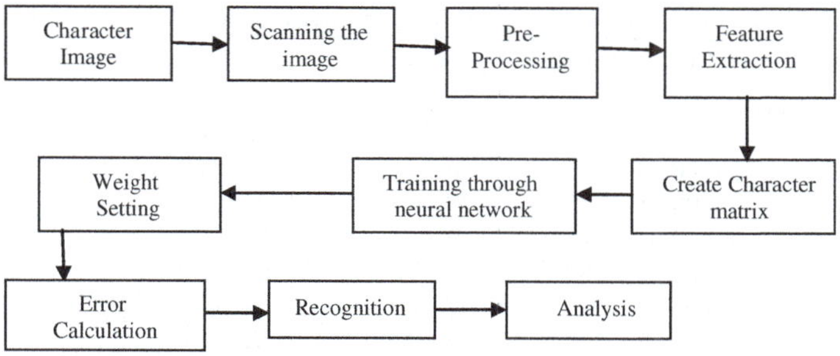

Fig. 15.1 Overall steps involved in the character recognition system

15.2.1 Character Matrixes

The character matrix is represented as an array of black and white pixels: the vector of 1 represented by black, and 0 by white. These are created manually, in whatever of size or font is basically imaginable. In addition, multiple fonts of the same alphabet may even be used under separate training sessions.

15.2.2 Neural Network

The network receives the 40 Boolean values as a 40-element input vector. It is then required to identify the letter by responding with a 36-element output vector. The 36 elements of the output vector each represent a letter. To operate correctly, the network have to respond with a 1 in the position of the letter being presented to the network. The neural network needs 40 inputs and 41 neurons in its output layer to identify the letters. The network is a two-layer log-sigmoid network. The log-sigmoid transfer function is picked because of its output range (0–1) is perfect for learning to output Boolean values. The hidden (first) layer has 10 neurons. This number is picked basically by guessing and experience. If the network has trouble learning, then neurons can be added to this layer for reducing the learning problem. The network is trained to output a 1 in the correct position of the output vector and to fill the rest of the output vector with 0's. Here, if the values of hidden layer and no of epochs varies it will varies the result also.

15.3 Simulated Results

Here, the performance analysis curve is shown where the relation between the epoch and the sse (sum squared error) is given. So, here the graph shows that if the no of epochs are increasing then the sse values are decreasing and at a certain limit

Table 15.1 Gradient descent training algorithm with momentum and adaptive learning

No. of hidden neurons	Time	Epoch (Max)	Performance (SSE)	Mean square error (MSE)	Training rate
10	0:00:02	260	35.9	0.13807	0.1028
12	0:00:02	174	34.2	0.19655	0.04787
13	0:00:05	525	33.4	0.06361	0.6445
14	0:00:02	280	32.8	0.11714	0.16533
15	0:00:04	328	32.2	0.09817	0.13029
16	0:00:07	878	31.9	0.03633	0.20283
17	0:00:04	405	30.3	0.074814	0.23955
18	0:00:03	392	29.2	0.07448	0.03167
20	0:00:04	545	27.8	0.05107	0.28481
22	0:00:07	812	26.2	0.03226	0.23024

Table 15.2 Batch training with weight and bias learning rules

No. of hidden neurons	Time	Epoch (Max)	Performance (SSE)	Mean square error (MSE)	Training rate
10	0:00:04	550	36.6	0.06654	0.05028
12	0:00:04	560	34.8	0.06214	0.06445
14	0:00:05	450	33.3	0.0744	0.02943
13	0:00:03	372	32.5	0.08733	0.30271
15	0:00:03	353	32.1	0.09093	0.13906
17	0:00:06	780	31.8	0.4076	0.32907
18	0:00:04	672	31.2	0.46422	0.13167
20	0:00:05	710	39.2	0.05523	0.39054
22	0:00:05	810	37.6	0.048412	0.29755

it will become constant. Here basically the no of hidden neurons values are varied on the basis of requirement. If these epoch and neurons values are changed then the training rate, its performance factor also changed. These are the simulation results shown using matlab. It also can be seen that Batch training with weight and bias learning rules trained the data faster than the Gradient Descent algorithm. Here the recognition rate is changed after changing the no of hidden neurons (Tables 15.1, 15.2, and 15.3).

15.3.1 Performance Characteristics

Here the characteristics graphs are shown below which show that the error value is decreasing after increasing the no of epochs. These two figure is used for two techniques as one is Gradient Descent Training Algorithm with momentum and adaptive Learning and other is Batch training with weight and bias learning rules (Figs. 15.2, 15.3, 15.4, and 15.5).

Table 15.3 Comparative results of different neural network model

No.	Different approach of OCR	Recognition rate (%)
1	Back propagation	84
2	Gradient descent training algorithm with momentum and adaptive learning	81
3	Batch training with weight and bias learning rules	84
4	Single layer perception learning	72
5	Pre-processing and feature extraction	80

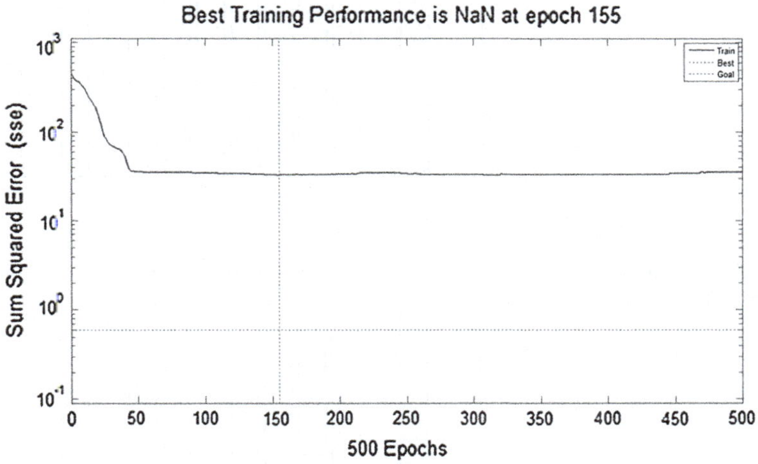

Fig. 15.2 Gradient descent training algorithm with momentum and adaptive learning

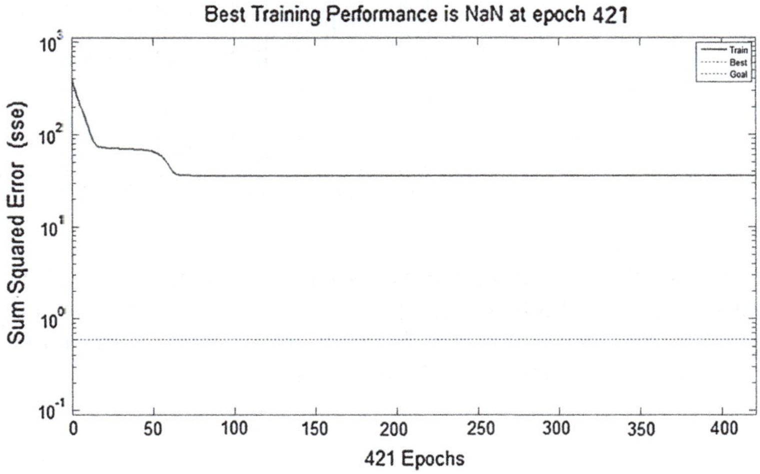

Fig. 15.3 Batch training with weight and bias learning rules

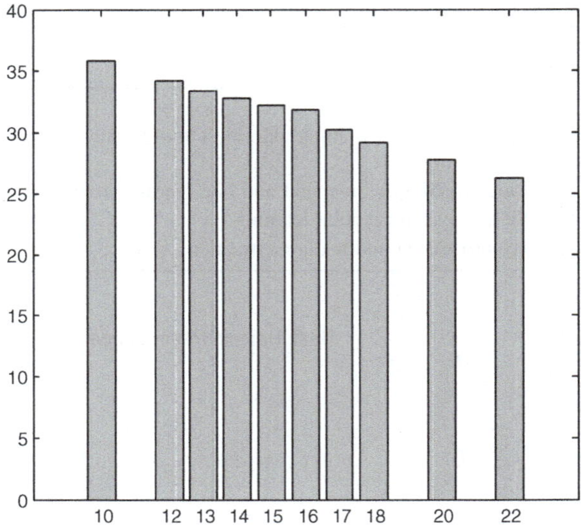

Fig. 15.4 Gradient descent (hidden neuron vs. SSE)

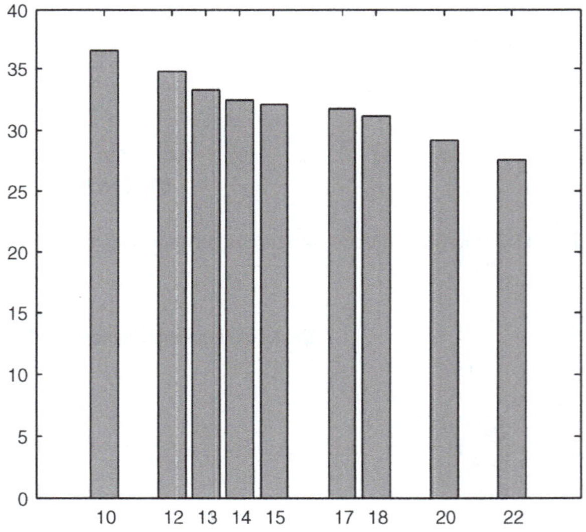

Fig. 15.5 Batch training (hidden neuron vs. SSE)

Here the two characteristics curves are used to show the characteristics between no of hidden layers and SSE values. If the no of hidden layer increases then the Sum Squared Error value (SSE) will decreases. The amount of error for Gradient descent is much lesser than the batch training approach.

15.4 Conclusion

The problem shows how much easily pattern recognition problem can be designed through neural network. In this paper the recognition of characters are studied through introducing templates separately by creating matrix and using back propagation neural network. Different types of training function are used for analyze the result properly. Here, an experimental result shows that back propagation network yields good recognition rate of more than 80 %. This problem gives partial solution to cope with the shift in position and the distortion in the shape of the different character patterns. So, after doing this hard work we get better recognition rate for the Batch training with weight and bias learning rules rather than image based or the gradient descent training approach.

Acknowledgments At first we firstly thanks to God for giving us the courage and blessing for doing this work. Also we thanks to our father's and mother's for giving us support. We should also like to be very much grateful to our college, because they have provided us an ideal working environment, facilities and also give moral support to complete this task.

References

1. Srihari SN (1993) Recognition of handwritten and machine printed text for postal address interpretation. Patterns Recogn Lett 14:291–302
2. Singh D, Dutta M, Singh SH (2012) Neural network based handwritten hindi character recognition. ACM Int J Mach Learn Comput 2(4)
3. Pal A, Singh D (2010) Handwritten English character recognition using neural network. India Int J Comput Sci Commun 1(2):141–144 (Department of Computer Science and Engineering, U.P. Technical University, Lucknow)
4. Mathur S, Aggarwal V, Joshi H, Ahlawat A (2008) Offline handwriting recognition using genetic algorithm. In: 6th international conference on information research and applications, Varna, Bulgaria, June–July 2008
5. Mamedov F, Hasna JFA Character recognition using neural network, IC-AI, CSREA Press, 2006, p 728–733
6. Devireddy SK, Rao SA (2005–2009) Handwritten character recognition using back propagation network. JATIT 5(3)
7. Gonzalez RC, Woods RE (2002) Digital image processing, 2nd edn. Prentice-Hall, USA
8. Singh R, Yadav CS, Verma P, Yadav V (2010) Optical character recognition (OCR) for printed devnagari script using artificial neural network. Int J Comput Sci Commun 1(1):91–95
9. Ganapathy V, Liew KL (2008) Handwritten character recognition using multiscale neural network training technique. World Acad Sci, Eng Technol 39:32–37
10. Ety Dey (2009) Recognition Bangla and English Text from the same Document. July 2009
11. Cheriet M, Kharma N, Liu C-L, Suen CY (2007) Character recognition systems, a guide for students and practitioners. Wiley, New Jersey
12. Pal U, Choudhuri BB (2004) Indian script character recognition: a survey. Pattern Recogn 37:1887–1899
13. Mori S, Suen CY, Kamamoto K (1992) Historical review of OCR research and development. Proc IEEE 80(7):1029–1058

14. Patil V, Shimpi S (2011) Handwritten English character recognition using neural network. Elixir Comput Sci Eng 41:5587–5591
15. Singh MP, Dhaka VS (2008) Handwritten character recognition using modified gradient descent technique of neural networks and representation of conjugate descent for training patterns. Database 5:20
16. http://mathworks.com/characterrecognition
17. Chanda PB, Datta S, Paul Choudhury J (2013) Analysis of character recognition using back propagation neural network algorithm. In: Proceedings of national conference on brain and consciousness, Sept 2013
18. Sivanandam SN, Deepa SN Principles of Soft Computing, John Wiley, 2007

Part III
Network Security and Cryptography

Chapter 16
Exploring Chaotic Neural Network for Cryptographic Hash Function

Prateek Singla, Payal Sachdeva and Musheer Ahmad

Abstract Due to the one way property of neural networks and high sensitivity of chaotic systems, chaotic neural networks make an ideal candidate for cryptographic hash function design. In this paper, a novel algorithm is proposed to construct an efficient cryptographic hash function using a four layer chaotic neural network. The proposed hash function satisfies the security requirements of confusion and diffusion, and the mechanism allows flexibility of the hash value length, which makes it resistant to birthday attack for hash lengths longer than 128 bits. Moreover, the running time of a neural network can be reduced with the help of parallel processing. The statistical analysis of the proposed algorithm proves it to be a promising choice for cryptographic hash function design.

Keywords Chaotic neural network · Hash function · Security · Chaotic systems

16.1 Introduction

Neural networks, by the virtue of their design, possess a strong one way property. If a neuron has multiple inputs and a single output, then the output can be obtained easily from the inputs but difficult to recover the inputs from the output. This property of neural networks makes them suitable for cryptographic hash function design [1–4]. Also, chaotic maps, being highly sensitive to initial conditions, have also been extensively used for hash algorithm designs [5, 6]. A hash function encodes a plaintext of variable length into a hash of fixed length, which is used in data signature or data authentication. A hash function has two major requirements: one way and high sensitivity. The hash function should also be secure against

P. Singla (✉) · P. Sachdeva · M. Ahmad
Department of Computer Engineering, Faculty of Engineering and Technology,
Jamia Millia Islamia, New Delhi 110025, India
e-mail: prateeks53@gmail.com

S. Sengupta et al. (eds.), *Emerging Trends in Computing and Communication*,
Lecture Notes in Electrical Engineering 298, DOI: 10.1007/978-81-322-1817-3_16,

collision and birthday attack, which makes it difficult to find two plaintexts having the same hash. It has recently been discovered that MD5 and SHA-1 algorithms are no longer secured. The collision attack on MD5 is very easy to carry out and the collision attack on SHA-1 is close to practical [7]. Therefore, it is significant and worthwhile to find some new and practical ways of constructing effective cryptographic hash functions. To explore the features of both chaos and neural networks, they are integrated to exhibit the properties suitable for constructing an effective hash function. In this paper, a four layer neural network having multiple inputs and single output is utilized. Furthermore, the structure of the neural network viz. the weights, biases and transfer function parameters are generated by using chaotic maps to make the system more random and sensitive to the plaintext. The proposed algorithm generates a secure hash function based on four-layer chaotic neural network which not only satisfies the security requirements, but also has high sensitivity, confusion and diffusion properties.

16.2 Proposed Hash Algorithm

The proposed hash function algorithm has the following subsections.

16.2.1 The Neural Network

In the proposed hash function, the neural network shown in Fig. 16.1 is used, which is composed of four layers: the input layer, the first hidden layer, the second hidden layer and the output layer. Let the inputs be a set of 8 characters, $P = [P_1 P_2 \ldots P_8]$, where P_i is an 8-bit character. The output of the input layer is defined as

$$C = F^{n0}(\Sigma W_0 P + B_0, Q_0)$$

where n_0 is a random number ($1 \leq n_0 \leq 10$) generated by the key generator and F is the transfer function, which is the piecewise linear chaotic map [8], defined as

$$x(k+1) = F(x(k), q) = \begin{cases} \frac{x(k)}{q} & 0 < x(k) \leq q \\ \frac{1-x(k)}{1-q} & q < x(k) < 1 \end{cases}$$

where q is the control parameter and satisfies $0 \leq q < 1$. $x(k)$ ranges from 0 to 1.

$W_O = [w_{0,0}, w_{0,1}, \ldots w_{0,7}, w_{1,0} \ldots w_{7,7}]$ is 8×8 set of input weights. B_O is the bias set of size 8×1, and Q_O is the control parameter set of size 8×1. Initially, W_O,B_O and Q_O are generated by the key generator. After C is computed, it becomes $Q_O(Q_O = C)$ for the next iteration, to maintain dependency with the message.

Similarly, D, E and Op are computed as:

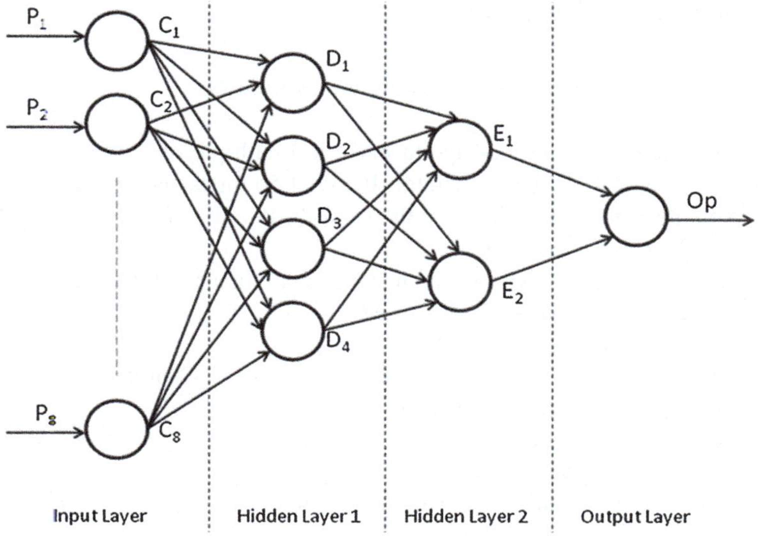

Fig. 16.1 The four layer chaotic neural network

$$D = F^{n1}(\Sigma W_1 C + B_1, Q_1)$$
$$E = F^{n2}(\Sigma W_2 D + B_2, Q_2)$$
$$Op = F^{n3}(\Sigma W_3 E + B_3, Q_3)$$

where W_1 is of size 4×8, W_2 of 2×4, W_3 of 1×2, B_1 of 4×1, B_2 of 2×1, B_3 of 1×1, Q_1 of 4×1, Q_2 of 2×1 and Q_3 of 1×1. After each iteration, the output of each layer becomes the control parameter for that layer for the next iteration ($Q_0 = C$, $Q_1 = D$, $Q_2 = E$, $Q_3 = Op$). Op is the output of the neural network and satisfies $0 \le Op < 1$ and n_1, n_2, n_3 are random numbers generated by the key generator ($1 \le n_1$, n_2, $n_3 \le 10$). The repeated iterations of the transfer functions improve the randomness of the relation between the input and output of each layer, and thus strengthens the hash function.

16.2.2 Key Generator

The key generator uses the 1-D cubic map and accepts a 32-bit key K (as the initial condition), and returns the generated value $y(n)$, defined as

$$y(n+1) = \lambda \cdot y(n) \cdot (1 - y(n) \cdot y(n))$$

where $\lambda = 2.59$, $y_0 = K/2^{32}$. The generated value $y(n)$ satisfies $0 \le y(n) \le 1$.

16.2.3 Proposed Algorithm

The steps of the proposed hash algorithm are as follows:

H.1. Add padding to the message to make it a multiple of 8 bytes
H.2. Input key K to the key generator. Iterate it 50 times and discard the values. Now iterate again to generate the sets W_0, W_1, W_2, W_3, B_0, B_1, B_2, B_3, Q_0, Q_1, Q_2, Q_3, n_0, n_1, n_2, n_3
H.3. Provide the first 8 bytes of the message P to the neural network and obtain the output. The output is discarded
H.4. Assign $Q_0 = C$, $Q_1 = D$, $Q_2 = E$, $Q_3 = Op$
H.5. Repeat step 3 for the next 8 bytes of the message. Continue to do this until the whole message is consumed
H.6. Generate a random number ω from the key generator, such that $0 \leq \omega \leq$ (message length-8)
H.7. Input the consecutive 8 characters of message P_{ω}, $P_{\omega+1} \ldots P_{\omega+7}$ to the neural network and obtain output Op
H.8. Multiply Op with 2^4 to get the first hexadecimal digit of the hash
H.9. Repeat steps 6–8 until a hash of the desired length is obtained.

16.3 Performance Analysis

16.3.1 Hash Results of the Messages

We follow the method used in [9] to analyze the proposed hash function. We have chosen a message of 1,024 null characters. This makes it easier to show the sensitivity of the hash function to the message.

Condition 1: The original message contains 1,024 null characters.
Condition 2: Change last character to 1.
Condition 3: Add one bit to the last character of the message.
Condition 4: Add a space to the end of the message.
Condition 5: Change λ (in Key Generator) = 2.59 to 2.590000000000001.

The corresponding 160-bit hash values in hexadecimals are given as follows:

Condition 1: 561E919CE69F62FE8E3BD3052CE23F27C7C057F6
Condition 2: 8B978E85552D28E6C6B153A717A8716C3079683E
Condition 3: 7CF5C0C060F0C953BB3F6604ECD758A0B428F5D7
Condition 4: D2216EF3694EF09E3EBE62716237FEE1652F786C
Condition 5: 1F304D20C3ECD4AD50102C713786AEF7950C3936

The results indicate that the one-way property is satisfied and the least change in the plaintext or key value causes huge changes in the final hash value.

16.3.2 *Statistical Analysis of Diffusion and Confusion*

Shannon pointed that 'It is possible to solve many kinds of ciphers by statistical analysis' [10]. Therefore, he suggested two methods of diffusion and confusion for the purpose of frustrating the powerful statistical analysis. In an ideal diffusion effect, tiny changes in the initial conditions should have a probability of 50 % of changing each bit of the hash value. A message is selected and the hash value for the message is generated; then, one bit of the message is changed randomly and a new hash value is generated. The two hash values are compared with each other, and the changed bits are counted and called B_i. This test is performed N times [9].

Minimum changed bit number $B_{min} = min(\{B_i\}_1^N)$
Maximum changed bit number $B_{max} = max(\{B_i\}_1^N)$
Mean changed bit number $\bar{B} = \sum_1^n \frac{Bi}{N}$
Mean changed probability $P = \frac{B}{128} * 100\,\%$

Standard variance of the changed bit number $\Delta B = \sqrt{\frac{1}{N-1}\sum_1^N (Bi - \bar{B})^2}$

Standard variance of probability $\Delta P = \sqrt{\frac{1}{N-1}\sum_1^N \left(\frac{Bi}{128-P}\right)^2} * 100\,\%$

Through the tests with N = 1,024, 2,048 and 10,000 respectively, the corresponding data are listed for 128-bit hashes and compared in Table 16.1. It is evident from the resultant statistical data that the mean changed bit number B and mean changed probability P are both very close to the ideal value 64 bits and 50 %. Also, ΔB and ΔP are small, which indicates the capability for diffusion and confusion is very strong and stable. The results obtained are consistent comparable to other hashes investigated in [9, 11].

16.3.3 *Resistance to Birthday Attack*

Birthday attack is a typical attack method used to break a hash function [2]. That is, to find a contradiction, is similar to find two persons with the same birthday. Thus for 64-length hash function, the attack difficulty is not 2^{64}, but much smaller (2^{32}). Considering the practical computing ability, the hash value's length should be at least 128 bits, which keeps the attack difficulty above 2^{64}. The proposed hash is 128 bits in length, and can be easily expanded to any greater length, with minimal alteration to the algorithm.

Table 16.1 Statistical number of changed bits B_i for N = 1,024, 2,048, 10,000

Parameter	In proposed			In Ref. [11]		
	1,024	2,048	10,000	1,024	2,048	10,000
B	63.965	64.087	63.976	63.834	63.895	63.919
$P(\%)$	49.972	50.068	49.981	49.870	49.918	49.937
ΔB	5.825	5.865	5.761	5.866	5.771	5.647
$\Delta P(\%)$	4.550	4.582	4.501	4.583	4.509	4.412
B_{min}	50	45	44	45	43	41
B_{max}	80	82	86	82	82	84

16.4 Conclusion

In this paper, a hash function based on chaotic neural networks is constructed. The chaotic maps provide the necessary sensitivity to the plaintext and key, while the neural network provides one-way property. The proposed hash function not only satisfies the security requirements, such as having a mean changed probability value very close to the ideal value of 50 %, but also has stable, consistent and strong diffusion and confusion capability. Furthermore, the proposed algorithm provides the flexibility to expand the hash length to any arbitrary length, which makes the system resistant to birthday attack for hashes greater than 128-bits. Besides, the proposed algorithm is simple to implement and can be executed much faster by using parallel processing.

References

1. Secure Hash Standard (2002) Federal information processing standards publications (FIPS PUBS) 180(2)
2. Vanstone SA, Menezes AJ, Oorschot PC (1996) Handbook of applied cryptography. CRC Press, Boca Raton, FL
3. Xiao D, Liao X, Wang Y (2009) Parallel keyed hash function construction based on chaotic neural network. Neurocomputing 72(10):2288–2296
4. Lian S, Sun J, Wang Z (2006) Secure hash function based on neural network. Neurocomputing 69(16):2346–2350
5. Wong KW (2003) A combined chaotic cryptographic and hashing scheme. Phys Lett A 307(5–6):292–298
6. Yang H, Wong KW, Liao X, Wang Y, Yang D (2009) One-way hash function construction based on chaotic map network. Chaos, Solitons Fractals 41:2566–2574
7. Wang X, Yin YL, Yu H (2005) Finding collisions in the full SHA-1. LNCS 3621:17–36
8. Li S, Chen G, Mou X (2005) On the dynamical degradation of digital piecewise linear chaotic maps. Int J Bifurcat Chaos 15(10):3119–3151
9. Akhshani A, Behnia S, Akhavan A, Jafarizadeh MA, Abu Hassan H, Hassan Z (2009) Hash function based on hierarchy of 2D piecewise nonlinear chaotic maps. Chaos, Solitons Fractals 42(4):2405–2412
10. Shannon CE (1949) Communication theory of secrecy systems. Bell Sys Tech J 28:656–715
11. Akhavan A, Samsudin A, Akhshani A (2013) A novel parallel hash function based on 3D chaotic map. EURASIP J Adv Signal Process 2013(126)

Chapter 17
Protocol to Authenticate the Objects Attached with Multiple RFID Tags

Subhasish Dhal and Indranil Sengupta

Abstract Use of multiple Radio Frequency Identification (RFID) tags in an object increases the detection rate in comparison to the single tagged object. Since, more than one tag is involved in the same object; the security such as authentication scheme needs to be revised. The authentication schemes that have been proposed so far are only for single tagged object. Modification of these schemes may be applicable to multi-tag environment. However, the advantage of having more resource (tags) can be utilized to enhance the security. We have used this advantage and proposed an authentication scheme which is not only applicable to multi tag environment but also enhance the security.

Keywords Multi tag RFID · Pair-wise secret · Authentication

17.1 Introduction

RFID technology is becoming a powerful tool for identification of any object uniquely. Single RFID tag in an object reveals less detection rate. This necessitates the use of more than one tag in each object. This technology is known as multi-tag RFID technology [1]. More than one tag increase the detection rate and reader-tag communication distance in the presence of metals, liquids, radio noise and adverse environmental conditions. Thus, it is highly applicable to those applications where reliability, availability, and safety are major requirements [1]. Above all these,

S. Dhal (✉) · I. Sengupta
Department of Computer Science and Engineering, Indian Institute of Technology Kharagpur, Kharagpur, India
e-mail: subhasis.rahul@gmail.com

I. Sengupta
e-mail: isg@iitkgp.ac.in

S. Sengupta et al. (eds.), *Emerging Trends in Computing and Communication*, Lecture Notes in Electrical Engineering 298, DOI: 10.1007/978-81-322-1817-3_17,

security is another major concern. RFID tags are easily accessible and hence the sensitive information within the tags is no longer secure. The security requirements such as privacy, authentication and integrity etc. are thus stringent requirements for RFID attached objects.

Our concern is about the authentication of RFID attached objects. Many authentication protocols [2–8] have been proposed for object with single RFID tag. These protocols may be applicable to multi tag RFID. We have tried to utilize the presence of multiple numbers of tags to enhance the security. In our proposed scheme, the secret between object (attached with multiple number of tags), and backend server is kept in the form of shares at object side. Shares are distributed to all the tags in a manner similar to [9]. Thus, the attacker needs to recover at least a threshold number of shares. This increases the security. Further, we have introduced the concept of Master Tag. For an object, a tag is selected randomly as master. This tag is responsible for checking the authentication or validity of received messages. In each session, the master tag releases its responsibility to another randomly selected tag. Thus, the new tag becomes the master tag. Distributing the master responsibility to each tag, we have tried to increase the difficulty for the adversary to guess the tag id of master tag. Further, it helps to reduce the power consumption of tags in case if we have the tags as active or semi passive type. This is because; only one tag is not responsible for computation in all session. For our proposed scheme, the tags attached in the same object need to have their own memory and also can communicate with each other. Thus, our scheme is applicable to active type of tags.

The remainder of the paper is organized as follows. In Sect. 17.2 we have briefly discussed the related schemes which have been proposed recently. In Sect. 17.3, we have described our proposed authentication scheme. In Sect. 17.4, we have analyzed the proposed scheme. We have concluded with the references in Sect. 17.5.

17.2 Related works

Many authentication protocols [2–8] have been proposed for RFID with single tag. We are revisiting a few of them.

Wies et al. [2] proposed hash based scheme. On requesting form reader, it will send the *meta-id* instead of the original *id*. The reader will check the validity of the same by consulting with the database through a secure channel. If that *meta-id* passes validity, the tag will be authenticated. This protocol is simple. However, this will suffer from location privacy problem, since the *meta-id* is not changing and attacker can easily track it. They also suggest *Randomized Access Control (RAC)*. This avoids the tracking attack as each session is having new random number. However, this scheme is suffering from impersonation attack. Because, the attacker can collect the hashed item along with the random number and may send the same to a legitimate reader. Thus, attacker can be able to get the id. Likewise, there are many hash based solutions [3–6].

Zhang et al. [7] proposed a scheme based on a session random number *TSN*, which has stored previously. This scheme suffers from synchronization problem. If the attacker chooses an arbitrary value and sends by pretending the same as the encrypted value of new *TSN*, the legitimate tag cannot be able to understand that this value is not valid. This is because, it still be successful to decrypt the new *TSN* using its keys. Therefore, it would calculate the *TSN* and would update accordingly. Now the *TSN* value in the tag and the backend database are not the same. Hence, tag would become unauthentic. Thus, it may suffer from blocking attack.

Lei et al. [8] proposed an authentication scheme which satisfies many security criteria. However, any of the intermediate value x get compromised, the successive x values for the same tag can be computed. Thus it is not providing forward security.

The above schemes for authentication in RFID are all for that environment where each object is attached with single tag. These schemes can be employed in multi-tag environment without any additional advantage. We have proposed an authentication scheme, which takes the advantage of multiple RFID tags.

17.3 Proposed Authentication Scheme

We assume that any given object would be attached by N number of tags, with each tag having its own memory. Also, they can communicate with each other. A tag is selected in each successful session and given the responsibility to do the authentication task and it does this task using the secret shares kept in other tags in the same object. It uses threshold number of shares to generate such pair-wise secret key. We also assume that the communication from reader and backend database is secure while the communication from reader and tag is insecure. The detail of our proposed scheme is as follows.

17.3.1 The Protocol

Figure 17.1 illustrates the proposed scheme.

Initialization: Before deployment of the RFID tags and readers a few parameters are initialized and preloaded.

1. Each tag is assigned an id *TID* and all the tags are divided into n groups where a group is assigned to a particular object.
2. Each object is assigned secrets from the set L_1, L_2, …, L_n. For simplicity, we denote that S is assigned to object G. Secret S for object G is divided into a number of shares S_1, S_2, …, S_m and distributed to tags for that particular object, using any scheme similar to [9]. Here m is the number of tags in object G.

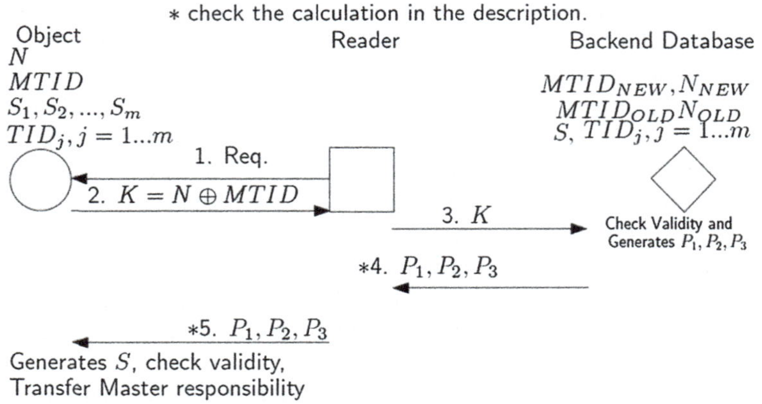

Fig. 17.1 Authentication protocol for multi-tag RFID

3. A tag T_{ij} from each object G_i is randomly selected as master tag for the corresponding object. Here T_{ij} means *jth* tag of object G_i. A different id $MTID_i$ is assigned to that tag. For simplicity, we denote *MTID* as the master *id* for object *G*. A session key *N* is also assigned to it.
4. All the relevant information for each object are kept in two sets, namely, *NEW* and *OLD* in backend database and loaded into corresponding object and hence into the tags. In this way, the objects are deployed.
5. Now, each reader R_t is assigned unique id TID_t along with the secret RS_t. For simplicity, we denote that reader *R* having *RID* as its *id* is assigned the secret *RS*. The relevant information for each reader is also kept in the backend database. In this way the readers are deployed.

Authentic communication steps: The following is the description of our proposed authentication scheme in step by step fashion.

Step 1: Reader *R* broadcasts request message.

Step 2: Master tag of object G calculates $K \leftarrow N \oplus MTID$ and sends back to the backend database via reader.

Step 3: In this step, backend server does the following:

(a) Searches for valid *MTID* from the set *NEW*.
Selects first record and calculates $N' = K \oplus MTID_{NEW}$.
Now, if N' is equals to the *N* of the corresponding record, then validity confirmed.
Otherwise, checks the next record.
Continue the search until a valid record found or exhaust the set *NEW* in the database.
If such record found, go to *Step 3.c*

(b) Searches the *OLD* set in the same way.
If any record found in the *OLD* set, then copy the relevant object information to the set *NEW*. However, if any of the set does not have any match, drops the message, and stops.
(c) Generates a new unique random number N_1.
Calculates $F = N \oplus N_1$, $P_1 = F \oplus S$. S is the secret key assigned to the corresponding object.
Selects another tag randomly from the same group as new master tag for the corresponding object. Suppose the *id* of the selected tag is TID_j.
Generates a new master tag id $MTID_1$.
(d) Calculates $P2 = TID_r \oplus N_1$, where TID_r is the id of the current master tag.
(e) Calculates $Q = TID_j \oplus N$.
Now, attaches $MTID_I$ to Q and get $M = MTID_1 \, || \, Q$.
Calculates $P3 = M \oplus N_1$.
(f) Updates the database as follows:
If K was validated from *NEW* set then copies the corresponding object information from *NEW* set to *OLD* set. Now in set *NEW*, make the following update. $N_{NEW} = N_1$ and $MTID_{NEW} = MTID_1$.
If K was validated then updates $N_{NEW} = N_1$ and $MTID_{NEW} = MTID_1$
(g) Sends P_1, P_2, *and* P_3 to tag via reader.

Step 4: In this step, master tag of object G does the following:

(a) Collects the shares from the other tags to generate S.
(b) Calculates $F = P_1 \oplus S$ and the new random $N_1 = F \oplus N$.
(c) Verifies the validity by calculating $TID_r = P_2 \oplus N_1$.
If TID_r is equals to its own *id*, then validity is confirmed and hence continues to *Step 6.d*, otherwise discards the message and stops.
(d) Determines the new master tag TID_j and master tag *id* $MTID_1$ from $M = P_3 \oplus N_1$.
(e) Updates the values as follows. $N = N_1$ and $MTID = MTID_1$.
(f) Handover the master responsibility to the newly selected tag having *id* TID_j.

In this way, a legitimate reader can read the information about a legitimate object.

17.4 Analysis of the Scheme

17.4.1 Authentication Analysis

We are now in the phase of analysis of our proposed scheme. In the proposed authentication scheme, the communication involves mainly three components namely reader, object, and backend server. The validity of master tag and hence object is confirmed at Step 4, where the backend database checks for a valid record. If any such record exists for which K is valid, then only the validity will be confirmed. Otherwise the backend database will simply drop the message. Authentication of backend server and reader is ensured at Step 6, where the master tag collects the shares from other tags to generate pair wise secret key S. Validity of new session key can be ensured only if the calculated session key (N_1) successfully decrypts the *id* of master tag *TID*. Again, valid N_1 can be recovered if F is valid. Validity of F is possible if the encryption is done by backend server with valid S.

17.4.2 Security Analysis

In our proposed scheme, the communication between reader and object is insecure and adversary may try to impose various kinds of attack.

Eavesdropping: All the messages transferred are encrypted by random key N. Therefore, attacker is unable to compromise the privacy. If she is able to capture any random secret at any time, she will able to capture the further random secret. However, any miss will break the chain.

Physical Attack: The objects we are considering are multi-tagged object. If an attacker wants to introduce a physical attack, she needs to compromise a certain number of tags which is a threshold value to recover the secret between the corresponding object and the database. This is because the secret between database and object lies distributive to all the tags in the same object in the form of shares, and therefore, it is quite difficult.

Location Privacy: The attacker may try to track the object. However, master tag *id MTID* and random secret N are changing in each authentic communication. Therefore, the attacker will be unable to track the object.

Man in the middle attack: Suppose the attacker is in between object and reader. She may block the original messages and send fake messages. However, fake messages will not be verified since the attacker does not have any secret key between object and backend server.

Blocking attack: If adversary blocks the signal at Step 5 (see Fig. 17.1) then the new information will not be updated in the object. Therefore, the information in *NEW* set of backend database and object are not same. However, we keep the old information in *OLD* set of backend database and hence future response from object will retrieve this information and able to authenticate the object. Thus, there will be no synchronization problem between object and backend server. It will also treat this fact as an anomaly. If the number of such events exceeds a certain threshold limit it will raise an alarm. If attacker blocks the message in communication 6 (see Fig. 17.1), it would merely successful to disturb the tag identification of the object under consideration, and however, it will be unable to get the sensitive information.

17.5 Conclusion

RFID tags are easily accessible to attacker and highly prone to attacks. The implementation of any application in these systems is very challenging since the resources are limited. Therefore, authentication as security requirement is also a very challenging task. Many schemes have been proposed for the RFID systems where single tag has been used in an object. These schemes may be applicable to those systems where multiple numbers of tags employed to a single object. However, we have taken the advantage of having multiple numbers of tags. A secret distributed to multiple tags is difficult to reconstruct until and unless a minimum number (threshold value) [9] of shares are collected. This strengthens the security to our proposed scheme. If the tags are active or semi passive type then the distributive computation will reduce the power consumption on a particular tag since the master responsibility is not fixed to that tag for all sessions. Hence, the power consumption is equal in average for all tags which is another advantage of our scheme. However, if successively two session key between object and reader is revealed, it may cause the further session key to be revealed. We have assumed that the attacker is unable to gather such information.

References

1. Bolotnyy L, Robins G (2007) Multi-tag RFID systems. J Internet Protoco Technol (IJIPT) 2(3/4):218–231. Special issue on RFID: technologies, applications, and trends
2. Weis SA, Sarma S, Rivest RL, Engels DW (2004) Security and privacy aspects of low-cost radio frequency identification systems. In: Proceedings of the first security in pervasive computing, LNCS-2802. Springer, New York
3. Tan C, Sheng B, Li Q (2007) Serverless search and authentication protocols for RFID. In: Proceedings of the fifth annual IEEE international conference on pervasive computing and communication (PerCom)
4. Tsudik G (2007) A family of dunces: trivial RFID identification and authentication protocols. In: Proceedings of PET
5. Burmester M, Van Le T, de Medeiros B (2006) Provably secure ubiquitous systems: universally composable RFID authentication protocols. In: Proceedings of SECURECOMM
6. Conti M, Di Pietro R, Vincenzo Mancini L (2007) RIPP-FS: an RFID identification, privacy preserving protocol with forward secrecy. In: Proceedings of PerCom'07, pp 229–234
7. Zhang Y, Li D, Zhu Z (2008) An efficient RFID tag-reader mutual authentication scheme. In: Proceedings of WiCOM'08, pp 1–4

8. Lei H, Yong G, Zeng-yu C, Na-na L (2009) An improved lightweight RFID protocol using substring. In: Proceedings of the 5th international conference on Wireless communications, networking and mobile computing, pp 3717–3720
9. Khalili A, Katz J, Arbaugh WA (2003) Toward secure key distribution in truly ad-hoc networks. In: Proceedings of symposium on applications and the internet workshops (SAINT 2003 Workshops), Saint-W, p 342

Chapter 18
ACO Based QoS Aware Routing for Wireless Sensor Networks with Heterogeneous Nodes

Sanjay Kumar, Mayank Dave and Surender Dahiya

Abstract Most of the existing routing protocols for Wireless Sensor Networks (WSN) consider homogeneous nodes wherein all sensor nodes have the same capabilities in terms of sensing, communication and computation capabilities. However, a homogeneous sensor network may suffer from poor performance and scalability. This paper presents an ant-based QoS routing protocol for Heterogeneous Wireless Sensor Networks (HWSN). The key feature of the protocol is its ability to meet diverse QoS requirements posed by different kinds of traffic generated due to heterogeneous nature of nodes thus maximizing network performance and its utilization. We have evaluated and compared the proposed novel solution with EEABR and AODV for environments of dynamic topology.

Keywords Wireless sensor networks · Wireless multimedia sensor networks · Heterogeneous wireless sensor networks · QoS · Ant routing

18.1 Introduction

The ability of wireless sensor devices to capture multimedia content from the environment has gradually shifted the paradigm from existing scalar sensor services (light, temperature, etc.) to a new world of real-time audio-visual applications and

S. Kumar (✉)
Hindu College of Engineering, Sonepat, 131001 Haryana, India
e-mail: skm_09@rediffmail.com

M. Dave
National Institute of Technology, Kurukshetra, Haryana, India
e-mail: mdave67@yahoo.com

S. Dahiya
Deenbandhu Chhotu Ram University of Science and Technology, Murthal, Sonepat, Haryana, India
e-mail: Dahiyasurender73@gmail.com

S. Sengupta et al. (eds.), *Emerging Trends in Computing and Communication*,
Lecture Notes in Electrical Engineering 298, DOI: 10.1007/978-81-322-1817-3_18,

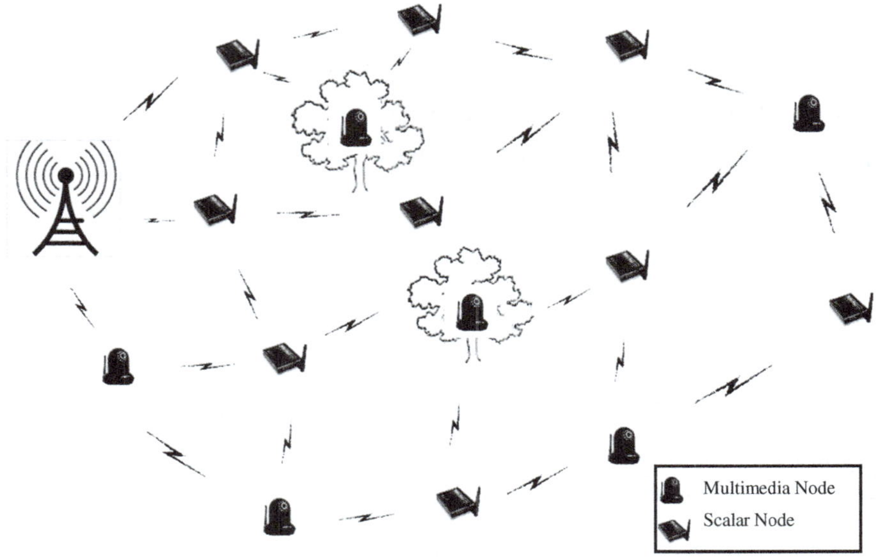

Fig. 18.1 Heterogeneous wireless sensor network

thus evolution of wireless multimedia sensor networks (WMSNs) [1, 2]. The WMSN may consist of heterogeneous nodes with moderated capabilities. Such a network is called Heterogeneous WSN (HWSN). HWSN nodes may be heterogeneous in terms of energy, wireless links, hardware or security. This helps in limiting the WSN cost as instead of using a full sensor set on a node, heterogeneous nodes may be deployed at different locations depending on the application and criticality of situation [3]. Due to the additional requirements of HWSNs it is important to consider the impact of node's heterogeneity in terms of energy, sensed data, and bandwidth requirement while designing routing algorithms for HWSNs so as to achieve optimal performance. This paper proposes an Ant based QoS routing protocol for HWSNs (AntQHSeN). The key feature of the protocol is its ability to meet diverse QoS requirements posed by different kinds of traffic generated by heterogeneous nodes thus maximizing network performance and its utilization. The routing decisions for control packets, scalar data packets and multimedia data packets are taken independently and in different manners satisfying their respective QoS requirements. Moreover, as some applications require minimum bandwidth support, which if not provided will make entire data useless, therefore, if minimum bandwidth requirement cannot be met for such applications, data should not be transmitted [4]. AntQHSeN addresses this issue by using admission control scheme. For all other applications which do not impose strict bandwidth constraints, the protocol determines the minimum bandwidth along the route from source to destination Figure 18.1 shows the heterogeneous sensor network.

The remainder of the paper is organized as follows: Sect. 18.2 provides a brief review of some of the closely related works. The proposed protocol is described in

Sect. 18.3. Then AntQHSeN is tested through a series of computer simulations presented in Sect. 18.4. Section 18.5 concludes the paper.

18.2 Related Work

The introduction of Wireless Multimedia Sensor Networks (WMSNs) has revolutionized the scope and applications of wireless sensor networks which require the delivery of multimedia content with a certain level of QoS. There has been a host of research works on QoS routing for WMSNs.

SPEED [5], a spatio-temporal, maintains a desired delivery speed across the network and provides soft real-time, end-to-end delay guarantees. MMSPEED [6] protocol is an integration of reinforcement learning based probabilistic multipath forwarding with soft real-time guarantee of SPEED. However, due to per packet route computation it consumes more energy and thus reduces network lifetime. A multi-path and multi-channel based QoS-aware protocol proposed by Hamid et al. [7] makes routing decision according to the dynamic adjustment of the required bandwidth and path-length-based proportional delay differentiation for real-time data. PEMuR [8] is an efficient protocol for video communication over WMSNs. It ensures low power consumption over all sensor nodes and high perceived video QoS by making a combined use of hierarchical routing and video packet scheduling models. SDRCS [9] provides soft real-time guarantees for event-based traffic in WSNs. It uses grouping approach to estimate end-to-end hop distance and to meet various application requirements. It performs distributed packet traversal speed estimation for traffic classification and admission control, and prioritized packet forwarding for local routing decisions. Sun et al. [10] proposed a new routing metric called Load Balanced Airtime with Energy Awareness (LBA-EA). EEQAR proposed by Lin et al. [11] adopts cellular topology to form the cluster structure and balances the energy consumption by structure movement resulting in enhanced performance.

Swarm intelligence techniques are also prominently used in solving routing issues in WSNs. An ant based protocol, ASAR [12] is aimed at periodically selecting QoS routing paths for each three types of services—event-driven service, data query service and stream query service. ACOLBR [13] is another hierarchical protocol which is based on the concept of constructing a minimum spanning tree rooted at the cluster-head for intra-cluster routing. ACOWMSN [14] for WMSN aims at finding an optimize path from source to sink node. Based on ant colony heuristics, it uses probabilistic approach to find next hop that can satisfy multiple QoS constraints. A notable related approach is AntSensNet [15], an ant-based multi-QoS routing protocol for multimedia sensor networks with heterogeneous nodes and proposes a biologically inspired clustering process. For providing QoS guarantee an ant, and mobile agent based protocol QR2A [16] extends the local as well as global pheromone deposition rules and updation rules of ant algorithm. The algorithm meets the QoS requirements and solves the problem of network load balancing effectively. BiO4SeL [17] is a distributed and autonomic ant-based routing protocol that aims to maximize sensor network lifetime. It uses battery power information to update the distributed routing tables as battery power is consumed. EEABR [18] is an improved version of the Ant based routing in WSN and is designed to extend the lifetime. However, EEABR is weak in terms of scalability and reliability as it lacks quality of service and increases excessive delay in packet delivery.

Based on above survey, is observed that the adaptive routing protocols are desirable for WSNs. Therefore, to meet the diverse requirements of HWSNs along with simplicity, it is desirable to have a multi-hop communication based protocol with a cross-layer support to select the best routes.

18.3 Ant Based QoS Routing Protocol for HWSNs

This section describes an Ant based QoS enabled routing protocol for Heterogeneous WSNs (AntQHSeN) with multifarious and inherently conflicting demands. The method is intended for networks with a single data sink. The network consists of multimedia sensor nodes, scalar nodes, and an access point. The multimedia nodes are capable of sensing multimedia data such as audio, video and photo. Scalar nodes are the nodes with simple sensing capabilities such as temperature sensor, humidity sensor etc., and the data gathered by these scalar nodes, hereinafter, is called scalar data.

AntQHSeN is a reactive routing protocol consisting of two operational phases—route discovery phase and route maintenance phase. The route discovery phase sets up paths when they are needed at the start of a session and no routing information for the destination node is available. Source node finds multiple paths to the destination by launching ant agents called Forward ants. These ants also carry the network information to be used for evaluating the quality of path such as available bandwidth and residual energy of the intermediate nodes lying on the path. Backward ants are then sent by the destination node to the source node, completing the reactive route setup phase. Once the routing path has been set, the source node starts sending data packets stochastically over different paths using pheromone values as well as a heuristic function taken together. Route maintenance phase starts when link failures are encountered.

AntQHSeN protocol considers residual bandwidth, minimum residual energy and route cost to compute the pheromone concentration. Residual bandwidth describes the bandwidth availability along the routing path and is a measure of residual channel capacity. If the source, demands for some minimum bandwidth, admission control scheme is used, else the minimum available bandwidth along the route is determined. In admission control scheme, if any node during route selection cannot satisfy the bandwidth constraints imposed by the source, the route selection procedure is terminated. Minimum residual energy identifies high energy paths and more pheromone is deposited along the path with high residual energy. Route cost is calculated in terms of end-to-end delay and hop count.

AntQHSeN uses Hello ants which are periodic messages and are used to find out immediate neighbor nodes and to detect link failures. Hello ant packet also contains the Bandwidth (b_i), Timestamp (T_j), Energy (e_i) and Pheromone (τ_i^d) information. When a source node has no routing information for a destination node, it generates Reactive Forward ants and initiates reactive route discovery process. These ants keep record of all intermediate nodes visited by it in reaching from source to the destination node. The information passed and accumulated through the Reactive Forward ants is used by the destination node to compute the pheromone values to be deposited on the route. In addition to source address, destination address, and sequence numbers following fields are required for making QoS based routing decisions:

Flag (*f*)—indicates whether the source is using the admission control scheme or not.

Requested Bandwidth (b_{rq})—denotes the minimum bandwidth as desired by the source node. This field is significant only if the flag is set and every intermediate node from source to destination receiving the forward ant compares the available bandwidth with the value stored in this field.

Minimum Bandwidth ($b_{\min}$)—indicates the minimum of bandwidth available with the nodes from source to the current node and is a measure of maximum bandwidth supported by the route.

Route Energy ($e_{\min}$)—stores the minimum of residual battery capacity of nodes from the source to the current node along the path traversed by the Forward ant and is an indication of the route's lifetime.

Each ant generated by source node s has a goal to determine a path to the destination d, which can satisfy given QoS requirements. Initially, when generated at source node, ant contains address of source node and that of destination node. On its way to the destination node, the ant keeps record of all the intermediate nodes visited by it. The source node broadcasts this *FA* to its neighboring nodes.

The intermediate node i when receives the ant, it first calculates its residual bandwidth b_i. Depending upon the status of flag bit and its residual bandwidth, the node either forwards the ant or drops it. If the flag bit is set $b_i > b_{rq}$, it forwards this *FA*. Otherwise, the node discards this ant. In case the flag bit is not set, the node compares its residual bandwidth with the minimum bandwidth field in the *FA*. If $b_i < b_{rq}$, the node replaces minimum bandwidth value in the ant with residual bandwidth. Otherwise, the node simply forwards the ant packet. Similarly, node updates route energy field in ant packet by comparing its own residual battery capacity e_i and value contained in the route energy $e_{\min}$ field of the packet. If former is greater than the latter, ant packet is forwarded else route energy field of the packet is updated using its residual battery capacity. Node n_i updates fields in the *FA* message as follows:

$$b_{\min} = \begin{cases} b_{\min} = \min(b_{\min}, b_i), & flag = 0 \\ b_{rq}, & flag = 1 \ \&\ b_i \geq b_{rq} \\ drop, & otherwise \end{cases} \tag{18.1}$$

$$e_{\min} = \min(e_{\min}, e_i) \tag{18.2}$$

Depending upon the availability of routing information for *d*, intermediate node either unicasts or broadcasts the ant packet. If routing information is available, the node makes probabilistic decision to select next hop for ant packet. The decision is based on the pheromone values associated with each next hop for *d*. The probability of selecting node *n* as next node by current node *i* is given as [19]:

$$P_{i\,n}^{d} = \frac{\left(\tau_{i\,n}^{d}\right)^{\beta_1}}{\sum\limits_{j \in N_i^d} \left(\tau_{ij}^{d}\right)^{\beta_1}}, \quad \beta_1 \geq 1, \tag{18.3}$$

where

$\tau_{i\,n}^{d}$ is pheromone value for next node n
N_i^d is the set of neighbors of i over which path to d is known
β_1 is a parameter that controls exploratory behavior of the ant.

If routing information for destination *d* is not available at the node, the node broadcasts the forward ant packet. The ant packet while traveling from source to destination collects status information of nodes along the route as per Eqs. (18.1) and (18.2). Therefore, when it reaches the destination, it has minimum bandwidth value that can be supported by the route and minimum energy of the route. This information is crucial in determining the quality of path in term of pheromone value.

On receiving a *FA*, destination node creates a *BA*. The status information of the route contained in *FA* is copied to *BA* in the following manner:

$$\begin{aligned} r_b &= b_{\min} \\ e_b &= e_{\min} \end{aligned} \tag{18.4}$$

The *BA* also contains the addresses of the forward ant's source node *s* and destination node *d*, as well as the full list of nodes that the forward ant has visited. The *BA* is unicast from destination *d* to source s along the same path the *FA* had traveled but in reverse direction.

The *BA* updates pheromone value $\tau_{i\,n}^{d}$ in the table for destination *d* on each intermediate node *i*, till it reaches source node. Here *n* is the node that the ant visited before *i* on its way back from *d*. The pheromone value to be deposited on a node is determined by the route status information carried by the ant. It also considers hop count and delay in reaching the current node from the destination node. It is interesting to note here that rather than relying completely on global information as provided by Forward ant, AntQHSeN combines this global information with the estimates calculated locally by the nodes for pheromone computation. h_{in}^{d} is the hop count and t_{in}^{d} is the delay incurred by a packet from d to i through n which is the node that the ant visited before i, t_{hop} is the time needed to take one hop in unloaded conditions. This is in order to improve reliability and provide better approximation of the measured values. The amount of pheromone released by Backward ant is given by $\tau_{in}^{d'}$ as follows:

$$\tau_{in}^{d'} = \alpha_r \times (R) + \alpha_e \times (E) + \alpha_t \times (T), \quad 0 \leq \alpha_r, \alpha_e, \alpha_t \leq 1 \tag{18.5}$$

where

$$R = \frac{r_b}{BW_{channel}},$$
$$E = e_b, \text{ and}$$
$$T = \left(\frac{t_{in}^{d} + h_{in}^{d} \times t_{hop}}{2}\right)^{-1}.$$

$BW_{channel}$ is bandwidth of wireless channel. Due to local burst of traffic or any other reason there may be large variations in time estimates gathered by the ants. To take into account these large oscillations, instead of considering time estimates gathered by the ants only, the average of estimated time and time under ideal circumstances has been taken. By doing so it also takes into account both end-to-end delay and number of hops. α_r, α_e, and α_t are weight factors of rate, energy and time respectively, and their values can be set as per QoS requirements. We can set value of α_r to minimum if the application is bandwidth insensitive, otherwise, higher value can be set and similar consideration can be made while choosing values for other weight factors.

The pheromone value τ_{in}^{d} in node i is updated as follows:

$$\tau_{in}^{d} = \gamma\, \tau_{in}^{d'} + (1 - \gamma)\, \tau_{in}^{d}, \quad 0 \leq \gamma \leq 1 \tag{18.6}$$

This updated pheromone value is diffused in the network by Hello ants. The source node starts data forwarding on receiving the Backward ant. Till that period the data packets are buffered in the source node. If no Backward ant is received within some stipulated interval, source node restarts the reactive path setup phase. However, if source node does not receive any Backward ant even after maximum number of retries, the source node discards the buffered data.

During the reactive path setup phase multiple paths are created between the source and destination pair. The algorithm does not determine a single better path out of available multiple paths for data transmission, rather data is forwarded stochastically. At each node there can be multiple next nodes for destination d and every intermediate node takes an independent decision to select next hop for data packet forwarding. The probability of selecting next hop for data forwarding is determined on the basis of pheromone value deposited on each node for destination d and the heuristic function. The probabilistic rule to determine the probability of moving from node i to node j for destination d is given as:

$$P_{in}^{d} = \frac{\left(\tau_{in}^{d}\right)^{\alpha}\left(\eta_{in}^{d}\right)^{\beta}}{\sum_{j \in N_i^d}\left(\tau_{ij}^{d}\right)^{\alpha}\left(\eta_{ij}^{d}\right)^{\beta}}, \quad \alpha, \beta \geq 1 \tag{18.7}$$

η_{ij}^{h} is the heuristic evaluation function. α and β are parameters that control the relative weight of the pheromone trail and heuristic value respectively. Pheromone value is an indication of global information, whereas, heuristic value is based on local status information. Therefore, both local as well as global status information contribute towards next hop selection.

As different applications pose different QoS requirements, therefore performing application specific data forwarding as per the QoS requirements is a major contribution of the paper. Here, we assume a heterogeneous WSN, in which we have multimedia nodes for sensing multimedia data and scalar nodes for sensing scalar data.

Multimedia traffic does not require 100 % reliability; rather it poses more strict requirements on minimum bandwidth, energy efficiency and bounded delay. Although, the pheromone value deposited on the nodes reflect global estimation of all these three factors, however, it is important to rely on local bandwidth estimation before making data forwarding decisions. High pheromone value does not necessarily account for high residual bandwidth, as it may be due to high energy or low delay. A low bandwidth link for multimedia data transmission can lead to high packet drop rate, thus resulting in frequent re-transmissions. To resolve this challenge, the proposed protocol gives due weightage to residual bandwidth of neighboring nodes along with value of deposited pheromone during data forwarding. Figure 18.3 shows the forwarding of data packets through high and low bandwidth paths as per their QoS requirements.

When a data packet is received by a source node or an intermediate node and routing path is available for destination d, the node first checks the type of received data packet. If it is multimedia data packet, the probability P_{ij}^{d} of selecting next node j is given as:

$$P_{ij}^{d} = \frac{\left(\tau_{ij}^{d}\right)^{\alpha}\left(b_j\right)^{\beta}}{\sum\limits_{k \in N_i^d}\left(\tau_{ik}^{d}\right)^{\alpha}\left(b_k\right)^{\beta}}, \quad \alpha, \beta \geq 1 \tag{18.8}$$

where b_j heuristic evaluation factor in considering j as next hop for destination d for multimedia data and is a measure of residual bandwidth of neighbor node j. The pheromone and heuristic value are controlled by α and β respectively. High value for β makes the algorithm greedy with respect to high bandwidth paths.

In heterogeneous WSNs, when both multimedia nodes and scalar nodes are sensing and transmitting data, scalar nodes should be more conservative in terms of energy while forwarding data. In the existing situation, where multimedia streams require high bandwidth routes, it becomes important to select high energy nodes for scalar data routing and hence enhance network lifetime. Considering this point of view, the proposed protocol takes residual energy of neighboring nodes into account while selecting next hop for data forwarding.

When a node receives scalar data for which routing information is available, the node chooses next hop with probability P_{ij}^{d} as:

$$P_{ij}^{d} = \frac{\left(\tau_{ij}^{d}\right)^{\alpha}\left(e_j\right)^{\beta}}{\sum\limits_{k \in N_i^d} \left(\tau_{ik}^{d}\right)^{\alpha}\left(e_k\right)^{\beta}}, \quad \alpha, \beta \geq 1 \tag{18.9}$$

where e_j is residual energy of neighbor node j and is heuristic evaluation factor in considering j as next hop for destination d for general data. Similar to Eq. 18.8. α and β control pheromone and heuristic value respectively. The protocol selects routes with higher residual energy with high β value.

Contrary, to conventional routing protocols in which a single best path is selected by source for data forwarding, the probabilistic routing strategy leads to automatic data load spreading according to the estimated quality of the paths. The protocol continuously senses the network status and adapts data traffic as per the QoS requirements and prevailing network conditions leading to enhanced performance.

18.4 Performance Evaluation

We use Mannasim [20] and Network Simulator ns-2.34 [21] to evaluate the performance and efficiency of AntQHSeN. We have considered two types of nodes scalars (S-sensor) and multimedia (M-sensor). Multimedia nodes have more energy and longer transmission range, than, the scalar nodes. But, at the same time they consume more energy in processing of multimedia data and its transmission. The radio range of scalar nodes spans 15 m while that of multimedia nodes spans up to 100 m. The data rate equals 1 Mbit/s. Each simulation run lasts for 600 s, and each result is averaged over five random network topologies.

Figure 18.2 shows the packet delivery fraction (PDF) of AODV, EEABR and AntQHSeN. We find that the PDF of AntQHSeN is significantly higher compared with AODV and EEABR. At the beginning, AntQHSeN lacks sufficient information in order to find appropriate routes, but after a certain period of time, when the algorithm converges and the ants have gathered much node and route information, the algorithm routes packets as per their desired QoS constraints and thus the quality of routes discovered for the AntQHSeN is superior to those found by AODV. The other important observation is that EEABR does not provide a consistent performance. The inconsistent behavior of EEABR is due to its proactive route discovery mechanism. As the number of control packets increase in the network, congestion occurs and collisions increase. As a result, forward ants start losing their way to the sink node and probability values do not stabilize which leads to the loss of data packets.

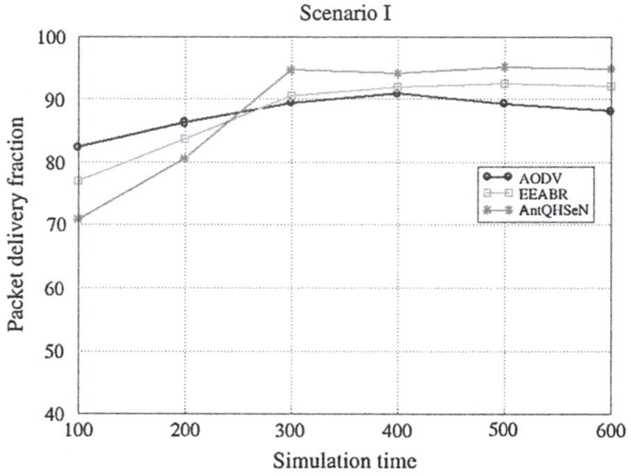

Fig. 18.2 Packet delivery fraction versus simulation time

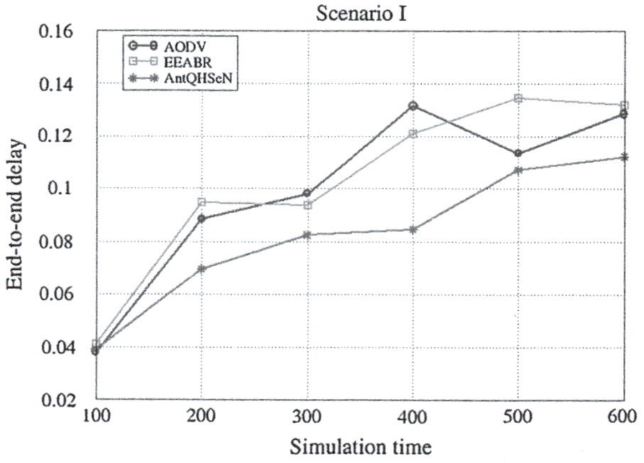

Fig. 18.3 End-to-end delay versus simulation time

In Fig. 18.3 the average end-to-end (EED) comparison between the protocols is depicted. Despite the fact that AODV selects shortest routing path, AntQHSeN protocol has considerably lower average EED than AODV. This is due to the discovery of multiple paths during route establishment, therefore, when a path to the destination breaks, packets could immediately continue to be forwarded using another paths without a new route discovery process. EEABR has approximately same average EED delay as that of AODV.

Routing overhead is shown in Fig. 18.4. Both AODV and AntQHSeN are reactive protocols, but, in AntQHSeN the size of control packets is larger than that of AODV due to extra control information required for network status information. Moreover, contrary to AODV in which any intermediate node having route to the destination can reply to the source node, in AntQHSeN only destination node can send backward ant which leads to extra overhead. Although, in AODV all nodes try to find the shortest path which may lead to congestion resulting in packet drops and re-transmissions still AntQHSeN has higher routing overhead than AODV due

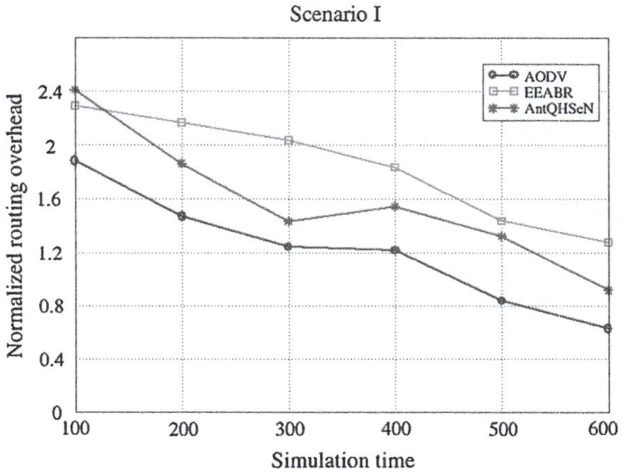

Fig. 18.4 Normalized routing overhead versus simulation time

to reasons cited above. Extra forward and backward ants required to maintain proactive paths in EEABR leads to still higher routing overheads in EEABR.

18.5 Conclusion

The pace of technological growth has led to the proliferation of multimedia sensor nodes and thus multiplicative enhancement in application areas of WSNs. Multimedia sensors as well as scalar sensors can be deployed in a region to monitor environmental data as well as to detect intrusion. Hence the application layer data can be categorized as scalar and multimedia with diverse QoS requirements. Given such motivation this paper proposes an ant based QoS routing protocol for heterogeneous WSNs—AntQHSeN. The routing algorithm categorizes entire traffic into routing traffic and data traffic. Data traffic is further categorized into multimedia traffic and scalar traffic. The routing decision is taken on the basis of traffic type as well as QoS constraints posed by that traffic. This paper proposes three different methods to handle three different types of data i.e. routing, multimedia and scalar data, thus improving network performance. Simulation results show that the performance of AntQHSeN outperforms the standard AODV and EEABR in terms of packet delivery fraction, end-to-end delay and routing overhead.

References

1. Gurses E, Akan OB (2005) Multimedia communication in wireless sensor networks. Ann Telecommun 60(7–8):799–827
2. Misra S, Dhurandher SK, Obaidat MS, Gupta P, Verma K, Narula P (2010) An ant swarm-inspired energy-aware routing protocol for wireless ad-hoc networks. J Syst Softw 83:2188–2199
3. Hadjidj A, Bouabdallah A, Challal Y (2011) HDMRP: an efficient fault-tolerant multipath routing protocol for heterogeneous wireless sensor networks. In: Proceedings of the 7th

international conference on heterogeneous networking for quality, reliability, security and robustness (Qshine), vol 74, Houston, USA, Nov 2010, Published in Springer LNICST, pp 469–482
4. Chen L, Heinzelman WB (2005) QoS-aware routing based on bandwidth estimation for mobile ad hoc networks. J Sel Areas Commun 23(3):561–572
5. He T, Stankovic JA, Lu C, Abdelzaher TF (2005) A spatiotemporal communication protocol for wireless sensor networks. IEEE Trans Parallel Distrib Syst 16(10):995–1006
6. Felemban E, Lee C, Ekici E (2006) MMSPEED: multipath multi-SPEED protocol for QoS guarantee of reliability and timeliness in wireless sensor networks. IEEE Trans Mobile Comput 5(6):738–754
7. Hamid M, Alam M, Seon HC (2008) Design of a QoS-aware routing mechanism for wireless multimedia sensor networks. In: Proceedings of the IEEE global telecommunications conference, pp 800–805
8. Kandris D, Tsagkaropoulos M, Politis I, Tzes A, Kotsopoulos S (2011) Energy efficient and perceived QoS aware video routing over wireless multimedia sensor networks. Ad Hoc Netw 9:591–607
9. Xue Y, Ramamurthy B, Vuran MC (2011) SDRCS: a service-differentiated real-time communication scheme for event sensing in wireless sensor networks. Comput Netw 55:3287–3302
10. Sun W, Song Y, Chen M (2010) A load-balanced and energy-aware routing metric for wireless multimedia sensor networks. In: Proceedings of the IET 3rd international conference on wireless, mobile and multimedia networks (ICWMMN 2010), Beijing, China, pp 21–24
11. Lin K, Rodrigues JJPC, Ge H, Xiong N, Liang X (2011) Energy efficiency QoS assurance routing in wireless multimedia sensor networks. IEEE Syst J 5(4):495–505
12. Bi J, Li Z, Wang R (2010) An ant colony optimization-based load balancing routing algorithm for wireless multimedia sensor networks. In: Proceedings of the 12th IEEE international conference on communication technology (ICCT'10), pp 584–587
13. Yu X, Luo J, Huang J (2011) An ant colony optimization-based QoS routing algorithm for wireless multimedia sensor networks. In: Proceedings of the IEEE 3rd international conference on communication software and networks (ICCSN), pp 37–41
14. Cobo L, Quintero A, Pierre S (2010) Ant-based routing for wireless multimedia sensor networks using multiple QoS metrics. Comput Netw 54:2991–3010
15. Huaihu C (2012) A QoS routing algorithm based on ant colony optimization and mobile agent. Procedia Eng 29:1208–1212. In: International workshop on information and electronics engineering (IWIEE)
16. Liu M, Xu S, Sun S (2012) An agent-assisted QoS-based routing algorithm for wireless sensor networks. J Netw Comput Appl 35:29–36
17. Ribeiro LB, Castro MF (2010) BiO4SeL: a bio-inspired routing algorithm for sensor network lifetime optimization. In: Proceedings of the 17th international conference on telecommunications, pp 728–724
18. Camilo T, Carreto C, Silva J, Boavida F (2006) An energy-efficient ant-based routing algorithm for wireless sensor networks. In: Proceedings of the ant colony optimization and swarm intelligence, Brussels, Belgium, pp 49–59
19. Di Caro G, Ducatelle F, Gambardella LM (2005) AntHocNet: an adaptive nature inspired algorithm for routing in mobile ad hoc networks. Eur Trans Telecommun 16(2):443–455
20. The Mannasim. http://www.mannasim.dcc.ufmg.br/
21. The Network Simulator ns2. http://www.isi.edu/nsnam/ns/

Chapter 19
RC4 Stream Cipher with a Modified Random KSA

Suman Das, Hemanta Dey and Ranjan Ghosh

Abstract The RC4 stream cipher has two components: KSA and PRGA. Though this simple and fast cipher proved itself as robust enough and it is trusted by many organizations, though a number of researchers claimed that RC4 has some weakness and bias in its internal states. Some researchers pointed to the *swap* function of RC4 as a main reason of weakness, especially in the KSA. The authors replaced the KSA randomly with a robust PRBG, BBS, to fill-up the internal state array, which they named as the KSA-R, to eliminate the swap function from KSA. The original RC4 and the modified RC4 are tested with NIST Statistical Test Suite, and it has been found that RC4 with KSA-R is giving a better security.

Keywords Modified RC4 · Random RC4 · Random KSA · Random S-Box · Modified KSA

19.1 Introduction

RC4, designed by Ron Rivest in 1987, is one of the most simple and fast stream ciphers. It contains an initialization routine and a random number generator. The random values are selected from a secret internal state array and two elements are swapped for every byte reported. Based on this table-shuffling principle, RC4 is designed for fast software and hardware implementation and widely used in many commercial products and standards. The RC4 cryptanalysis has been mainly

S. Das (✉)
Department of Computer Science and Engineering, University of Calcutta, Kolkata, India
e-mail: aami.suman@gmail.com

H. Dey · R. Ghosh
Institute of Radio Physics and Electronics, University of Calcutta, Kolkata, India
e-mail: hemantadey13@gmail.com

S. Sengupta et al. (eds.), *Emerging Trends in Computing and Communication*, Lecture Notes in Electrical Engineering 298, DOI: 10.1007/978-81-322-1817-3_19,

Table 19.1 The RC4 Stream Cipher

KSA	PRGA
Input: Secret Key K	Input: S-Box S—The o/p of KSA
for $i = 0, \ldots N - 1$	
$S[i] = i;$	$i = 0; j = 0;$
next i	while *TRUE*
	$\{i = i + 1$
$j = 0;$	$j = j + S[i]$
for $i = 0, \ldots, N - 1$	swap($S[i]$, $S[j]$);
$\{j = j + S[i] + K[i]$	$z = S[S[i] + S[j]];$
swap($S[i]$, $S[j]$);	}
}	
next i	
Output: S-Box S generated by K	Output: Random Stream Z

devoted to the statistical analysis of the output sequence, or to the initialization weaknesses.

RC4 contains a secret array S of size N (generally, 256), in which integers (0 to $N - 1$) are *swapped*, depending upon two index pointers i and j in a deterministic (for i) and pseudo-random (for j) way. There are two components of the cipher: The Key-Scheduling Algorithm (KSA), and the Pseudo-Random Generation Algorithm (PRGA).

The KSA turns an identity permutation into a random-looking permutation and the PRGA generates the keystream bytes, which are XORed with the plaintext bytes to generate ciphertext bytes. All additions in both the KSA and the PRGA are additions modulo N (Table 19.1).

There are several works on strength and weakness of RC4. It has been argued that there are many biases in the PRGA due to the propagation of biases in the KSA, which shows that there is significant interest in the cryptographic community for RC4. In this paper, we have modified the KSA in a random way and compared and analyze this variant of RC4 (RC4-R, as we named it) statistically with the original RC4, following the guidelines given by NIST (National Institute of Standards and Technology), USA in their Statistical Test Suite. It has been found that though RC4 itself is quite secured to use, even after so many years of its primary design, the new variant is able to prove itself more efficient.

19.2 Motivation

RC4 has gone through tremendous analysis since it has become public. Roos [1] showed some weakness in KSA and defined weak keys for RC4 with some important technical results. He showed strong correlation between the secret key bytes and the final key stream generated. He also identified several classes of weak keys.

Paul and Preneel [2] presented new statistical bias in the distribution of the first two output bytes of the RC4 key stream generator and also proposed a new key stream generator namely RC4A with much less operations per output byte. They also described a new statistical weakness in the first two output bytes of RC4 key stream and recommended to drop at least the initial 2 N bytes, where N is the size of the internal S-Box. They proposed to introduce more random variables in PRGA to reduce the correlation between the internal and the external states.

Maitra and Paul [3] revolved the non-uniformity in RC4 KSA and proposed for additional layers over the KSA and PRGA. They named the modified cipher as RC4+, which avoids existing weaknesses of RC4. They presented a three-layer architecture in a scrambling phase after the initialization to remove weaknesses of KSA (KSA+). They also introduced some extra phases to improve the PRGA (PRGA+).

Akgün et al. [4] detected a new bias in the RC4 KSA and proposed a new algorithm to retrieve the RC4 key in a faster way. Their framework significantly increases the success rate of key retrieval attack. They showed that KSA leaks information about the secret key if the initial state table is known.

Noman et al. [5] presented efficient network implementation of RC4A to achieve high data throughput, parameterized to support variable key lengths. They proposed RC4A as a more suitable alternative in respect of cost and security.

Tomasevic and Bojanic [6] introduced an abstraction in the form of general conditions about the current state of RC4. Strategy has been used to favor more promising values that should be assigned to unknown entries in the RC4 table. They proposed a new technique to improve cryptanalytic attack on RC4, which is based on new information from the tree representation of RC4.

Nawaz et al. [7] introduced a new 32-bit RC4 like faster key stream generator. It has a huge internal state and offers higher resistance against state recovery attacks. This is suitable for high speed software encryption.

19.3 A Modified Random KSA of RC4 (KSA-R)

Roos and others strongly discussed about the weakness of RC4 KSA and weak keys in RC4. In this paper, we will give our attention on the weakness of KSA only. Roos [1] argued that in KSA, only the line of *swap* directly affects the state table S while exchanging two elements and hence the previous line $j = j + S[i] + K[i]$ is responsible for calculating the indexes. Here the variable i is deterministic and j is pseudo-random. Therefore, the swap between two elements may happen once, more than once, or may not happen at all—thus brings a weakness in the KSA. He argued that there will be a high probability of about 37 % for an element not to be swapped at all.

In this paper, we propose to introduce a new type of KSA (KSA-R) in place of the original one, where all the cells of the internal state array S will be filled up by a strong suitable PRNG/PRBG (Pseudo-Random Number/Bit Generator). By this way we are able to remove the swap function and its complexities. As a PRBG, we have used the BBS (Blum–Blum-Shub), which is considered to be a secured PRBG by many researchers. Junod [8] discussed about the strength of BBS as a generator that it is cryptographically strong and robust—the proof of security of BBS is treated in details by him. The cryptographic security of the Blum–Blum-Shub PRBG follows from an assumption on a number-theoretic problem, which is known as the quadratic residuosity problem [9]. We put here some definitions and theorems to claim the robustness of BBS to be used in place of RC4 KSA without compromising with the security of RC4.

19.3.1 The BBS Algorithm

The format of BBS is as follows:

$$x_{n+1} = x_n^2 \bmod M = x_n^2 \bmod (p.q),$$

where M is the product of two large prime numbers p and q. The output is generally the least significant bit of x_{n+1}; though sometimes more least significant bits or the bit parity of x_{n+1}; are used. Here, $p, q \equiv 3 \pmod 4$, which confirms that each quadratic residue has one square root, which is also quadratic residue. The initial seed x_0 is generally an integer co-prime to M.

To prove that the generator is secure, "modulo" the quadratic residuosity assumption, Blum et al. [9] showed first how an advantage in guessing the parity of an element to the left of the sequence can be converted in an advantage for determining quadratic residuosity.

Definition 1 (*Quadratic Residues*) If $n \in N$, then $a \in Z_n^*$ is called a quadratic residue modulo n if there exists $b \in Z_n^*$ such that

$$a \equiv b^2 (\bmod n)$$

The set of quadratic residues modulo n is denoted by QR_n and

$$QNR_n = Z_n^* \backslash QR_n$$

is called the set of quadratic non-residues.

Theorem 1 *Let p be an odd prime, and let* $a \in Z_p^*$, *then*

$$\left(\frac{a}{p}\right) \equiv a^{\frac{p-1}{2}} (\bmod p)$$

where $\left(\frac{a}{p}\right)$ *is the Legendre symbol.*

Proof Let a $\in QR_p$, i.e., $a = b^2$ in Z_p^* for some b $\in Z_p^*$. Then

$$a^{\frac{p-1}{2}} \equiv \left(b^2\right)^{\frac{p-1}{2}} \equiv b^{p-1} = 1\,(\mathrm{mod}\,p)$$

as per Fermat's Little Theorem.

If $a \in QNR_p$ and g be a generator of Z_p^*, then $a = g^t$ for some odd $t = 2\ s + 1$, and

$$a^{\frac{p-1}{2}} \equiv \left(g^t\right)^{\frac{p-1}{2}} \equiv \left(g^{2s}\right)^{\frac{p-1}{2}} .\ g^{\frac{p-1}{2}} \equiv\ g^{\frac{p-1}{2}}(\mathrm{mod}\,p)$$

Now $\left(g^{\frac{p-1}{2}}\right)^2 = 1$, so $g^{\frac{p-1}{2}} \in \{-1, 1\}$. As g is a generator of Z_p^*, the order of g is equal to $p - 1$ and $g^{\frac{p-1}{2}} = -1$.

Definition 2 (*Blum Prime Number*) A prime number p with $p \equiv 3$ (mod 4) is called a Blum prime number. An important property of Blum primes is the following:

Theorem 2 *Let p be an odd prime number,*

$$-1\ \in QNR_p \Leftrightarrow p \text{ is a } \textit{Blum prime}$$

Proof
By theorem 1, $\left(\frac{-1}{p}\right) \equiv (-1)^{\frac{p-1}{2}} \mathrm{mod}\,p$.

As p is odd by assumption, it must be congruent to 1 or to 3 modulo 4. But $\frac{p-1}{2}$ is odd if and only if p $\equiv$ 3 (mod 4), hence proved.

It can be predicted that the BBS PRBG is an unpredictable (cryptographically secured) generator, i.e., for each probabilistic polynomial-time predicting algorithm $A(n, x)$, and positive integer t, A has at most an $1/s^t$-advantage for n in predicting sequences to the left, s being the length of n, for sufficiently large n and for all but $1/s^t$ prescribed numbers n of length s.

The outputs of RC4 (original and modified), have been tested statistically using the guidance of NIST, by the NIST Statistical Test Suite. For the original RC4, a text file has been encrypted 300 times by using 300 encryption keys, generating 300 ciphertexts, each of which is of at least 1,342,500 bits, as recommended by NIST. For the modified RC4, RC4-R (with KSA-R), the same text file has been encrypted 300 times by using 300 initial seeds of BBS to generate the internal state array S, generating 300 ciphertexts with size as mentioned above. The ciphertexts are then tested statistically to find out if the security varies depending upon the original KSA and KSA-R.

19.4 The NIST Statistical Test Suite

NIST developed a Statistical Test Suite, which is an excellent and exhaustive document consisting of 15 tests developed to test various aspects of randomness in long binary sequences produced by RNGs and PRNGs [10, 11]. The tests are listed as follows:

1. *Frequency (Monobit) Test*: Number of 1's and 0's in a sequence should be approximately the same, i.e., with probability ½.
2. *Frequency Test within a Block*: Whether frequency of 1's in an M-bit block is approximately M/2.
3. *Runs Test*: Whether number of runs of 1's and 0's of various lengths is as expected for a random sequence.
4. *Test for Longest-Run-of-Ones in a Block*: Whether the length of the longest run of 1's within the tested sequence (M-bit blocks) is consistent with the length of the longest run of 1's as expected.
5. *Binary Matrix Rank Test*: Checks for linear dependence among fixed length sub-strings of the original sequence, by finding the rank of disjoint sub-matrices of the entire sequence.
6. *Discrete Fourier Transform Test*: Detects periodic features in the sequence by focusing on the peak heights in the DFT of the sequence.
7. *Non-overlapping Template Matching Test*: Occurrences of a non-periodic pattern in a sequence, using a non-overlapping m-bit sliding window.
8. *Overlapping Template Matching Test*: Occurrences of a non-periodic pattern in a sequence, using an overlapping m-bit sliding window.
9. *Maurer's Universal Statistical Test*: Whether or not the sequence can be significantly compressed without loss of information, by focusing on the number of bits between matching patterns.
10. *Linear Complexity Test*: Finds the length of a Linear Feedback Shift Register (LFSR) to generate the sequence—longer LFSRs imply better randomness.
11. *Serial Test*: Determines number of occurrences of the 2^m m-bit overlapping patterns across the entire sequence to find uniformity—every pattern has the same chance of appearing as of others.
12. *Approximate Entropy Test*: Compares the frequency of all possible overlapping blocks of two consecutive/adjacent lengths (m and $m + 1$).
13. *Cumulative Sums Test*: Finds if the cumulative sum of a sequence is too large or small. Focuses on maximal excursion (from 0) of random walks defined, which should be near 0.
14. *Random Excursions Test*: Finds if number of visits to a state within a cycle deviates from expected value, calculates the no. of cycles having exactly K visits in a cumulative sum random walk.
15. *Random Excursions Variant Test*: Deviations from the expected visits to various states in the random walk, calculates the number of times that a state is visited in a cumulative sum random walk.

In each test, for a bit sequence, NIST adopted different procedures to calculate the P-values of different tests from the observed and expected values under the assumption of randomness. The Test Suite has been coded by us and had been used to study the randomness features of AES with different S-Boxes.

Table 19.2 Comparison of POP status and uniformity distribution generated by the 15 NIST tests for RC4 and RC4-R

Test↓	POP status		Uniformity distribution	
	RC4-R	RC4	RC4-R	RC4
1	Successful	Successful	*4.186050^{-01}*	3.187653^{-01}
2	Successful	Successful	6.965042^{-02}	*4.368401^{-01}*
3	Successful	Successful	7.110673^{-02}	*2.846729^{-01}*
4	Successful	Successful	*7.660667^{-01}*	9.269018^{-02}
5	Successful	Successful	*7.979853^{-01}*	5.004468^{-01}
6	Successful	Successful	*2.093654^{-01}*	5.530616^{-02}
7	Successful	Successful	*4.126253^{-01}*	6.965042^{-02}
8	Successful	Successful	2.406654^{-01}	*5.882242^{-01}*
9	Successful	Successful	8.550180^{-02}	*4.745147^{-01}*
10	Successful	Successful	2.245991^{-01}	*9.780716^{-01}*
11	Successful	Successful	9.511961^{-01}	*9.573122^{-01}*
12	Successful	Successful	*7.463362^{-01}*	5.403653^{-01}
13	Successful	Successful	*5.103033^{-01}*	2.731592^{-01}
14	Successful	Unsuccessful	*7.186485^{-01}*	1.414488^{-01}
15	Successful	Unsuccessful	*4.119953^{-02}*	2.095199^{-04}
Total:	*15*	13	*9*	6

19.5 Results and Discussions

After analyzing the outputs of the original RC4 and modified RC4, using the NIST Statistical Test Suite, as described above, it has been found that KSA-R creates a tweak in RC4 to increase its security. The final analysis and comparison is displayed in Table 19.2, where the POP status and uniformity distribution of the NIST tests for these two algorithms are displayed and compared. The best values of a particular test for each S-Box are emphasised (in rows) and then the numbers of emphasised cells for each S-box are counted (in columns). The highest count (here RC4-R) gives the best result for a particular algorithm, which shows that this algorithm has a better security than the other, at least for this particular data-set.

POPs and uniformity distribution generated by these 2 algorithms for the 15 tests, compared to the expected values, are displayed in Table 19.3a, b. Distribution of POPs generated by the algorithms for the 15 tests are displayed in Table 19.4a, b. Histograms on distribution of POP values of two tests (5 & 10) for the 15 tests are displayed in Figs. 19.1, 19.2. Scattered Graphs on the POP Status of the 15 tests are displayed in are displayed in Fig. 19.3a, b.

Finally, it has been observed that besides using the original KSA, a suitable PRNG/PRBG can also be used to generate a secured internal initial S-array for RC4, which may give even better randomization in ciphertexts.

Table 19.3a POP status and uniformity distribution generated for RC4-R

Test↓	Expected POP	Observed POP	Status	Uniformity distribution	Status
1	0.972766	0.990000	Successful	4.186050^{-01}	Uniform
2	0.972766	0.996667	Successful	6.965042^{-02}	Uniform
3	0.972766	0.986667	Successful	7.110673^{-02}	Uniform
4	0.972766	0.993333	Successful	7.660667^{-01}	Uniform
5	0.972766	0.990000	Successful	7.979853^{-01}	Uniform
6	0.972766	0.993333	Successful	2.093654^{-01}	Uniform
7	0.972766	0.990000	Successful	4.126253^{-01}	Uniform
8	0.972766	0.990000	Successful	2.406654^{-01}	Uniform
9	0.972766	0.980000	Successful	8.550180^{-02}	Uniform
10	0.972766	0.986667	Successful	2.245991^{-01}	Uniform
11	0.977814	0.988333	Successful	9.511961^{-01}	Uniform
12	0.972766	0.986667	Successful	7.463362^{-01}	Uniform
13	0.977814	0.985000	Successful	5.103033^{-01}	Uniform
14	0.983907	0.985833	Successful	7.186485^{-01}	Uniform
15	0.985938	0.992037	Successful	4.119953^{-02}	Uniform

Table 19.3b POP status and uniformity distribution generated for RC4

Test↓	Expected POP	Observed POP	Status	Uniformity distribution	Status
1	0.972766	0.983333	Successful	3.187653^{-01}	Uniform
2	0.972766	0.993333	Successful	4.368401^{-01}	Uniform
3	0.972766	1.000000	Successful	2.846729^{-01}	Uniform
4	0.972766	0.990000	Successful	9.269018^{-02}	Uniform
5	0.972766	0.983333	Successful	5.004468^{-01}	Uniform
6	0.972766	0.990000	Successful	5.530616^{-02}	Uniform
7	0.972766	0.996667	Successful	6.965042^{-02}	Uniform
8	0.972766	0.996667	Successful	5.882242^{-01}	Uniform
9	0.972766	0.986667	Successful	4.745147^{-01}	Uniform
10	0.972766	0.990000	Successful	9.780716^{-01}	Uniform
11	0.977814	0.991667	Successful	9.573122^{-01}	Uniform
12	0.972766	0.990000	Successful	5.403653^{-01}	Uniform
13	0.977814	0.993333	Successful	2.731592^{-01}	Uniform
14	0.983907	0.980417	Unsuccessful	1.414488^{-01}	Uniform
15	0.985938	0.982778	Unsuccessful	2.095199^{-04}	Uniform

Table 19.4a POP distribution generated for RC4-R

Test↓	0	1	2	3	4	5	6	7	8	9	10
1	3	27	40	28	32	23	22	31	37	29	28
2	1	33	30	26	17	23	44	31	33	34	28
3	4	26	37	29	33	16	29	20	33	40	33
4	2	33	27	25	26	33	30	38	28	26	32
5	3	22	34	29	29	34	30	23	28	37	31
6	2	19	42	30	24	27	29	35	26	37	29
7	3	35	27	29	40	30	33	29	24	27	23
8	3	27	32	21	33	42	29	25	33	22	33
9	6	27	44	36	24	26	22	31	27	22	35
10	4	25	33	36	32	23	17	32	38	27	33
11	7	54	69	62	54	61	60	63	53	56	61
12	4	26	36	35	22	28	32	32	26	26	33
13	9	57	57	51	57	50	57	69	72	57	64
14	34	205	239	240	219	236	231	249	230	252	265
15	43	420	526	546	517	544	541	570	559	563	571

Table 19.4b POP distribution generated for RC4

Test↓	0	1	2	3	4	5	6	7	8	9	10
1	5	39	25	35	27	29	31	28	23	29	29
2	2	23	22	31	31	28	32	41	35	25	30
3	0	38	32	33	18	27	29	33	36	31	23
4	3	23	30	35	30	30	44	29	34	25	17
5	5	32	34	37	27	31	20	28	32	27	27
6	3	18	43	25	25	38	26	38	33	24	27
7	1	27	35	30	26	29	31	25	26	48	22
8	1	32	31	30	27	32	31	22	36	36	22
9	4	27	29	24	35	34	27	19	36	33	32
10	3	24	33	25	28	28	30	33	33	30	33
11	5	52	61	65	60	56	51	62	58	65	65
12	3	25	36	35	35	24	27	27	27	37	24
13	4	49	66	65	48	63	70	63	68	59	45
14	47	221	240	245	214	223	225	219	250	253	263
15	93	516	511	490	499	492	526	577	569	535	592

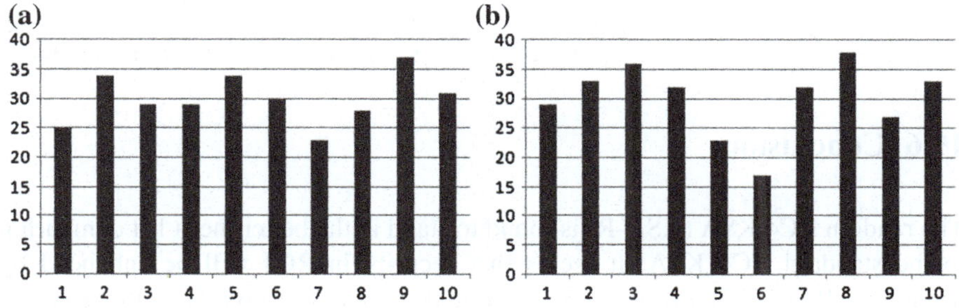

Fig. 19.1 **a**, **b** Histograms for POP Distribution of Test 5 & 10 for RC4-R

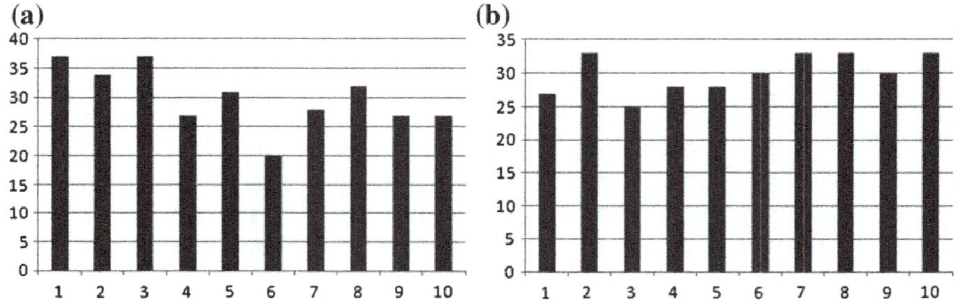

Fig. 19.2 **a**, **b** Histograms for POP Distribution of Test 5 & 10 for RC4

Fig. 19.3a Scattered Graph on POP Status on 15 NIST Tests for RC4-R

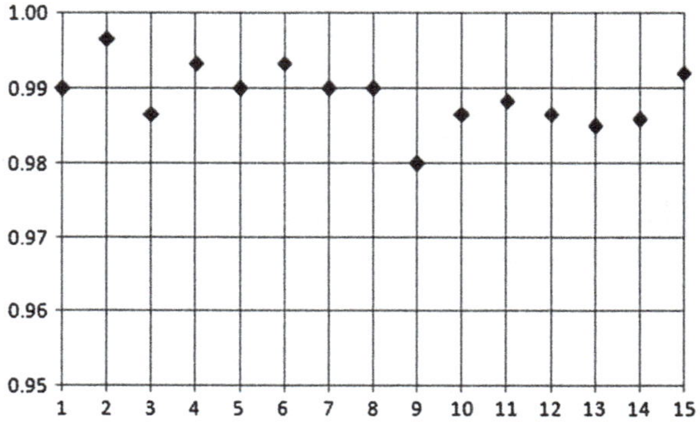

Fig. 19.3b Scattered Graph on POP Status on 15 NIST Tests for RC4

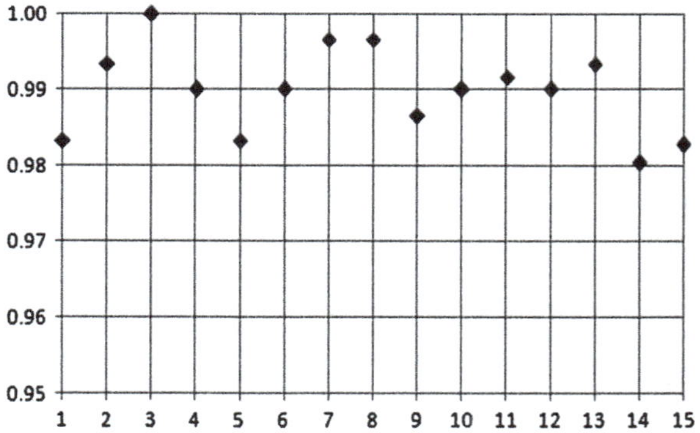

19.6 Conclusion

The random RC4 KSA (KSA-R) is found to stand in the better merit list comparing to the standard RC4 KSA. It seems that security in RC4 will be enhanced by driving a PRNG/PRBG to generate the initial internal S-array. Also, the user can choose and generate any random state array S according to his/her own choice of initial seeds from a large set of options. In the case of suspicion of a trapdoor in the ciphertext, an S-array might be replaced by another one by the user. Other PRNGs/PRBGs may also be used to create random RC4 KSA and studies on them are required to find better opportunities to generate secured RC4 internal state arrays.

References

1. Roos A (1995) A class of weak keys in the RC4 stream cipher. Post in sci crypt, message-id 43u1eh$1j3@hermes.is.co.za (1995), http://marcel.wanda.ch/Archive/WeakKeys
2. Paul S, Preneel B (2004) A new weakness in the RC4 keystream generator and an approach to improve the security of the cipher. In: FSE 2004, LNCS, vol 3017. Springer, Heidelberg, pp 245–259. http://www.iacr.org/archive/fse2004/30170244/30170244.pdf (Last accessed on: 22 July 2013)
3. Maitra S, Paul G (2008) Analysis of RC4 and proposal of additional layers for better security margin. In: Lecture notes in computer science, INDOCRYPT, 5365. Springer, Berlin, pp 40–52. http://eprint.iacr.org/2008/396.pdf (Last accessed on: 22 July 2013)
4. Akgün M, Kavak P, Demicri H (2008) New results on the key scheduling algorithm of RC4. In: Lecture notes in computer science, INDOCRYPT, 5365, pp 40–52. Springer. http://link.springer.com/content/pdf/10.1007/978-3-540-9754-5_4.pdf (Last accessed on: 22 July 2013)
5. Noman AA, Sidek RS, Ramli AD (2009) Hardware implementation of RC4A stream cipher. Int J Cryptology Res 1(2):225–223. http://www.mscr.org.my/V1(2)/PP%225-233.pdf (Last accessed on: 22 July 2013)
6. Tomašević V, Bojanić S (2004) Reducung the state space of RC4 stream cipher. In: Bubak M et al (eds) ICCS 2004, LNCS 3036, Springer, Berlin, pp 644–647 http://link.springer.com/chapter/10.1007%2F978-3-540-24685-5_110#page-1 (Last accessed on: 22 July 2013)
7. Nawaz Y, Gupta KC, Gong G (2013) A 32-bit RC4-like keystream generator, IACR Eprint archive. 2005. eprint.iacr.org/2005/175.pdf (Last accessed on: 22 July 2013)
8. Junod P (1999) Cryptographic secure pseudo-random bits generation: the Blum–Blum-Shub generator. http://www.cs.miami.edu/~burt/learning/Csc609.062/docs/bbs.pdf (Last accessed on: 22 July 2013)
9. Blum L, Blum M, Shub M (1983) Comparison of two pseudo-random number generators. In: Rivest RL, Sherman A, Chaum D (eds) In: Proceedings of CRYPTO 82. Plenum Press, New York, pp 61–78
10. National Institute of Standards & Technology (NIST) Technology Administration, U.S. Dept. of Commerce, A Statistical Test Suite for RNGs and PRNGs for Cryptographic Applications, April, 2010, http://csrc.nist.gov/publications/nistpubs800/22rec1SP800-22red1.pdf
11. Kim SJ, Umeno K, Hasegawa A (2004) Corrections of the NIST statistical test suite for randomness. Communications Research Lab., Inc. Admin. Agency, Tokyo

Chapter 20
Design of a Novel Power Efficient Routing Scheme for Mobile Ad-Hoc Network

Koushik Majumder, Samrat Sarkar and Joydeep Kundu

Abstract The Mobile Ad-Hoc network is composed of a group of autonomous wireless node without any centralized administration. Energy awareness is one of the main challenges in MANET due to its constrained power resources. To overcome the problem of energy scarcity, several routing protocols have been developed in recent years. In this paper, we propose a new novel energy aware routing scheme named Power Aware Routing using Standard Deviation (PARSD) for Mobile Ad-Hoc network using standard deviation which ensures an even distribution of the battery power dissipation for the individual nodes. As a consequence, this protocol increases the network lifetime and also tries to minimize the total transmission cost of the network. After analyzing the protocol, it has been seen that there are several advantages of this protocol over the other basic power and battery aware routing protocols.

Keywords Power aware routing · Mobile ad-hoc network · Network lifetime

20.1 Introduction

Mobile Ad-Hoc Network is a self-organizing network without wired backbone or a centralized control, in which the node forwards the packets to each other in a multi-hop manner. Typical applications include military, cell phone, laptop, personal conferences, meeting rooms etc.

As the Mobile Ad-Hoc network is infrastructure less, so it faces different challenges [1] that are not present in wired network. Topology changes occur in MANET both rapidly and unexpectedly. Working with limited bandwidth,

K. Majumder (✉) · S. Sarkar · J. Kundu
Department of Computer Science and Engineering, West Bengal University of Technology, Kolkata, India
e-mail: koushik@ieee.org

S. Sengupta et al. (eds.), *Emerging Trends in Computing and Communication*,
Lecture Notes in Electrical Engineering 298, DOI: 10.1007/978-81-322-1817-3_20,

maintaining quality, and assuring security—these are considered as important challenges in MANET. Power efficient routing is very important in MANET because the transmission power is expensive in wireless network when the nodes have limited battery energy.

Mobile devices in MANET use lithium batteries with average lifetime of one day. Conventional MANET routing protocols search routing path based on the delay which mainly result in the shortest path. But the nodes in the selected path will 'die' soon since they are overused. A single node in the routing path going to 'dead' condition can cause the entire network to fail.

In recent years, a large number of researchers are trying to solve the problem of energy-efficient data transfer in the context of Mobile Ad-Hoc networks [2]. The different existing protocols can be classified in the following two categories:

The protocols in the first category are based on minimum-power routing algorithms. A standard protocol in this category selects the route by minimizing the total power consumption of the entire network. The second category of protocols is based on battery aware routing algorithms that take care about battery life of individual nodes.

This paper is organized as follows- Sect. 20.2 gives a classification of different protocols of MANET based on power aware routing. Section 20.3 contains reviews of some existing research on power aware routing algorithm. Section 20.4 describes the details of the proposed power aware routing protocol using standard deviation. Section 20.4.1 elaborates the case study with a network scenario which compares the proposed protocol with some existing battery aware routing protocols.

20.2 Classification of Power Aware Ad-Hoc Routing Protocol

Several power aware routing protocols have been proposed in recent years. A classification [3] of these protocols has been shown in Fig. 20.1. The power-aware routing protocols can be classified into two categories: activity-based and connectivity-based. The activity based protocols deal with the power consumption based on network activity and connectivity based protocols maintain an effective connectivity.

20.2.1 Multicasting/Broadcasting

Multicasting/broadcasting protocols deal with power consumption efficiency when a single data or packet is sent to multiple destinations. Some examples are: Energy-efficient multicasting routing protocol (E2MRP) [4], On demand multicasting routing protocol (ODMRP) [5].

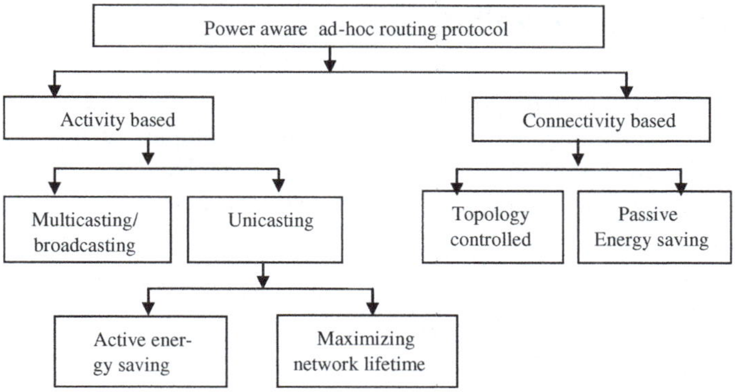

Fig. 20.1 Classification of power aware ad-hoc routing protocol

20.2.2 Active Energy Saving

Active energy saving protocol deals with minimizing the total energy consumed for a single packet. When a source n_i has some data to send to the destination n_k, the energy consumed per packet e is formulated by,

$$e = \sum_{i=0}^{k-1} T(n_i, n_{i+1})$$

where $T(n_i, n_{i+1})$ represents the power consumption to transfer packet from node n_i to n_{i+1}. This protocol selects the route by minimizing the energy consumed per packet. Some examples of this type of protocols are Power-Aware Routing Optimization (PARO) [6], Location-Aided Power-Aware Routing Protocol (LAPAR) [7].

20.2.3 Maximizing Network Lifetime

The maximizing network lifetime protocol focuses on the energy consumption for the individual nodes to ensure that it is used in a balanced manner. Overuse of single node can cause network failure. So this protocol selects the route through the node which has sufficient battery remaining. Some examples of this type of protocols are Minimal Battery Cost Routing (MBCR) [8], Power-Aware Source Routing (PSR) [9].

20.2.4 Topology Controlled

The protocols in topology controlled category adjust the node's transmission power while maintaining desired topology. Some examples of this type of protocols are: Small Minimum-Energy Communication Network (SMECN) [10], Minimum-Energy Communication Network (MECN) [11].

20.2.5 Passive Energy Saving

Passive energy saving protocols deal with some passive mechanism by allowing a set of nodes to go to sleep mode at different period of time when the nodes are idle while maintaining the desired connectivity. Some examples are: PAMAS [12], Adaptive Fidelity Energy-Conserving Algorithm (AFECA) [13].

20.3 Review of Existing Power Aware Protocols

Several researches on power aware routing have been done in recent years. In this part we have presented a short review on these power aware routing protocols for MANET.

The MTPR [12] is a basic power aware routing protocol which always tries to minimize the total transmission power. In MTTPR, first it calculates the total transmission power for all possible routes between source *0* and destination *D*. Calculation of the total transmission power $P(n_i, n_j)$ between two hosts n_i and n_j can be used as a metric. The total transmission power $P_l = \sum_{i=0}^{D-1} P(n_i, n_{i+1})$.

Finally it selects the route with minimum total transmission power to transfer a packet from source to destination. The MTTPR algorithm does not take care of battery life of every individual node, so MBCR algorithm is proposed by introducing an extra battery cost function [12] which is the inverse of remaining battery capacity. This algorithm minimizes the total summation of the inverse of remaining battery capacities for all routing paths. The algorithm cares about the individual nodes but still it does not protect those critical nodes which have very low remaining battery capacity. MMBCR [12] is a modification of the MBCR algorithm in such a way that no critical node will be overused. Without summing the battery cost functions of all nodes of every individual route, MMBCR finds the maximum battery cost among all nodes of different routes to find the critical nodes. Then it selects the route based on the minimum battery cost among those critical nodes. So this algorithm gives balanced use of the battery capacity of the nodes in the network. A hybrid approach- CMMBCR, was devised by C.K Toh [12], that tries to minimize the total transmission power and also avoid the battery having low remaining capacity by adding an extra threshold to each battery node. This algorithm first finds the minimum battery capacity (R_i) for all nodes of each route. If $R_i \geq Y$ (chosen threshold value) is true for some or all routes between a source and destination, then the MTPR scheme is applied to select the route among all possible paths which satisfy the previous condition. If any path does not satisfy the condition then the route is selected with the help of maximum battery remaining capacity by using the protocol MMBCR. Protocol MRPC [14] is a power aware routing algorithm that increases the lifetime of wireless network. It works in the same way as basic protocol MMBCR however MRPC chooses the nodes, not just by their remaining battery capacity but also with the energy required in network transmission for forwarding a packet over a specific link. MRPC first selects the node having smallest residual packet transmission capacity as 'critical' node among different possible paths. Then it selects the path having largest packet capacity among those 'critical' nodes. Another protocol PSR [9] tries to extend the lifetime of every node because if any one node dies there is a possibility of network partition. The algorithm finds a route π at route discovery time t such that the following cost function is minimized.

$$C(\pi, \mathrm{t}) = \sum_{i \in \pi} \rho i \left(\frac{F_i}{R_i(t)} \right)^{\alpha}$$

where, ρ_i is the transmit power of node i, F_i: full-charge battery capacity of node i, R_i: remaining battery capacity of node i at time t and α is a positive weighting factor. Maleki et al. [15] proposed an on-demand routing protocol named Lifetime Prediction Routing that is based upon the prediction of battery lifetime of individual nodes. The main objective of this protocol is to extend the lifetime of the MANET. The objective function of LPR is as follows: $\max_{\pi} T_{\pi}(t) = \min_{i \in \pi}(T_i(t))$, where $T_{\pi}(t)$ is lifetime of path π and $T_i(t)$ is the predicted lifetime of the node i in path π. LPR uses dynamic distributed load balancing approaches to avoid the power-congested nodes and to use the path that is lightly loaded.

20.4 Proposed Work on Power Aware Routing Protocol Based on Standard Deviation

The main objectives of power aware routing are maintaining the minimum transmission power and increasing the network lifetime. There should be a balance between these two goals.

The optimal route from the battery life point of view will be such a route where all the nodes in the route lie in an average manner. For building such a protocol for power aware routing the mean can be used. But the mean cannot give a proper route selection if the battery costs are highly spread out from its mean. So the formulation of the standard deviation [16] can be applied for optimal path selection. Standard deviation actually shows a given set, how much spread out from its mean or average. After calculating the possible routes' standard deviation from different node's battery cost function, the optimal route can be selected in such a way that its standard deviation value is minimized. Another approach is introduced by multiplying a weight factor. This weight factor is calculated based on the number of nodes and nodes below mean value which gives more accurate results from the total link cost point of view.

20.4.1 Assumption

1. Every node is capable of calculating its remaining battery life.
2. Every node maintains a Remaining Energy Table which contains the details about battery capacity of that node and its neighbor nodes.
3. Link cost is dependent directly upon the hop count.

20.4.2 Algorithm Description

Our proposed power aware routing protocol PARSD is based on two phases: route discovery and route establishment.

In the first phase the protocol tries to find the possible paths and their remaining battery capacity by sending a Route Discovery Request packet. If the sender has some data to send, it sends the RDR packet to its neighbor nodes. The neighbor nodes attach its IP address and its remaining battery capacity in the RDR packet and forward the packet to its neighbor nodes except from where it was received, until the packet reaches its destination. After receiving the packet, the destination starts a timer (T_r) and waits for other RDR packets from the same source. When the timer (T_r) expires, it sends Reply RDR packet containing all the possible path list and its nodes remaining battery capacity details to the source. The RRDR packet is sent to the sender via the route through which the first RDR packet was received, because that path contains the minimum hop count. When the sender receives the RRDR packet, it stores the list of possible paths with its remaining battery capacity details.

The second phase does the route establishment by calculating the weight factor and standard deviation of the possible routes. First, it calculates the battery cost function of the nodes in the possible routes, which is the inverse of its remaining battery capacity. Then it calculates the mean of the battery cost functions for every possible route. The weight factor is calculated as the ratio between the total number of nodes on the selected path and the number of nodes whose battery cost function is below the mean in that path. The standard deviation is calculated by taking the battery cost functions of the nodes as population for the possible routes. The optimal value is calculated by multiplying the summation of standard deviation and mean. The final route is selected with the minimum optimal value among possible routes.

Algorithm 1: Route discovery

Step 1. The sender node floods RDR packets to initiate a route discovery to all its neighbor nodes.

Step 2. After receiving RDR packet every node checks whether it is destination node or not

2.1. If the node is not the destination, then

2.1.1. Attach its IP address in the RDR packet.

2.1.2. Attach its remaining battery capacity to the RDR packet.

2.1.3. Forward the packet to its neighbor nodes except the node from where the packet was received.

2.2. Else, the node is the destination node

2.2.1. Starts a timer (T_r) for a specific period of time.

2.2.2. Collects all the arriving packets of the same source until the timer (T_r) expires.

2.2.3. When the timer expires, attach the possible path list and the remaining battery capacity of the nodes in a Reply RRD packet.

2.2.4. Sends the packet to source node by the path from where the first RDR packet was received.

Algorithm 2: Route establishment

Step 1. The source node receives RRDR packet.

Step 2. The source node collects the possible path list and the remaining battery capacities of the nodes belong to that path from the packet.

Step 3. The source node calculates the battery cost function of the nodes in the possible route. The battery cost function is the inverse of battery remaining which can be achieved from the following equation.

$$f_i\left(c_i^t\right) = \frac{1}{c_i^t}$$

where c_i^t is the battery capacity of a host n_i at time t and $f_i(c_i^t)$ is a battery cost function of node n_i at time t.

Step 4. The source node calculates the mean μ of those battery cost functions for every possible route. If N is the total number of nodes and x is the number from the population set then,

$$\mu = \frac{1}{N}\sum_{i=1}^{n} x$$

Step 5. The algorithm then finds out the total number of nodes having a lower battery cost function than mean on every selected path. Then it calculates the weight factor by dividing the total number of nodes on the selected path and the number of nodes with battery cost function below the mean.

$$weight\,factor = \frac{Total\,number\,of\,nodes}{Number\,of\,nodes\,below\,mean}$$

Step 6. Then it finds the standard deviation for each possible route by using each nodes cost function as a population. The standard deviation σ can be found using the following equation,

$$\sigma = \sqrt{\frac{1}{N}\sum_{i=1}^{N}(x_i - \mu)^2}$$

where x_i is the variable of population set and μ is the mean value. The standard deviation is calculated in following steps.

6.1. Take the battery cost function of all nodes for a single path as population set.
6.2. Calculate mean μ for the selected population set of the path.
6.3. Find list of deviations set by subtracting the mean from every population set
6.4. Square the deviation values.
6.5. Find the variance by calculating the average of all the square values of deviations.
6.6. Find standard deviation by calculating the square root of the average value.

Step 7. For every route an optimal value is calculated from the following equation,

$$Optimal\,value = (Standard\,Deviation + mean) * weight\,factor$$

Step 8. The final route is selected based on the minimum optimal value among all possible paths.

20.4.3 Case Study

Let's take a network scenario with 6 possible paths between source and destination. A single path consists of minimum 2 nodes and maximum 8 nodes. The values in the nodes are the battery cost function of the nodes (Fig. 20.2; Tables 20.1, 20.2, 20.3).

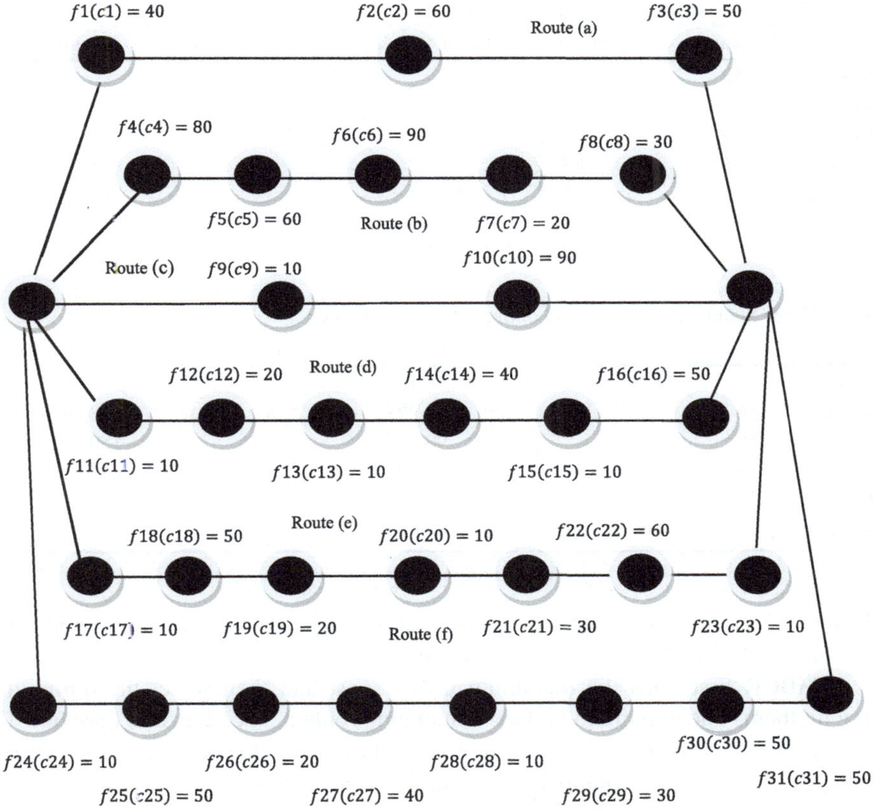

Fig. 20.2 Example network scenario

Table 20.1 Battery cost function

	Battery cost function of all nodes in the selected route							
Route a	40	60	50					
Route b	80	60	90	20	30			
Route c	10	90						
Route d	10	20	10	40	10	60		
Route e	10	50	20	10	30	60	10	
Route f	10	50	20	40	10	30	50	50

MTTPR selects the route by minimizing the total transmission cost. So, MTTPR scheme will select route *c*. The selection is not desirable from the network longevity point of view because the node 10 has very low remaining battery capacity. The path will be disconnected as soon as the node will die.

MBCR selects the route by minimizing the total battery cost function for a single path. The total battery cost of route a, b, c, d, e, and f are 150, 280, 100, 150,

Table 20.2 Weight factor calculation

	Mean	Node less than mean	Total number of nodes	Weight factor
Route a	50	1	3	3
Route b	56	2	5	2.5
Route c	50	1	2	2
Route d	25	4	6	1.5
Route e	27.1	4	7	1.75
Route f	32.5	4	8	2

Table 20.3 Standard deviation and optimal value calculation

	Mean	Standard deviation	SD + mean	Weight factor	Optimal value
Route a	50	8.16	58.16	3	174.48
Route b	56	27.27	83.27	2.5	208.175
Route c	50	40	90	2	180
Route d	25	18.92	43.92	1.5	65.88
Route e	27.1	19.05	46.15	1.75	80.76
Route f	32.5	16.39	48.89	2	97.78

190, and 260. So it will select the route *c*. But route *c* contains a critical node 10. So the selected path is not an optimal route.

MMBCR first finds the maximum battery cost and then select the minimum among them. The maximum battery cost among route a, b, c, d, e, and f are 60, 90, 90, 60, 60, and 50. Route *f* (50) has minimum among them. So, route *f* will be selected. The route *f* contains 8 hop counts which increase the total transmission cost. With respect to total battery cost (which is 260), it is greater than other four routes. The route selection is not optimal from both the battery cost and total transmission power point of view.

Our proposed protocol PARSD selects the route based on minimum optimal value which can be derived by multiplying the weight factor with the summation of standard deviation and mean value for every possible route. Here route *d* has minimum optimal value. So route *d* will be selected. Route d contains node 11, 12, 13, 15 which has enough battery remaining to transfer data. The total battery cost 150 is also less and the route does not contain any critical node. This ensures the even dissipation of power by all the nodes and consequently longer network lifetime.

20.5 Conclusion

Designing an efficient power aware routing protocol is a very challenging task which requires meeting both the goals of minimizing the total transmission cost and increasing the network lifetime. Our proposed protocol PARSD tries to achieve both these goals. The technique of using standard deviation enables us to select an optimal route where the difference in the energy level between the highest and lowest battery capacity nodes is minimum. This proper power optimization actually ensures an even distribution of the battery power dissipation by the nodes thereby increasing the overall lifetime of the network. The weight factor gives the result more accuracy as it increases the optimal value for those routes where maximum nodes lie above the mean. From the link cost point of view this weight factor tries to select the minimum hop count route thereby reducing the total transmission power.

References

1. Singh A, Tiwari H, Vajpayee A, Prakash S (2010) A survey of energy efficient routing protocols for mobile ad-hoc networks. Int J Comput Sci Eng 02(09):3111–3119
2. Goldsmith AJ, Wicker SB (2002) Design challenges for energy-constrained ad hoc wireless networks. IEEE Wirel Commun 9(4):8–27
3. Li J, Cordes D, Zhang J (2005) Power-aware routing protocols in ad hoc wireless networks. IEEE Wirel Commun 12:69–81
4. Jiang H, Cheng S, He Y, Sun B (2002) Multicasting along energy-efficient meshes in mobile ad hoc networks: IEEE wireless communication and net. In: Conference, vol 2. pp 807–811
5. Lee SJ, Su W, Gerla M (2000) On-demand multicast routing protocol in multihop wireless mobile networks. Mobile Netw Appl 7(6):441–453
6. Gomez J, Campbell A (2001) Power-aware routing optimization for wireless ad hoc networks. In: High speed network workshop. pp 27–32
7. XueY, Li B (2001) A Location-aided Power-aware Routing Protocol in Mobile Ad Hoc Networks. In: Proceedings of IEEE global telecommunication conference, vol 5. pp 2837–2841
8. Toh CK (2001) Maximum battery life routing to support ubiquitous mobile computing in wireless ad hoc networks. IEEE Commun Mag 39(6):138–147
9. Maleki M, Dantu K, Pedram M (2002) Power aware source routing protocol for mobile ad hoc networks. In: Proceedings of 2002 international symposium low power electronics and design. pp 72–75
10. Li L, Halpern JY (2001) Minimum energy mobile wireless networks revisited. Proc IEEE ICC 1:278–283
11. Rodoplu V, Meng TH (1999) Minimum energy mobile wireless networks. IEEE JSAC 17(8):1333–1344
12. Singh S, Woo M, Raghavendra CS (1998) Power aware routing in mobile ad hoc networks. In: Proceedings of 4th annual international conference on mobile computing and network. pp 181–90
13. Xu Y, Heidemann J, Estrin D (2000) Adaptive energy conserving routing for multihop ad hoc networks: Technical report 527, USC/Information Sciences Institute
14. Misra A, Banerjee S (2002) MRPC: maximizing network lifetime for reliable routing in wireless environments. In: IEEE wireless communications and networking conference (WCNC), pp 800–806
15. Maleki M, Dantu K, Pedram M (2003) Lifetime prediction routing in mobile ad hoc networks: In: Proceedings of IEEE wireless communications and networking conference. pp 1185–1190
16. Al-Saleh MF, Yousif AE (2009) Properties of the standard deviation that are rarely mentioned in classrooms. Austrian J Stat 38(3):193–202

Chapter 21
Trust Based Network Layer Attacks Prevention in MANET

Mousumi Sardar, Subhashis Banerjee, Kishore Majhi and Koushik Majumder

Abstract MANET is a collection of wireless nodes that cooperate with each other for relaying packets during the communication. Due to the absence of centralized authority and frequent nodes mobility ad-hoc networks become more attractive towards attackers. For this reason security becomes an important issue in MANET. So protocols are made secure by using trust as a parameter. Trust parameter can efficiently handle the secure route finding procedure. Several trust based protocols already exist in MANET. In this paper we also use trust as a tool for mitigating the severe network layer attacks in MANET. Our proposed scheme uses both the direct and indirect trust. Trust percentage is calculated dynamically during route establishment process. We have also considered delay parameter to avoid end-to-end delay which degrades the performance of MANET. Our scheme efficiently finds secure and reliable route based on trust by mitigating several network layer attacks.

Keywords Worm-hole attack · Black-hole attack · Grey-hole attack · Jellyfish attack · Byzantine attack · Trust · Reputation

21.1 Introduction

Mobile ad-hoc networks (MANETs) are formed by the collection of wireless nodes that communicates with each other in the absence of any centralized authority. MANET does not need any predefined network structure as it is able to configure by itself. It has dynamic network topology because the nodes frequently move in the network. As a result the communication between any two nodes in the

M. Sardar · S. Banerjee · K. Majhi · K. Majumder (✉)
Department of Computer Science and Engineering, West Bengal University of Technology, Kolkata, India
e-mail: koushik@ieee.org

S. Sengupta et al. (eds.), *Emerging Trends in Computing and Communication*, Lecture Notes in Electrical Engineering 298, DOI: 10.1007/978-81-322-1817-3_21,

network depends on the other nodes' cooperation. For this reason the vulnerability of attack increases in ad-hoc network than the wired network. Several attacks are happened in different layers of the network. Attacks in this network are broadly classified into two categories: active and passive attack. In active attacks attacker modifies the original message to disrupt the normal routing flow but in passive attacks attacker only tries to know the content of message without doing any modification. Several attacks are happened in different layers of the network [1, 2]. In this paper we address some of the severe network layer attacks in MANET, which are: Worm-hole attack, Black-hole attack, Grey-hole attack, Jellyfish attack and Byzantine attack. In Worm-hole attack attacker captures packet and tunnels it to another distant node. Black-hole attack causes continuous packet dropping whereas Grey-hole attack causes selective packet dropping which is more difficult to detect than Black-hole attack. In case of Jellyfish attack attacker tries to degrade the performance of the network by introducing delay in packet forwarding. Due to all these attacks the normal routing function is disrupted. As a result communication between nodes hampers. So network layer security becomes a main concern in ad-hoc networks. For secured and reliable communication routing protocols are made secure by using many traditional techniques like cryptographic mechanism. But cryptographic mechanisms increases overhead for huge computation thus increases cost as well as in many cases we cannot apply this type of solution in MANET. So to provide cheap secure and reliable communication trust mechanism is introduced in routing protocols. Trust depends on the behaviour of nodes while relaying packets. In this paper we proposed a new trust based approach to mitigate several network layer attacks simultaneously without incurring heavy overhead. Depending on nodes packet forwarding nature trust on the nodes are calculated. And also nodes' behaviour towards the other nodes in the network is also taken into consideration which helps us to detect byzantine behaviour of a malicious node. In this way our method provides a secure and reliable communication between two nodes in the network.

The rest of the paper is organized as follows: in Sect. 21.2 different network layer intrusions are discussed. Section 21.3 represents a brief idea on trust mechanism. In Sect. 21.4 a survey is done on different trust based protocols in MANET and in Sect. 21.5 our proposed trust based approach is discussed and finally we conclude in Sect. 21.6.

21.2 Network Layer Intrusions in MANET

The connection between wireless nodes relies on the cooperation among the nodes. Network layer basically provides the hop-to-hop connection. The protocols in the network layer in MANET suffer from various types of attack during communication. Some of the mostly happened network layer attacks in MANET are discussed in the following:

21.2.1 Worm-Hole Attack [3]

This network layer attack is easy to launch but difficult to detect. In this attack, attacker sends reply of having route with minimum hop-count towards destination via worm-hole link in response of sender's RREQ packet. Next when sender selects that route malicious node captures the data packet and tunnels it to another colluding malicious node which is located nearest to the destination.

21.2.2 Black-Hole Attack [4]

One malicious node or set of colluding nodes can launch this type of attack. First it includes itself in routing path and then drops all packets without forwarding it to the next node in that path. In case of Single Black-hole attack, a single malicious node creates this attack whereas in Cooperative Black-hole attack a set of nodes act in a group.

21.2.3 Grey-Hole Attack [5]

This attack is a special case of Black-hole attack where a malicious node drops the packet selectively without dropping all packets unlike Black-hole attack. The malicious node does this to avoid the detection. It is more difficult to detect than the Black-hole attack.

21.2.4 Jellyfish Attack [5]

In Jellyfish attack the main aim of attacker is to degrade the performance of the network by increasing end-to-end delay. When the malicious node receives the packet it intentionally waits for some time before forwarding it to the next node which results a great delay in reaching of packets to the destination.

21.2.5 Byzantine Attack [5]

In this type of attack malicious nodes try to create loop in routing path or send the packet through the path that is non-optimal or selectively drop packets. This type of attack is difficult to detect. Figure 21.1 shows different types of network attack.

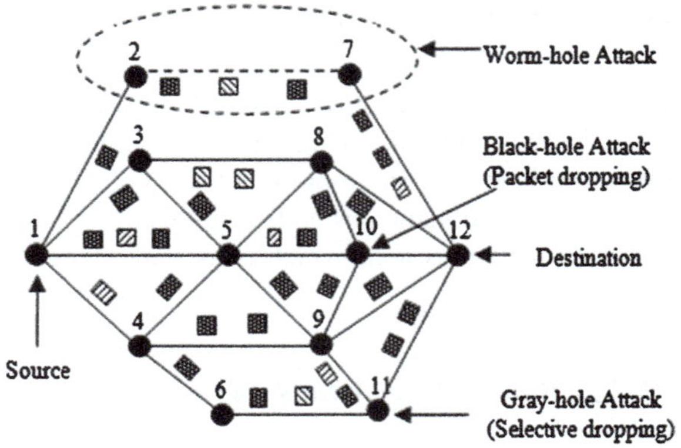

Fig. 21.1 Example of different network layer attacks

21.3 Trust Mechanism

Trust Mechanism is basically depends on the fact that how much a node behaves correctly towards a particular node or the other nodes in the network. It is very similar to our real life where a person becomes trusted to someone when the person behaves properly. Similarly trust on a node in the network depends on the node's action i.e. whether the node forwards the packet correctly or not. Depending on the nodes behaviour trust can be classified in two categories: Direct trust and Indirect Trust. Direct trust is defined from the history of direct communication with a node. Indirect trust is a reputation that other nodes in the network give about a particular node depending upon the direct experience the nodes have with that node. Using this trust parameter the malicious nodes in the network can easily be detected without incurring heavy computational overhead.

21.4 Literature Survey

A. M. Pushpa et al. proposed a trust based mechanism based on AODV protocol i.e. TAODV [6] in which trust on a node and route is used to choose a reliable and secure path for packet forwarding. For node trust calculation trust value from the other neighbour nodes i.e. recommendation is collected using TREQ and TREP packet and the route trust between two nodes is determined on the basis of direct communication between them. Route trust and node are determined by direct trust and indirect trust respectively. TAODV can detect selfish node in network but can't detect Worm-hole attack and Black-hole attack and Jellyfish attack.

Ad Hoc On-Demand Trusted Path Distance Vector (AOTDV) protocol [7] is another trust based approach in which trust is evaluated using packet forwarding ratio. Node on a trust depends on number of control and data packets forwarded by that node. Depending on node trust path trust is calculated which is minimum node trust value among all node along that path. The path with high trust value is chosen for communication. This protocol provides multiple loop free paths and shows better performance in compared to AODV. Grey-hole, black-hole attacks are reduced but this algorithm can't prevent Worm-hole attack as the Worm-hole nodes don't drops packet.

Secure Routing Protocol (SRT) based on both NTP and AODV protocol is presented in [8]. By flooding beacon frames trust is evaluated and then based on this trust value nodes are categorized in three different lists: ally, acquaintance and associate list. In this protocol if destination and all the intermediate nodes are not in same level with source node trust is compromised. This protocol provides better throughput, packet delivery ratio, average routing load, and average path length, in terms of mobile mobility only in presence of Black-hole node. In the presence of Worm-hole, Grey-hole and Jellyfish attacks the performance degrades.

T. Eissa et al. proposed Friendship Based Ad Hoc on Demand Distance Vector (FrAODV) protocol [9] in which two separate algorithms are used to establish forward and reverse connection between nodes. The friendship values are assigned to every node ranging from 0 to 100. The friendship values are calculated by summing all the friendship values of nodes along a path. The nodes accept RREQ and RREP packets if it is coming from a friend node otherwise reject. Depending on friendship value secure route is established. In terms of QoS it gives better performance than AODV. In this protocol as the end-to-end delay is not considered Jellyfish attack can be easily launched. As a result, the performance of the network will be degraded in case of real time application.

A Distributed Trust Management Framework for detecting malicious packet dropping nodes in a Mobile ad-hoc network is proposed in [10] where every node needs to run some security module. Monitor module is used to observe nodes' behaviour whether they drop any packets or tampering with data. If any malicious activity is observed trust on that node is calculated from the reputation values of other neighbour nodes of that node. This protocol can detect Black-hole nodes and Grey-hole nodes. But this mechanism is failed to detect the Worm-hole nodes as it don't drop packets. Also in this scheme delay is not considered in performance measurement so it is not able to detect Jellyfish attack.

N. Bhalaji et al. includes trust mechanism in DSR protocol in [11] to improve the performance of DSR protocol by finding the more secure route. This scheme evaluates trust on the basis of direct interaction with a node. The trust on a route is assigned by doing average on all node trust values along that path. This scheme improves the performance of DSR can detect Black-hole and Grey-hole nodes. But this protocol can't detect Jellyfish and Worm-hole node as the trust value depends on amount of packet forwarding.

After going through the existing various types of trust based protocols it has been seen that no protocols are alone capable of detecting more than two network layer attacks. But our scheme is able to detect maximum of the network layer attacks without incurring heavy overhead. In our scheme two types of trust are taken into consideration: direct trust and indirect trust. Direct trust is based on direct communication with a node and on the basis of a delay parameter. The direct communication with a node helps to detect the black-hole and Grey-hole node. If the malicious nodes drop packets direct trust will be low. The delay parameter is also included in calculation of direct trust for detecting that kind of malicious node which tries to degrade the performance of the network by introducing end-to-end delay. The indirect trust i.e. the reputation from other nodes about a particular node is used to detect the nodes that don't drop packets but breaks the confidentiality of the message. In this way our scheme mitigates most of the severe network layer attack.

21.5 Proposed Trust Based Routing for Preventing Network Layer Attacks in MANET

Now we discuss our proposed reliable routing method that used DSR [12] as a base routing protocol with some extra phases like false packet transmission, trust calculation and trust exchange, then the sender select the most trusted path that is attack free. In the following sections the assumptions, data structures, packet format and the proposed algorithm will be discussed.

21.5.1 Assumptions

In the next section we outline the assumptions that we make regarding the properties of the physical and the network layer and the node characteristics of the ad-hoc network.

1. We assume that nodes in the ad-hoc network use the DSR as the underlying routing protocol with an extra false packet sending phase during which each node calculates the direct trust value of its neighbour nodes.
2. We introduces two new packet format one is Trust request packet (TREQ) another is Trust reply packet (TREP) which are used for getting the recommendation values from all the intermediate nodes in the path.
3. Route selection is done based on the route trust calculated by the sender not based on the minimum hop count of a route.
4. We assume that there is some asymmetric key cryptography used by the nodes. Each node has a public and private key pair. For a node X it keeps its private key secret and distributes the public key among other node in the network.
5. Nodes encrypt packets using the destination's public key while sending.

IP Header	DSR Header	DSR Option	Transport Layer Data

Fig. 21.2 IP packet with DSR information

Option Type	Opt Data Len	Identification
Target Address		
Address[1]		
...		
Address[n]		

Fig. 21.3 DSR Route Request Option (RREQ) format

	Option Type	Option Data Length		Reserved
Address[1]				
Address[2]				
...				
Address[n]				

Fig. 21.4 DSR Route Reply Option (RREP) format

Option Type	Opt Data Len	Identification
Target Address		
Direct Trust value [1]		
Address[1]		
...		
Direct Trust value [n]		
Address[n]		

Fig. 21.5 TREQ and TREP header format

21.5.2 Packet Formats Used by Our Algorithm

Here we give standard DSR packet format with two additional packet format Trust Request packet (TREQ) and Trust Reply packet (TREP).

DSR RREQ and RREP packet format (Figs. 21.2, 21.3, 21.4).

Trust Request (TREQ) and Trust Reply (TREP) pack format (Fig. 21.5).

Direct Trust Value [1] is assigned by the sender node, and next node sets its neighbour trust value and so on up to the destination. When the destination receives the TREQ packet it changes its option field corresponding to TREP packet, and sends back it to the source.

Fig. 21.6 Direct trust table

Node ip address	Direct trust value
172.192.125.182	67%
...	...
172.192.158.169	42%

Fig. 21.7 Indirect trust table

Node ip address	Indirect trust value
172.192.125.182	67%
...	...
172.192.158.169	42%

21.5.3 Data Structure Used by Our Algorithm

Three additional data structures DTT, ITT and NTAL are used by our algorithm for storing and calculating the trust value of the nodes in the network. In the following section we give details of all the data structures.

Direct Trust Table

It has two fields Node's IP address and corresponding direct trust value. Each node maintains this table and initializes it during false packet forwarding. An example of DTT is given next.

Example: (Fig. 21.6)

Indirect Trust Table

It has two fields Node's IP address and corresponding indirect trust value. The originator node only maintains this table and initializes it after the originator receives the TREP packets from the intermediate nodes. We give an example of ITT next.

Example: (Fig. 21.7)

Neighbour Trust Adjacency List

Each intermediate node along a path will maintain an adjacency list called Neighbour Trust Adjacency List (NTAL) for storing the packet forwarding delay by its next neighbour node. Each cell in the main array is divided in two parts, the first part used for storing the total no. of packets that the node forwarded to the particular neighbour corresponding to its index. And the second part stores the pointer to a linked list, in which every node stores the delay that the corresponding neighbour node of it has done while forwarding a packet. We give an example of NTAL next.

Example: (Fig. 21.8)

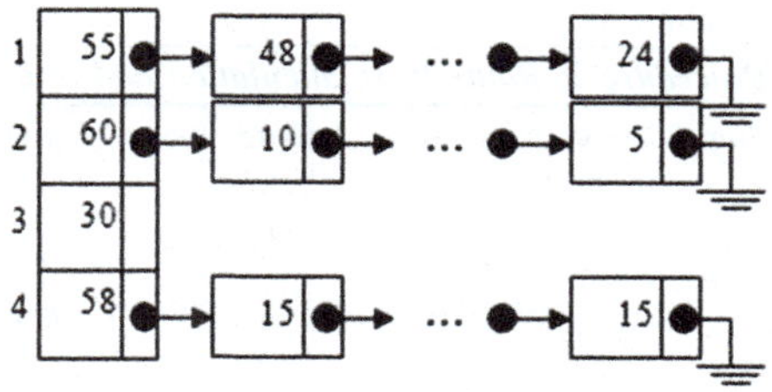

Fig. 21.8 Neighbour trust adjacency list

21.5.4 Proposed Algorithm

Procedure 1: Route discovery

Step 1: Sender node initiates the route discovery by flooding the RREQ packets in the network

Step 2: As we are considering the source routing protocol (DSR), each intermediate node that receives the RREQ packet forwards it towards the destination

Step 3: Then destination sends the RREP packets back to the sender corresponding to every RREQ it receives

Step 4: After receiving a RREP packet the sender extracts all the intermediate node ids from a RREP packet that forms a route to the destination

Step 5: Next the sender calculates the route trust of that *route by using the Route trust calculation procedure* described next

Step 6: Repeat step 4–5 until there are no RREP packet

Step 7: End

Procedure 2: Route trust calculation and Route establishment

Step 1: After the route discovery phase the sender node initiates a *trust exchange* phase

Step 2: The sender node creates false data packets of different type. Some of them are TCP packet and some are UDP packets. Then sends them via different routes, that it already discovered during the route discovery phase

Step 3: During the false packet propagation each intermediate node calculates the direct trust value for its entire neighbour using the *Direct trust calculation procedure*, described next

Step 4: After a certain time period the sender stop sending the false data packets and generate Trust Request Packet (TREQ), which was given in the packet structures section

Step 5: Sender creates TREQ packet for each route. The sender assigns sequence of intermediate node's addresses in Address[i] to Address[n] field in the TREQ packet that it got from a valid RREP packet

Step 6: During the propagation of the TREQ packet intermediate nodes along the path assign their pre calculated trust value about their neighbour towards the destination and also encrypt the trust value with the help of public key that it has corresponding to the particular source

Step 7: When the destination node receives the packet change the option type of the TREQ packet to TREP packet and unicast the RREP packet to the sender node

Step 8: Next the sender initializes the Direct Trust Table (DTT) and Indirect Trust Table (ITT) by using the direct trust calculation and DTT creation and ITT creation and initialization procedure given next

---

(continued)

(continued)

Procedure 2: Route trust calculation and Route establishment
Step 9: Next the sender calculates the route trust of each route by the help of DTT and ITT by using the formula given below: $RT(P) = \sum_{\substack{i \in P \wedge \\ \text{Neighbour of} \\ \text{Originator}}} DTT(i) + \sum_{\substack{i \in P \wedge \\ \text{Other node} \\ \text{on the path}}} ITT(i)$ $[P \text{ is the path and } i \text{ is a node on the path}]$
Step 10: Now the source selects the route that has the maximum RT value and starts sending packets via this route
Step 11: End

Procedure 3: Direct trust calculation and DTT creation
Step 1: When an intermediate node receives a false data packet it forwards it to the next hop (its neighbour node) on the path towards the destination
Step 2: Before forwarding the data packet the node sets two counters C1 and C2 and initializes both of them by its current timer value
Step 3: Now the forwarding node goes into the promiscuous mode to overhear that its neighbour further forwards the packet or not
Step 4: If the neighbour forwards the packet then the forwarding node records its current time when it overhears its neighbour nodes' transmission in the second counter C2
Step 5: Next the forwarding node calculates the packet forwarding delay (PFD). For the packet i the delay generated by neighbour node j is: $PFD_j(i) = C2 - C1$ //so if the neighbour does not re forwards the packet PFD should be 0
Step 6: Then the sender stores the delay in the next node of the linked list that indexed at *j*th row in the Neighbour Trust Adjacency List (NTAL), given in the data structures section
Step 7: Each intermediate node repeats the steps 2–6 until the originator node sends the TREQ packet
Step 8: After an intermediate node receives the TREQ packet it calculates the percentage of the packet forwarding delay for each of its neighbour. PFD percentage for neighbour node j like this: $PFD(j)\% = \frac{\sum PFDj(i)}{\text{Total no. of packets forwarded to node } j} \times 100$
Step 9: Also each intermediate node while forwarding the false data packets calculate Packet forwarding percentage like this: $PF\%(j) = \frac{\text{No. of packet correctly forwarded by the neighbour node } j}{\text{Total no. of packet sent}} \times 100$
Step 10: Now the forwarding node calculates the direct trust value for a neighbour node j like this: $DT(j) = \frac{\alpha \text{ X } PFD\%(j) + \beta \text{ X } PF\%(j)}{2}$ [*Both percentages are weighted through the values* α *and* β, *which can range from 0 to 1*]
Step 11: Each intermediate node calculates the direct trust value for all of its neighbours and stores it in the direct trust table maintained by the node
Step 12: The originator node also calculates the direct trust value for all of its neighbours and creates the DTT using the same process described in this procedure while it sends the false data packets
Step 13: End

Procedure 4: ITT creation and initialization

Step 1: After the originator receiver all the TREP packets it creates an indirect trust table (ITT)
Step 2: If the sender wants to calculate the IT for the node k, it calculates as follows:

2.1. It first extracts the direct trust vales for k from all the TREP packets that contains a DT value for node k

2.2. After extracting all the DT values it uses the formula given below for calculating the IT value for node k like this:

$IT(k) = \sum_{i \in neighbour(k)} DT(i,k)X\frac{1}{|N(k)|}$

N(k) is the no. of neighbour node that k has
Step 3: End

21.6 Conclusion

MANET is prone to many different types of security attacks, if we consider a layered model like TCP/IP there are many dangerous attacks can be occurred in different layer. Among seven layer network layer has the maximum responsibility like secure and efficient routing. In the case of MANET many times we compromise the efficiency in routing because some time the shortest route does not has the maximum remaining power, but we cannot compromise the security at any cost.

In this paper we try to prevent all types of network layer attacks under one umbrella called *trust*. There are many research articles available that prevent one or two of the particular network layer attacks, but according to our survey we do not find any protocol or method that can deal with all types of network layer attacks.

In our proposed method we use the logic of trust for preventing those network layer attacks. Trust parameter has the inbuilt capability for preventing the attacks like black-hole and gray-hole if trust is calculated on the basis of packet dropping ration. And attacks like jelly-fish can be detectable if we are also consider the packet forwarding delay for calculating the trust. And simple attack like worm-hole can be ignored if we are use some sort of asymmetric key cryptography technique, as well since the worm-hole link is a low latency link it is the optimal path and we can use that for fast traffic transmission. So if we use cryptography then worm-hole attack is good for the routing.

In our trust calculation we consider all the parameter like delay, packet dropping percentage and also the implementation of this algorithm uses an asymmetric key cryptography the proposed key distribution mechanism is described in the algorithm. So all the network layer attacks like packet dropping, looping, end-to-end delay and snooping can efficiently be handled by our trust calculation procedure and selecting the routing path based on the trust evaluated by our algorithm.

Now consider the most severe issues like power consumption, overhead that are the main concern of this type of low resource network (MANET). Yes we do not claim that our method does not introduce any type of overheads, but we think the overhead is so minimal. It only uses an extra phase and sends some false packets for calculating the trust values of the whole network. But nodes do not need this very frequently because after one time calculation the nodes can use the calculated trust value up to a fixed time period before it becomes stale, the only extra phase is needed is to flooding the TRQE and TREP packets for receive the trust value of the routes before route establishment.

So we can claim that this is the first trust base technique that can efficiently prevent all types of network layer attacks in MANET with a minimal overhead that we think permissible at the present stage of Mobile Ad-Hoc Networks.

References

1. Nguyen HL, Nguyen UT (2008) A study of different types of attacks on multicast in mobile ad hoc networks. Ad Hoc Netw 6(1):32–46
2. Karmore P, Bodkhe S (2011) A survey on intrusion in ad hoc networks and its detection measures. Int J Comput Sci Eng IJCSE 3:1896–1903
3. Banerjee S, Majumder K (2012) A comparative study on wormhole attack prevention schemes in mobile ad-hoc network. In: Proceedings of recent trends in computer networks and distributed systems security, vol 335. pp 372–384
4. Banerjee S, Majumder K (2012) A survey of blackhole attacks and countermeasures in wireless mobile ad-hoc networks. In: Proceedings of recent trends in computer networks and distributed systems security, vol 335. pp 396–407
5. Rai AK, Tewari RR, Upadhyay SK (2010) Different types of attacks on integrated MANET-internet communication. Int J Comput Sci Secur IJCSS 4(3):265–274
6. Pushpa AM (2009) Trust based secure routing in AODV routing protocol. In: Proceedings of international conference on internet multimedia services architecture and applications (IMSAA). IEEE Press, USA, pp 1–6
7. Li X, Jia Z, Wang L, Wang H (2010) Trust-based on demand multipath routing in mobile ad hoc networks. In: Proceedings of IET information security, vol 4. pp 212–232
8. Edua EN, Radha S, Priyadarshini S, Jayasree S, Naga SK (2012) SRT-secure routing using trust levels in MANETs. Eur J Sci Res 75(3):409–422 ISSN 1450-216X
9. Essia T, Razak A, Khokhar RS, Samian N (2011) Trust-based routing mechanism in MANET: design and implementation. Springer, Berlin
10. Sen J (2010) A distributed trust management framework for detecting malicious packet dropping nodes in a mobile ad hoc network. Int J Netw Secur Appl (IJNSA) 2(4):92–104
11. Bhalaji N, Mukherjee D, Banerjee N, Shanmugam A (2008) Direct trust estimated on demand protocol for secured routing in mobile ad-hoc networks. Int J Comput Sci Secur 1–5
12. Johnson D, Hu Y, Maltz D (2007) The dynamic source routing protocol (DSR). RFC 4728, The internet engineering task force, Network working group http://www.ietf.org/rfc/rfc4728.txt

Part IV
Software Engineering and Soft Computing

Chapter 22
Analysis on Food Web Structure, Interaction, Strength and Stability of Different Mathematical Models of Prey and Predator

Paritosh Bhattacharya, Susmita Paul and K. S. Choudhury

Abstract This paper deals with the dynamics of a predator–prey model. In this paper, we put some models where the parameters of the biological growth model systematically change over time. The densities of both prey and predator populations are obtained as functions of time. We will be concerned with time intervals of the control process and time dependence of the control functions. Here we have discussed about two important growth models.

Keywords Predator · Prey · Time delay · Allee effect · Nonlinear system

22.1 Introduction

The fundamental building blocks of any ecosystem, the food webs, which are assemblages of species through various interconnections, provide a central concept in ecology. Interactions between species in a food web can be of many types, such as predation, competition, mutualism, commensalism, and ammensalism, which are described in below. Volterra (1926) first proposed a simple model for the predation of one species by another to explain the oscillatory levels of certain fish catches in the Adriatic, which is described in model-2. These models primarily concerned with the robustness of the food web structure against modifications (i.e., removal and addition) of vertices and links. The increasing study of realistic and practically useful mathematical models in population biology, whether we are dealing with a human population with or without its age distribution, population of

P. Bhattacharya (✉) · S. Paul
Mathematics Department, NIT Agartala, Agartala, India
e-mail: P_bhattacharya2001@yahoo.co.in

K. S. Choudhury
Mathematics Department, Jadavpur University, Kolkata, India

S. Sengupta et al. (eds.), *Emerging Trends in Computing and Communication*, Lecture Notes in Electrical Engineering 298, DOI: 10.1007/978-81-322-1817-3_22,

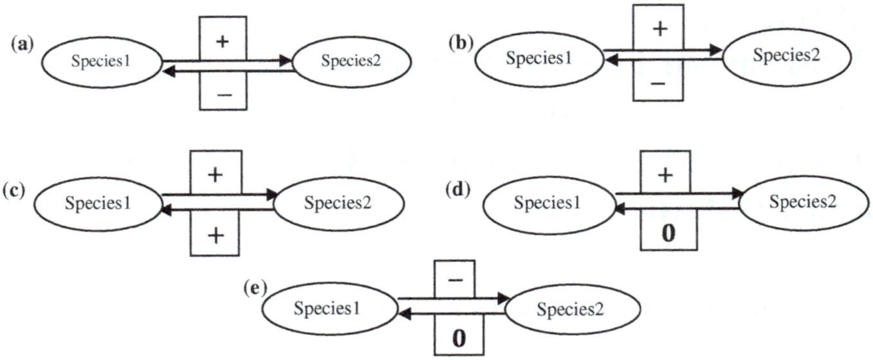

Fig. 22.1 Types of ecological interactions: **a** predation **b** competition **c** mutualism **d** commensalism **e** ammensalism

an endangered species, bacterial or viral growth and so on, is a reflection of their use in helping to understand the dynamic process involved and in making practical prediction.

22.2 Some Basic Definitions

A community in ecology comprises all species populations interacting in an area. An example of a community is a coral reef, where numerous populations of fishes, crustaceans, and corals coexist and interact.

A food web in an ecosystem is an assemblage of various organisms that are interconnected with each other through their different life history processes, such as feeding and shelter [1, 2].

A trophic level in a food web consists of all the species that prey on the same species and are also preyed upon by the same species [3].

Ecological interactions are the relationships between two species in an ecosystem [4]. Based on either effects, or on mechanisms, these relationships can be categorized into many different classes of interactions as shown in Fig. 22.1.

1. **Predation** is a biological interaction in which one species feeds on another. Most of the interactions in a food web are predatory. Figure 22.1a shows the network for this interaction, where species 2 preys on species 1. This interaction enhances the fitness of predators (indicated by "+"), but reduces the fitness of the prey species (shown by "−").
2. **Parasitism** is similar to predation by mechanism, as it enhances the fitness of the parasite, but impairs the host (Figs. 22.2, 22.3).
3. **Competition** between two species occurs when they share a limited resource and each tends to prevent the other from accessing it. This reduces the fitness of

Fig. 22.2 Example of predation. Here the predator cat eats the prey rabbits

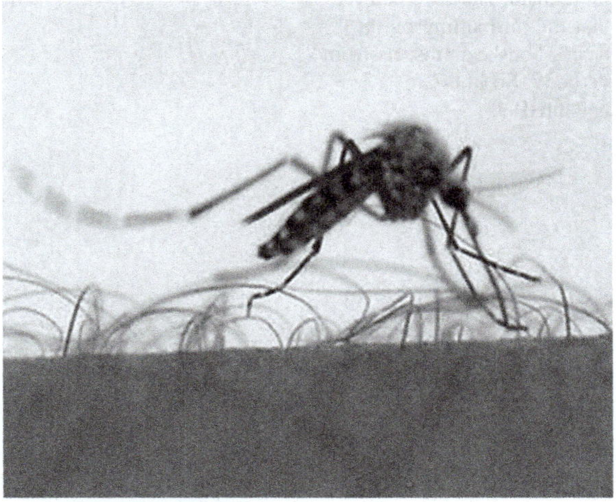

Fig. 22.3 Example of parasitism. In this picture mosquito is parasite and human body is the host. Here mosquito decreases the body's immunity by seeking our blood

one or both species. This reduces the fitness of one or both species, as is shown by "−" in Fig. 22.1b (Fig. 22.4).

4. In **mutualism** or **symbiosis**, two species provide resources or services to each other. This enhances the fitness of both species (shown by "+" in Fig. 22.1c) (Fig. 22.5).
5. **Commensalism** is an interaction, where one species receives a benefit from another species. This enhances the fitness of one species without any effect on fitness of the other species. (Shown by "0" in Fig. 22.1d) (Fig. 22.6).
6. In **ammensalism**, one species impedes or restricts the success of the other without being affected positively or negatively by its presence. (Shown by "−" and "0" respectively in Fig. 22.1e) (Fig. 22.7).

Fig. 22.4 Example of competition. Here in the picture we can see deadly hyena and the tiger fighting out for a limited resource (food)

Fig. 22.5 Example of mutualism. Here we can see that the bird sitting on the dear's body eats insects from its body. So both are benefited

Fig. 22.6 Example of commensalism. Here in the picture we see cows eating grasses meanwhile birds eating insects sharing the same spaces. That means common sharing and interaction of one species with another without harming each other

Fig. 22.7 Example of ammensalism. In this picture we see that the species lion restricts the growth of buffalo by eating it where as both species can live without harming the other

22.3 Food Web Structure, Interaction, Strength and Stability

Food web models are extensions of bioenergetics consumer-resource models, which by definition focus exclusively on trophic interactions. In a recent study, it was shown that predation is the most important process determining the community structure and dynamics [5, 6]. There are a few important related factors that regulate the strength of this process, such as metabolic efficiencies, handling times, foraging strategies, and frequencies of encounters. In the following section, we consider one simple ecological networks of the prey-predator interaction and discuss how the emergence of new functional components that alter interaction strengths can regulate the stability in food web dynamics.

22.3.1 Model 1

The figure bellow shows a prey-predator system where the predator species is commensally on the prey species [7]. This model is a ratio-dependent prey–predator system, which is termed as Volterra's principle, can be found in Lotka-Volterra model (Legovic 2008). In this simple model, in the absence of the predator (Y), the prey species (X) follows a density-dependent logistic growth with r as its intrinsic growth rate, and K is the carrying capacity of the environment. However, in the presence of the predator, the growth of the prey is reduced due to predation of Y on X. This interaction follows a hyperbolic function with Υ denoting the half saturation coefficient of predation and α deciding the strength of interaction, i.e., the per capita consumption rate. In the absence of prey, the predator species dies out exponentially at a rate d. On predation, the rate at which this food adds to the growth of the predator population is given by the conversion rate β. The rate of change of prey $\left(\frac{dX}{dt}\right)$ and predator $\left(\frac{dY}{dt}\right)$ populations with time is governed by the following equations (Fig. 22.8).

$$\frac{dX}{dt} = rX\left(1 - \frac{X}{k}\right) - \frac{\alpha XY}{\gamma + X}$$

$$\frac{dY}{dt} = Y(-d + \frac{\beta\alpha X}{\gamma + X})$$

22.3.2 Model 2

If *N(t)* is the prey population and *P*(t) that of the predator at time *t* then

$$\frac{dN}{dt} = N(a - bP) \tag{22.1}$$

$$\frac{dP}{dt} = P(cN - d) \tag{22.2}$$

where *a*, *b*, *c* and *d* are positive constants. This is known as Lotka-Volterra model [8, 9].

The assumptions in the model are: (1) The prey in the absence of any predation grows unboundedly in a Malthusian way; this is the *aN* term in Eq. (22.1). (2) The effect of the predation is to reduce the prey's per capita growth rate by a term proportional to the prey and predator populations; this is the *–bNP* term. (3) In the absence of any prey for sustenance the predator's death rate results in exponential decay, that is, the *–dP* term in Eq. (22.2). (4) The prey's contribution to the predators' growth rate is *cNP*; that is, it is proportional to the available prey as well as to the size of the predator population.

22.4 Results and Discussions

22.4.1 For This Study, In Model 1

The parameter values are taken as r = 0.5, K = 5, d = 3, $\Upsilon = 0.8$, and $\beta = 0.7$. The main parameter, α, which indicates the interaction strength of predation, is varied from 5.5 to 6.5 to describe the change in dynamics in this model network. Here let, X = 13, Y = 5 and t is varied from 0 to 5. Now for $\alpha = 5.5$, we plot the Fig. 22.9. We can see from the Fig. 22.9. When time t = 0, the number of prey is maximum & it is 13. On the other hand the number of predator at time t = 0.4 is maximum and it is becoming approximately 5.6. This shows that initially number

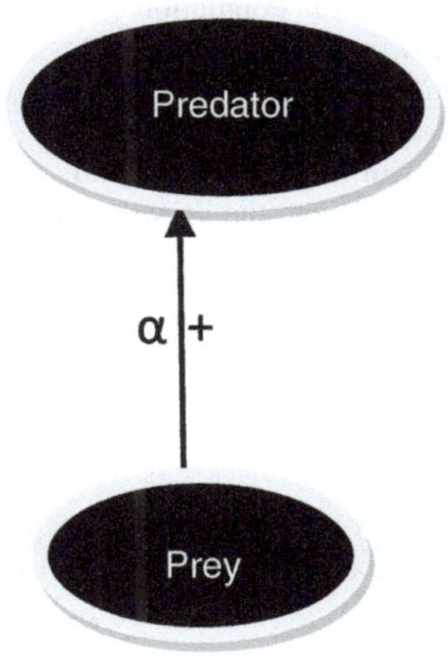

Fig. 22.8 Food web configurations of Model1

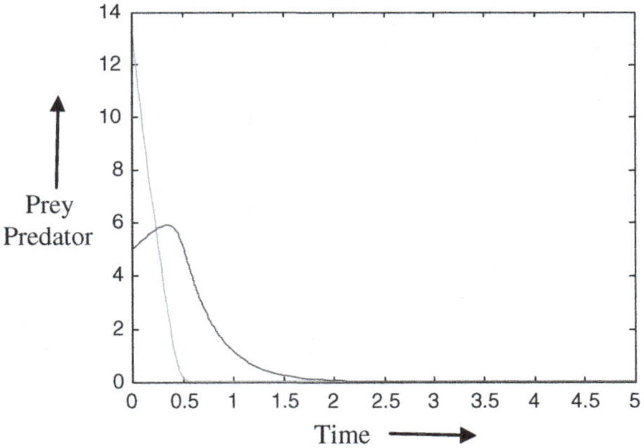

Fig. 22.9 Population growth rate forms for prey and predator in case of $\alpha = 5.5$. Here *blue color* denotes the growth rate of prey(X) and *green* denote the growth rate of predator(Y)

of prey is higher than predator but gradually it decreases than number of predators. Moreover predator is highest when prey is nearly minimum and gradually predator becomes less with the decrease in number of prey, and ultimately predator finishes as there is no prey left.

Now, for $\alpha = 6.4$ we plot the graph Fig. 22.10 and other values remaining same. Here in Fig. 22.10 the number of predator is maximum at time t = 0.4 and it is becoming approximately 6.8. Otherwise Fig. 22.1a, b almost similar. The only difference between the highest point of predator.

Now let, X = 1; Y = 0.5 and t is varied from 0 to 10. Now for $\alpha = 5.5$ we plot the Fig. 22.11 and we can see from the Fig. 22.11 that at time t = 10 the number of prey is maximum and it is becoming approximately 4.2. On the other hand the number of Predator at time t = 0 is maximum and it is 0.5. This shows that there is competition between predator and prey.

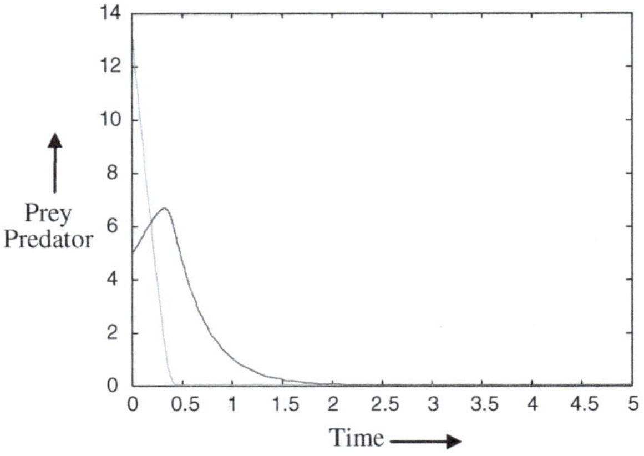

Fig. 22.10 Population growth forms for prey and predator in case of $\alpha = 6.4$. Here blue color denotes the growth rate of prey(X) and green denote the growth rate of predator(Y)

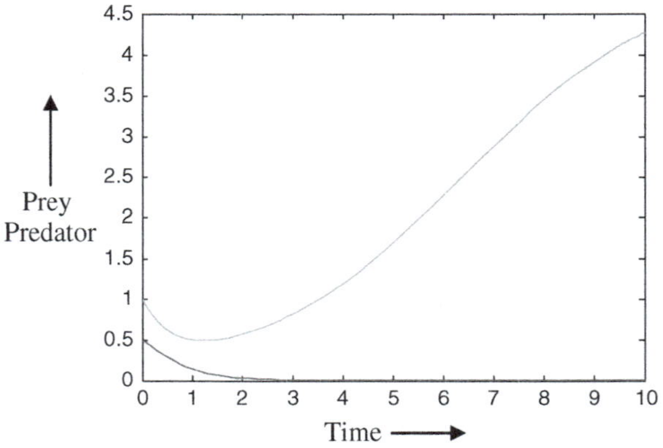

Fig. 22.11 Population growth forms for prey and predator in case of $\alpha = 5.5$. Here *blue color* denotes the growth rate of prey (X) and *green* denote the growth rate of predator(Y)

Now for $\alpha = 6.4$ we plot Fig. 22.12 and other values are remaining same. We can see from the Fig. 22.12, that at time t = 10 the number of prey is maximum and it is becoming approximately 3.8. On the other hand the number of Predator at time t = 0 is maximum and it is 0.5. This shows that there is competition between predator and prey (Figs. 22.13, 22.14).

22.4.2 In Model 2

The parameter values are taken as, a = 0.5471; b = 0.0281; c = 0.0266; d = 0.8439. Here let t = 13 and P = 21 and for these values we plot the Fig. 22.15. We can see from the Fig. 22.15 that at time t = 6 the number of prey is maximum and it is becoming approximately 63. On the other hand the number of Predator at time t = 8 is maximum and it is becoming approximately 42. This shows that there is competition between predator and prey.

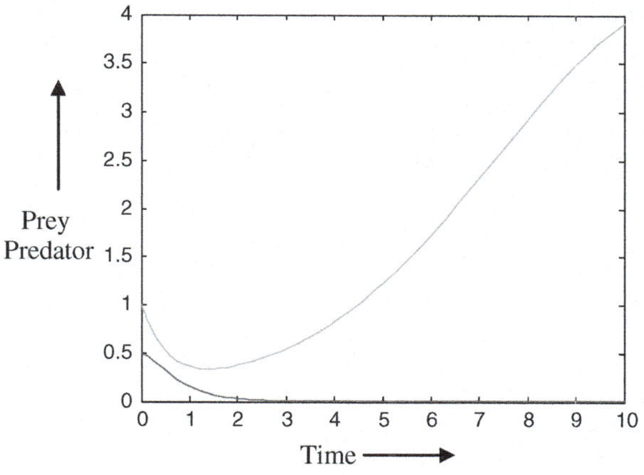

Fig. 22.12 Population growth forms for prey and predator in case of $\alpha = 6.4$. Here *blue color* denotes the growth rate of prey(X) and *green* denote the growth rate of predator(Y)

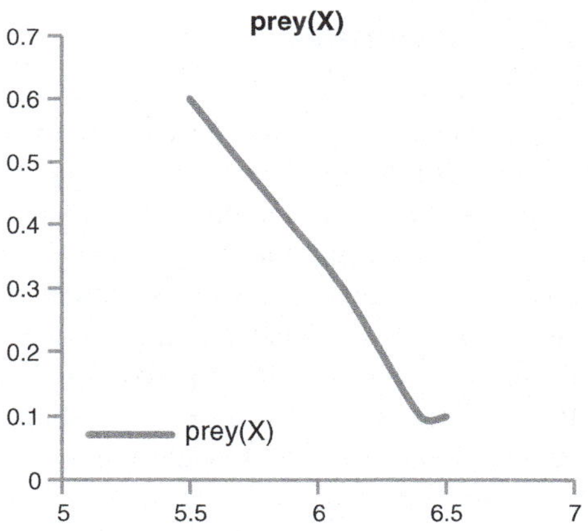

Fig. 22.13 Diagram of the prey with increasing predation strength α

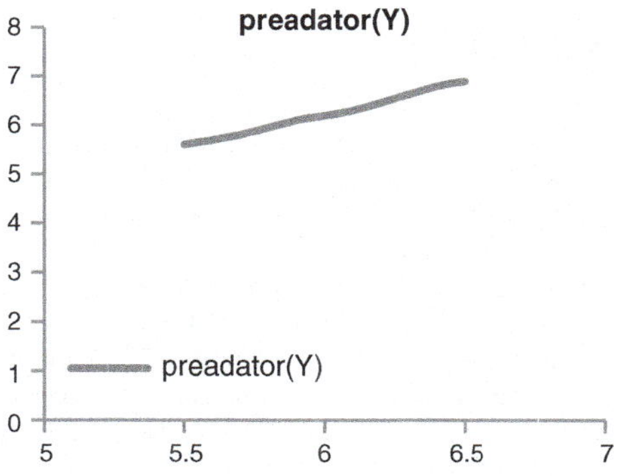

Fig. 22.14 Diagram of the predator with increasing predation strength α

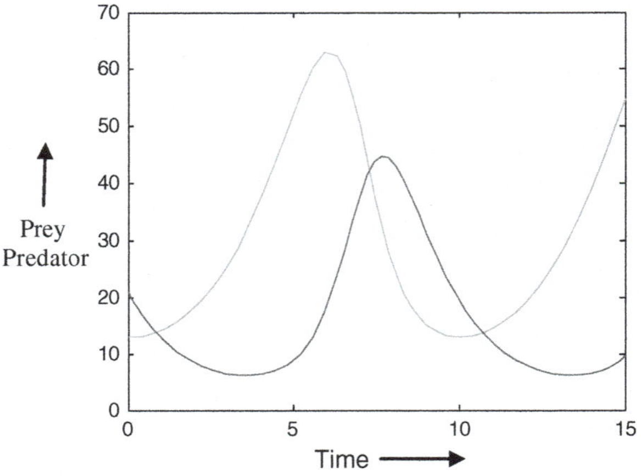

Fig. 22.15 Population growth forms for prey and predator. Here *blue color* denotes the growth rate of prey and *green* denote the growth rate of predator

22.5 Conclusions

In this paper we formulated two basic mathematical models which are based on the existence of different Prey–Predator system. Although we have mainly considered 2-species interactions in this paper, in nature, and in the sea in particular, there are many species or trophic levels where energy, in the form of food, flows from one species to another. Most of the recent research on food web theory in ecology centres around the local dynamics of a community, but the evolution of food web dynamics across different spatial scales has also received considerable attention [6, 10]. Here the densities of the populations are obtained as functions of time. Refined Lotka-Volterra models appear to be the appropriate level of mathematical sophistication to describe simple predator-prey models.

References

1. Bascompte J, Melian CJ (2005) Simple trophic modules for complex food webs. Ecology 86:2868–2873
2. Camacho J et al (2007) Quantitative analysis of the local structure of food webs. J Theor Biol 246:260–268
3. Berlow EL, Brose U, Martinez ND (2008) The "Goldilocks factor" in foodwebs. Proc Natl Acad Sci USA 105:4079–4080
4. Bastolla U, Lassig M, Manrubia SC, Valleriani A (2001) Diversity patterns from ecological models at dynamical equilibrium. J Theor Biol 212:11–34
5. Peterson EE, Theobald DM, VerHoef JM (2007) Geostatistical modeling on stream networks: developing valid covariance matrices based on hydrologic distance and stream flow. Freshw Biol 52:267–279
6. Memmott J et al (2006) Biodiversity loss and ecological network structure. In: Pascualand M, Dunne JA (eds) Ecological networks: linking structure to dynamics in food webs. Oxford University Press, Oxford

7. Saez E, Gonzalez-Olivares E (1999) Dynamics on a predator–prey model. SIAM J Appl Math 59(5):1867–1878
8. Freedman HI (1980) Deterministic mathematical models in population ecology. Marcel Dekker, New York
9. Ross R (1911) The prevention of malaria. Murray, London
10. Wang G, Liang X-G, Wang F-Z (1999) The competitive dynamics of populations subject to an Allee effect. Ecol Model 124:183–192

Chapter 23
Eigen Value and It's Comparison with Different RBF Methods by Using MATLAB

Abhisek Paul, Paritosh Bhattacharya and Santi Prasad Maity

Abstract Neural network is being used in various research areas in recent time. In this paper we have introduced Radial Basis Function (RBF) of neural network for the analysis of Eigen value. Eigen value is the characteristic value of any given system. We have incorporated various radial basis functions such as Gaussian RBF, Multi-Quadratic RBF and Inverse-Multi-Quadratic RBF in matrix for the calculation of Eigen value. Comparative analysis and simulation results show that Gaussian RBF gives better result compared to the other relevant radial basis functions.

Keywords Neural network · Eigen value · Radial basis function

23.1 Introduction

In recent time of research, neural networks are being used in various areas such as pattern recognition, optimization techniques, image processing, and classification. Various neural networks are there, out of that we have introduced Radial Basis Function (RBF) neural network for the computation process.

As we know Eigen value is the characteristic value of an equation which is modelled by that system. In this paper we have incorporated some RBF neural

A. Paul · P. Bhattacharya (✉)
Department of Computer Science and Engineering, National Institute of Technology, Agartala, India
e-mail: p_bhattacharya2001@yahoo.co.in

A. Paul
e-mail: abhisekpaul13@gmail.com

S. P. Maity
Department of Information Technology, Bengal Engineering and Science University, Shibpur, India

S. Sengupta et al. (eds.), *Emerging Trends in Computing and Communication*, Lecture Notes in Electrical Engineering 298, DOI: 10.1007/978-81-322-1817-3_23,

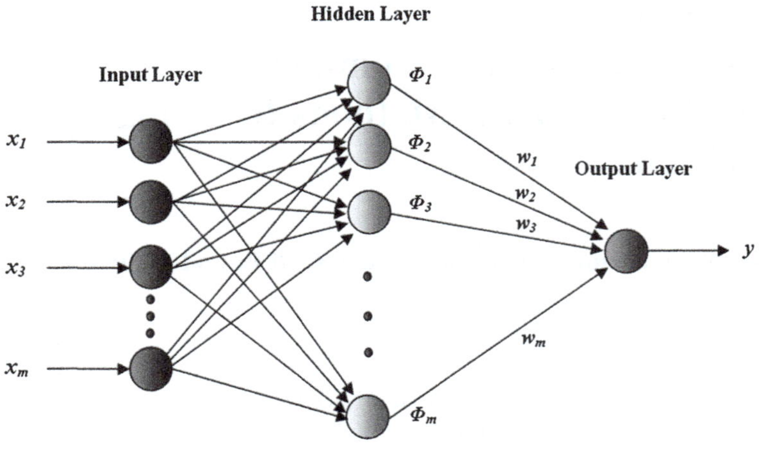

Fig. 23.1 Architecture of radial basis function neural network

network methods to compute Eigen values of some small matrixes and some matrixes of image. Radial basis functions like Gaussian RBF, Multi-Quadratic RBF and Inverse-Multi-Quadratic RBF are introduce to compute Eigen value and compared with normal methods [1–3].

The reminder of the sections is as follows. In Sect. 23.2 overview of RBF neural network is introduced. Section 23.3 comprises the mathematical analysis. Section 23.4 depicts simulation and experiment. Lastly, conclusion is given in Sect. 23.5.

23.2 Radial Basis Function

In Fig. 23.1 an overview of RBF neural network architecture is given. In this neural network there are three layers, such as input layer, hidden layer and the output layer. In x_1, x_2, x_3,...x_m are inputs which are fed into the input layer. Computational units are in the hidden layer which are called radial centres and represented by c_1, c_2,...,c_m vectors. Here, dimension of each centre for m input network is $m \times 1$. The output of each centre which is Φ_i is the function of the Euclidian distance between c_i and x. Output y is obtain by proper choice of $w_{j.}$, which is the weight of jth canter. The output is simply the summation of $\Phi_i w_i$. [1, 2].

$$y = \sum_{j=1}^{m} \phi_j w_j \tag{23.1}$$

$$\phi_j(x) = (||x - x_j||) \tag{23.2}$$

Various radial basis function equations are given below. Here z is the Euclidian distance; σ is the maximum distance from centre.

- Gaussian RBF:

$$\phi(z) = e^{-z^2/2\sigma^2} \tag{23.3}$$

- Multi-Quadratic RBF:

$$\phi(z) = (z^2 + r^2)^{1/2} \tag{23.4}$$

- Inverse-Multi-Quadratic RBF:

$$\phi(z) = (z^2 + r^2)^{-1/2} \tag{23.5}$$

23.3 Mathematical Analysis

RBF neural network needs optimal selections of some parameters such as the weight and the centres. We have introduced pseudo inverse technique to update the required parameters.

$$\phi = [\phi_1, \phi_2, \phi_3 \ldots, \phi_m] \tag{23.6}$$

$$w = [w_1, w_2, w_3 \ldots, w_m]^T \tag{23.7}$$

$$\phi w = y^d \tag{23.8}$$

Desired output is y^d and the weight vector can be computed as below. Here, the pseudo inverse of Φ is Φ' [4].

$$w = (\phi^T \phi)^{-1} \phi^T y^d \tag{23.9}$$

$$w = \phi' y^d \tag{23.10}$$

We have shown some example of small matrixes namely matrix A and matrix B which are of 3 × 3 and 4 × 4 dimensions respectively. We have also included image of size 128 × 128 pixels. We extract the red channel, green channel, and blue channel components of the image and also calculate the Y channel component [5] which is shown in Eq. 23.11.

$$Y = 0.299R + 0.587G + 0.114B \tag{23.11}$$

$$\mathrm{A}_{3\times3} = \begin{bmatrix} 2 & 1 & 3 \\ 1 & 2 & 1 \\ 3 & 1 & 2 \end{bmatrix} \quad \mathrm{B}_{4\times4} = \begin{bmatrix} 5 & 11 & 6 & 3 \\ 2 & 4 & 6 & 7 \\ 6 & 8 & 9 & 3 \\ 2 & 8 & 4 & 6 \end{bmatrix}$$

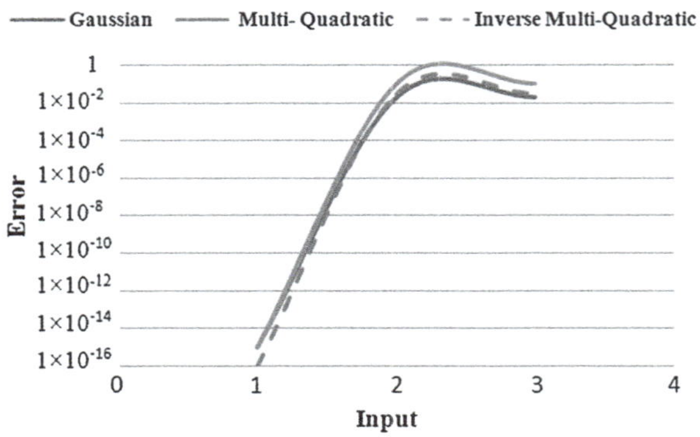

Fig. 23.2 Comparison of error of Eigen values of matrix A with different RBF methods

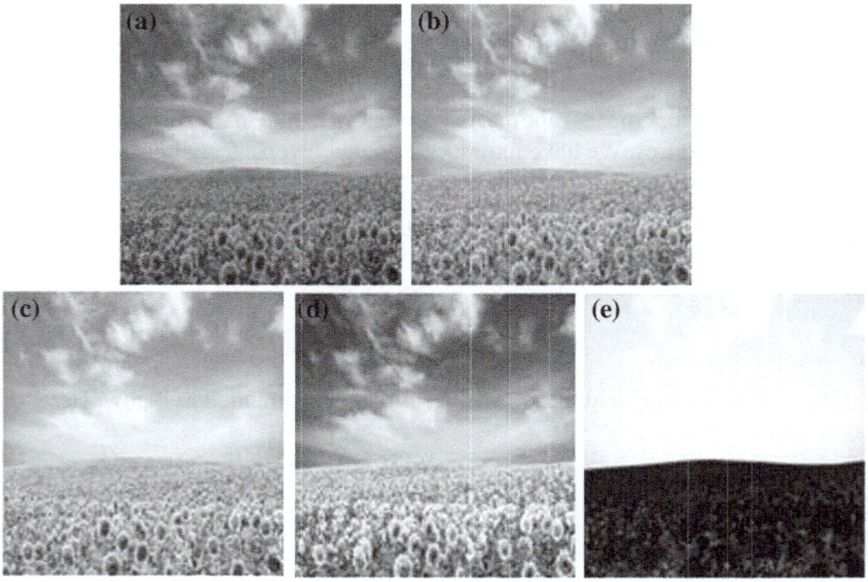

Fig. 23.3 Landscape image **a** Original color image, **b** Y channel component, **c** Red channel component, **d** Green channel component, **e** Blue channel component

23.4 Simulation and Experiment

Matrix A and B are experimented for Eigen value calculations with RBF methods. Comparison of errors is given in Tables 23.2, 23.4. Figure 23.2 also shows the same thing for matrix A. In Fig. 23.3 we have shown image Landscape and it's

Table 23.1 Eigen values of matrix A with different RBF methods

Normal method	Gaussian RBF	Multi quadratic RBF	Inverse multi quadratic RBF
−1.000000000000000	−1.000000000000001	−0.980392156862739	−0.980392156862743
1.438447187191170	1.438447187191170	1.410242340383501	1.410242340383501
5.561552812808831	5.561552812808832	5.452502757655707	5.452502757655715

Table 23.2 Comparison of error of Eigen values of matrix A in different RBF methods

Error in Gaussian RBF	Error in multi quadratic RBF	Error in inverse multi quadratic RBF
0.000000000000001	0.019607843137261	0.019607843137257
0.000000000000000	0.028204846807669	0.028204846807669
0.000000000000001	0.109050055153124	0.109050055153116

Table 23.3 Eigen values of matrix B with different RBF methods

Normal method	Gaussian RBF	Multi quadratic RBF	Inverse multi quadratic RBF
22.229382345925909	22.229382345925906	22.229382345925899	22.229382345925913
4.161391961254129	4.161391961254127	4.161391961254126	4.161391961254131
0.160982370085429	0.160982370085432	0.160982370085437	0.160982370085433
−2.551756677265459	−2.551756677265459	−2.551756677265464	−2.551756677265460

Table 23.4 Comparison of error of Eigen values of matrix B in different RBF methods

Error in Gaussian RBF	Error in multi quadratic RBF	Error in inverse multi quadratic RBF
0.000000000000003	0.000000000000010	0.000000000000004
0.000000000000002	0.000000000000003	0.000000000000002
0.000000000000003	0.000000000000008	0.000000000000004
0.000000000000000	0.000000000000005	0.000000000000001

Table 23.5 Mean or average Eigen value of landscape image with different RBF methods

Input (landscape image)	Gaussian RBF	Multi quadratic RBF	Inverse multi quadratic RBF
Red channel	144.1718750000003	144.1718749995045	144.1718750000001
Green channel	170.0703125000001	170.0703125000763	170.0703124999997
Blue channel	147.6562499999998	147.6562499993812	147.6562499999996
Y channel	159.8749999999999	159.8750000001236	159.8749999999998

Table 23.6 Comparison of error in Eigen value of landscape image in different RBF methods

Input (landscape image)	Error in Gaussian RBF	Error in multi quadratic RBF	Error in inverse multi quadratic RBF
Red channel	0.0000000000000	0.0000000004948	0.0000000000002
Green channel	0.0000000000001	0.00000000000763	0.0000000000003
Blue channel	0.0000000000001	0.0000000006188	0.0000000000003
Y channel	0.0000000000002	0.0000000001235	0.0000000000003

red, green, blue and Y channel component images. Here, in Table 23.5 the average or mean Eigen values are given and in Table 23.6 the corresponding errors are also shown. All the simulations are being done in MATLAB 7.6.0 software tools [6] (Tables 23.1 and 23.3).

23.5 Conclusion

In this paper, Gaussian RBF, Multi-Quadratic RBF and Inverse-Multi-Quadratic RBF methods are used for experiment. Experimental results and simulations show that Gaussian RBF method gives lesser error for the computation of Eigen values of small size matrix as well as for large size matrix. So, we can conclude Gaussian RBF can be used in neural network for better result compared to the other relevant neural network methods.

Acknowledgments The authors are grateful to the anonymous referee for a careful checking of the details and for helpful comments that improve this paper.

References

1. Mao KZ, Huang G-B (2005) Neuron selection for RBF neural network classifier based on data structure preserving criterion. IEEE Trans Neural Netw 16(6):1531–1540
2. Schölkopf B, Sung KK, Burges CJC, Girosi F, Niyogi P, Poggio T, Vapnik V (1997) Comparing support vector machines with Gaussian kernels to radial basis function classifiers. IEEE Trans Signal Process 45:2758–2765
3. Luo FL, Li YD (1994) Real-time computation of the eigenvector corresponding to the smallest Eigen value of a positive definite matrix. IEEE Trans Circuits Syst 41:550–553
4. Klein CA, Huang CH (1983) Review of pseudo-inverse control for use with kinematically redundant manipulators. IEEE Trans Syst Man Cybern 13(3):245–250
5. Noda H, Niimi M (2007) Colorization in YCbCr color space and its application to JPEG images. Pattern Recogn 40(12):3714–3720
6. Math Works. MATLAB 7.6.0 (R2008a) (2008)

Chapter 24
Effectiveness of Test-Driven Development as an SDLC Model: A Case Study of an Elevator Controller Design

Sayani Mondal and Partha Pratim Das

Abstract Test-driven development (TDD) is a new software development model where codes are written to meet the tests as specified from the specs. It is an agile method and claims to be more effective and efficient than the traditional waterfall (and other derivative) SDLC models. In this paper we use the development of an elevator controller as a target system and compare TDD against waterfall through independent development. Using three progressive "versions" of elevator system, we show the advantages of TDD over Waterfall.

Keywords Test-driven development · Agile methods · CppUnit framework

24.1 Introduction

A Software Development Life Cycle (SDLC) model like Waterfall, Prototype, and Spiral is a roadmap for a software product from conception till retirement. In these models we gather requirements, analyze, design, implement and finally test the code. Bugs found during test need a lot of rework here. Agile Methods [1, 2] attempt to change this workflow to minimize rework. Test-Driven Development (TDD) is one such method where the design and coding is driven from the test plan and test suites. This is expected to reduce unnecessary codes and ensure that any code that goes in is indeed already tested. Hence TDD is expected to be more efficient and effective.

S. Mondal (✉) · P. P. Das
Department of Computer Science and Engineering, Indian Institute of Technology, Kharagpur 721302, India
e-mail: sayani_mon46@yahoo.co.in

P. P. Das
e-mail: partha.p.das@gmail.com

S. Sengupta et al. (eds.), *Emerging Trends in Computing and Communication*, Lecture Notes in Electrical Engineering 298, DOI: 10.1007/978-81-322-1817-3_24,

The objective of the paper is to compare TDD against Waterfall using the development of an *Elevator Controller System* as a case-study. The system is implemented with both TDD and Waterfall, starting separately from an initial specification. Incrementally, the system is improved through changes at three phases to completion.

The paper is organized as follows: Sect. 24.2 briefly reviews the different SDLC models. Section 24.3 explains the research methodology for the development of the system. In Sect. 24.4, the results are discussed. Section 24.5 summarizes our findings.

24.2 Review of SDLC Models

Common SDLC models [2, 3] include Waterfall, Spiral, and Prototype. These models have feasibility study, requirement analysis, design, coding, and testing phases. The workflow is unidirectional and any re-designing or re-coding affects many phases. Prototype Model of a product has limited functionalities, low reliability and inefficient performance compared to the final one; but it helps to build the final product. The Spiral Model [4] is divided into 4 quadrants with loops. Each quadrant performs different activities like product validation, risk examination, and iteration planning. Recently, various agile models like TDD have started becoming popular.

24.2.1 TDD Model

Test-driven development (TDD) [5, 6] is a basic agile development process, where tests are created before production code is designed. A number of software development methods such as extreme programming (XP), feature-driven development, scrum, dynamic systems development, and adaptive software development fall into this category. In TDD first the test code is written according to the test cases. The production code is then developed to pass the test cases. Hence Tests here Drive the Development. TDD's development process for each test follows a cycle [7] as: (1) Write the test code, (2) Compile the test code, (3) Implement just enough to compile without errors, (4) Run the test and see if it fails, (5) Implement just enough to make the test pass, (6) Run the test and see if it passes, (7) Refactor for clarity and to eliminate duplication, and (8) Repeat the steps.

24.3 Research Methodology

To compare TDD against Waterfall, we develop the controller of an elevator using both approaches. The development is done in 3 phases—a basic phase where the elevator has only 1 cab, a change request phase where another cab is introduced and a final phase with arbitrary number of cabs. The phases are planned to simulate the develop-test-use-update life-cycle of software.

24.3.1 Phase 1: Basic Phase (# Cabs = 1)

System Specification: Initially a simple system is developed with one cab:

Number of Buttons: Cab Button, Floor Button, Open door button, Close door button

Number of Floors (assumed): 10 (Ten)

TDD—Test Plan:

1. Whether lifts exists or not will be tested first.
2. Whether cab can move 1 floor up/down is tested.
3. For multiple floors, check if the floor button requests are accepted or not.

TDD—Design: Sequence diagram of cab movement shown in Fig. 24.1a.

Waterfall—Design: Flow chart of cab movement is shown in Fig. 24.1b.

24.3.2 Phase 2: Change Request Phase (# Cabs = 2)

System Specification: Number of cabs is increased to 2. We formulate a movement matrix of rules for cab selection on basis of the direction of floor requests.

TDD—Test Plan: As according to the movement matrix:

1. Whether cabs can move 1 floor up/down is tested.
2. Whether the elevator processes the signal from any floor and changes its destination floor or not is tested.
3. When the same floor no is pressed on which the lift is already there.

TDD—Design: Sequence diagram in Fig. 24.2a, shows movement of two cabs according to the movement matrix for better efficiency and less cost.

Waterfall—Design: This does not consider any test plans before implementation. Hence, bugs are likely to occur. Our results show a list of bugs on a set of inputs. Figure 24.2b shows the flowchart of cab movement.

24.3.3 Phase 3: Final Phase (# Cabs = n, n > 2)

System Specification: Number of cabs is increased to n (generalized, as per requirement). The rule matrix is followed for cab selection and movement. Based on the time of request we use the rule matrix. Increase test floors to more than 10.

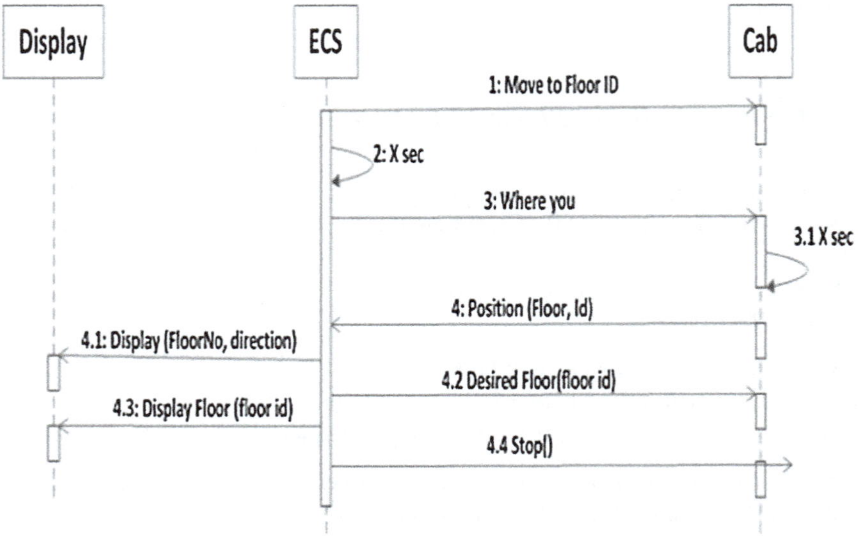

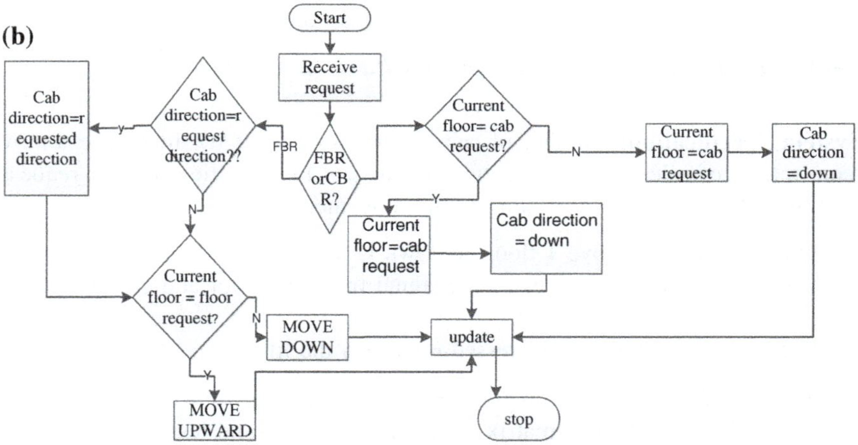

Fig. 24.1 **a** Sequence diagram for movement of 1 Cab. **b** Flow chart of the movement

TDD—Test Plan: According to the generalized lift structure:

1. Test cabs can move 1 floor up and down direction with single passenger.
2. Any idle cab will always accept a floor request, at any time.
3. If a cab is selected for any particular floor request, and if it is moving in reverse direction. Then the request will be stored in its queue.

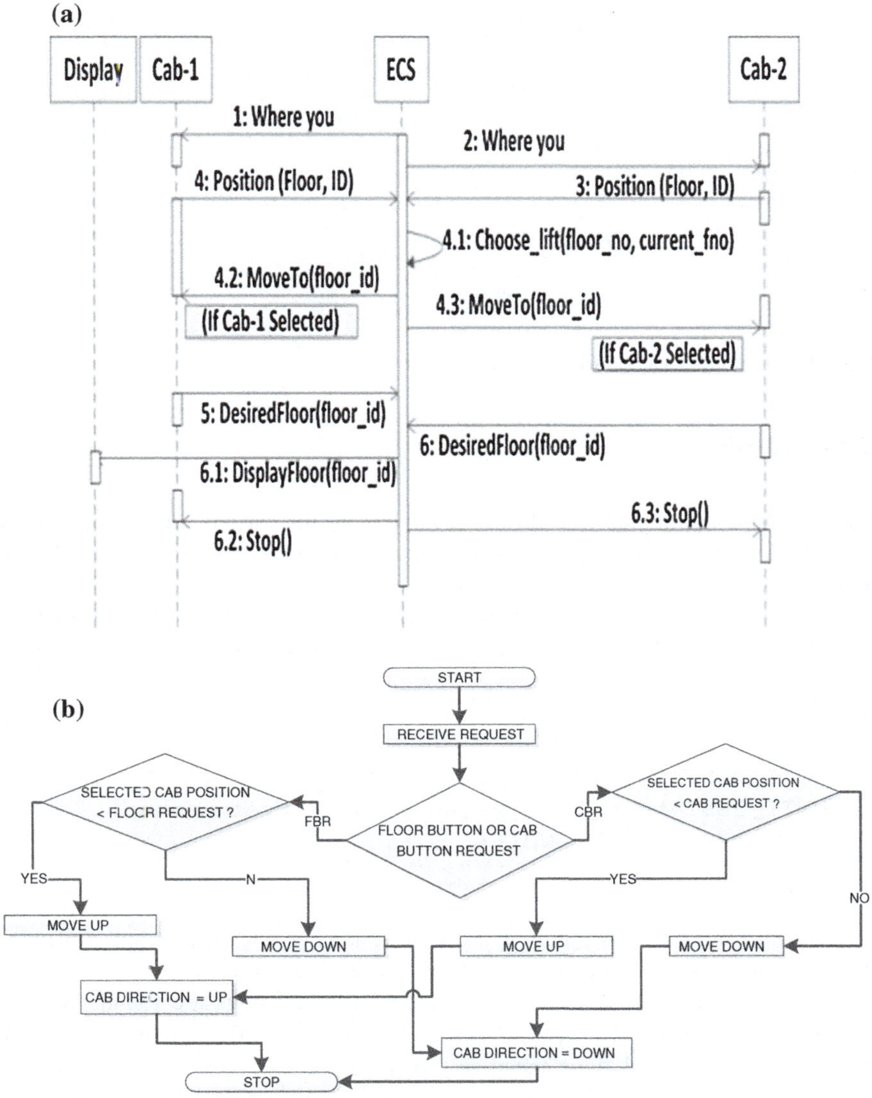

Fig. 24.2 **a** Sequence diagram for 2 Cabs. **b** Flow chart for 2 cabs

TDD—Design: As number cabs are not fixed, but as soon as the number of cabs is fixed, the number is immediately sent to the controller. The rest of the movements and selection of the cabs are same as that of the previous cases.

Waterfall—Design: Traditional flowchart in Fig. 24.3.

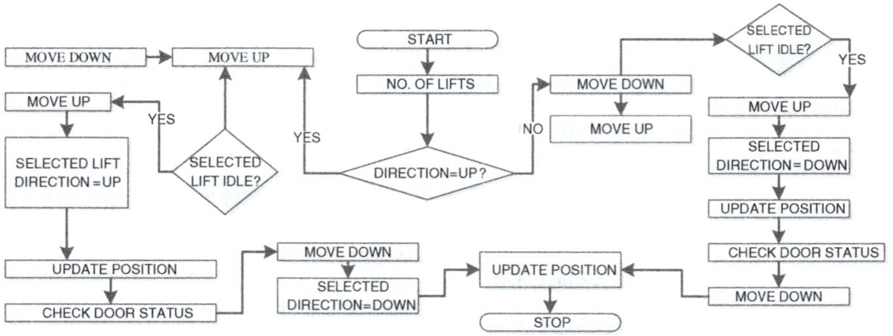

Fig. 24.3 The flow chart showing the movement of n-cabs

24.4 Results

For the above-mentioned test plans, we implement for various set of inputs. Here we provide some of the snapshots, which depict how a test case is countered by the TDD as well as Waterfall model approach.

Phase-1: TDD Test Plan-1: Using CppUnit framework [8, 9], the test methods are written. The code to be included for the test cases is shown below.

```
class test_lift_ecs : public CPPUNIT_NS :: TestCase
{

CPPUNIT_TEST_SUITE(test_lift_ecs);
CPPUNIT_TEST(test_lift_exists);//Set test method
CPPUNIT_TEST_SUITE_END();
}
```

Here an example shows how a test case fails (Fig. 24.4a) and how on introduction of production code (Fig. 24.4b) it passes (Fig. 24.4c).

Waterfall: The code is developed without tests and bugs are found. When a sample input {{3, 0}, {2, 1}, {5, 0}, {4, 1}} is tested, instead of stopping at desired floor, the cab stopped 1 floor ahead as shown in Fig. 24.5.

Phase-2: TDD Test Plan: Some portions of production code for modified system is developed and the previous test cases are run. Some of them failed (Fig. 24.6a) leading to development of separate ones. For rest of the test cases, test methods are generated as in phase 1. Figure 24.6b shows a sample testing whether right cabs are note selected.

Waterfall: Fig. 24.7 shows an example where bug is found in selecting the cabs, as same cab is selected repeatedly, which is rectified later after testing for a set of inputs.

Phase 3: TDD Test plan: Amongst the many test cases, when FBR and CBR become same, test case will fail as it will not be accepted as shown in Fig. 24.8.

(a) (b) (c)

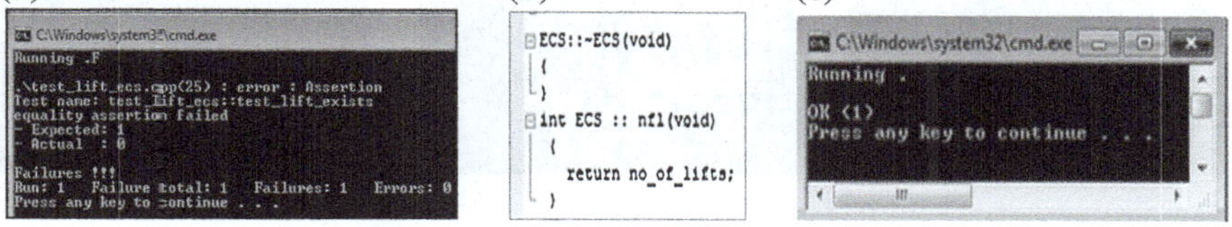

Fig. 24.4 **a** Test case fails. **b** Production code is added. **c** Test case passes

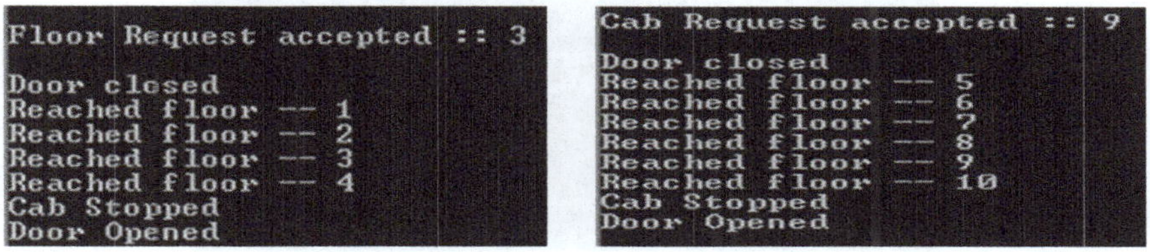

Fig. 24.5 Sample output where actual floor no is not served

(a) (b)

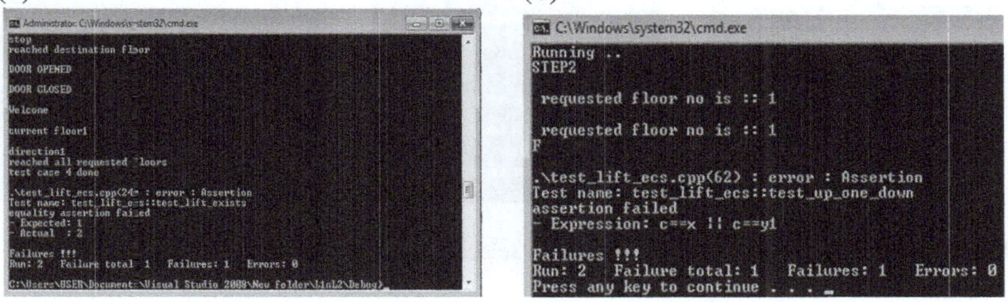

Fig. 24.6 **a** Test case fail for no of cab = 2. **b** Test cases fails as correct ones not selected

Waterfall: A bug was detected in selection of cabs as shown in Fig. 24.9.

We analyze and compare the development using software metrics in Table 24.1. On the basis of the lines of code and cyclomatic complexity, it is observed that initially it is expensive to write the test cases with TDD. As a result quantity of the code and complexity increases. Later due to better understanding of TDD, the test cases and the code are generated as "enough code" and the measures reduce. The number of bugs detected is also less. But with Waterfall, the number of bugs found is much more than TDD, as well as the coverage of different features is not that easy, which requires much effort to complete the development of the system.

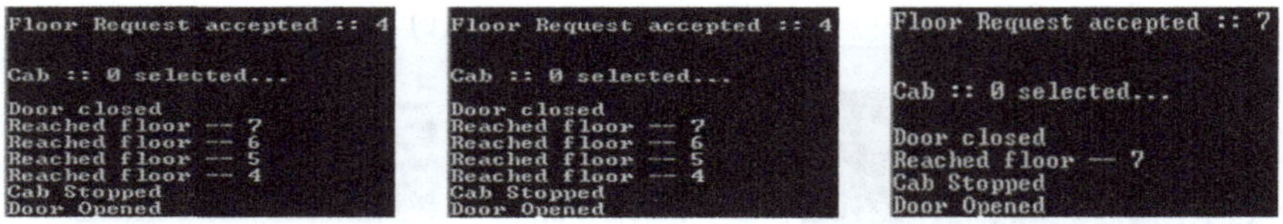

Fig. 24.7 Wrong cab selected

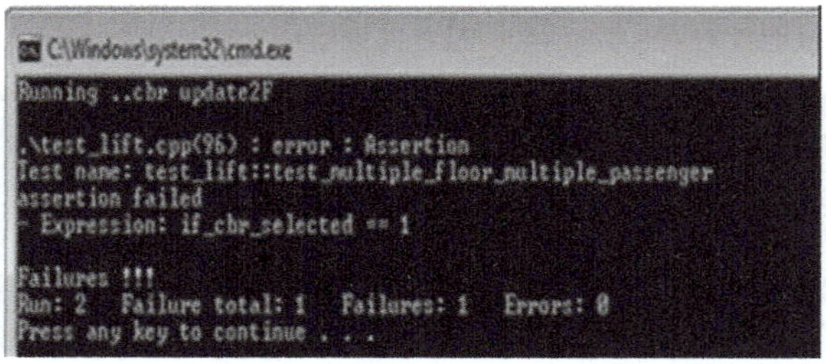

Fig. 24.8 The floor button request (*FBR*) and cab button request (CBR) are same—test failed

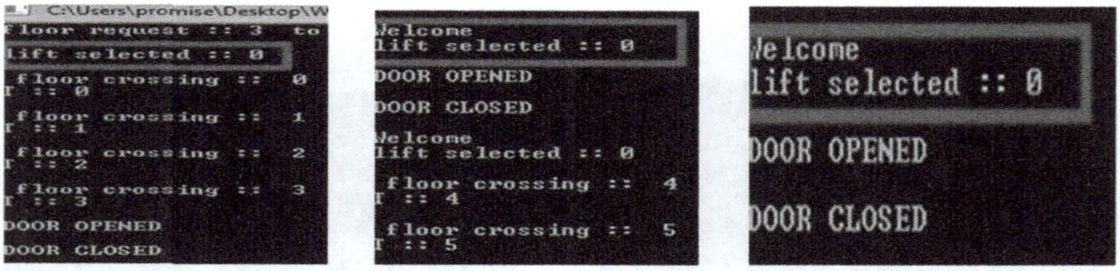

Fig. 24.9 The *marked lines* show the wrong selection of the cabs

Table 24.1 Result of the code metrics

Metric or measure	Phase-1		Phase-2		Phase-3	
	Waterfall	TDD	Waterfall	TDD	Waterfall	TDD
Source files	1	2	1	3	1	3
Lines of code	129	183	189	638	548	600
Physical executable lines of code	101	160	168	468	439	312
Logical executable lines of code	71	86	125	244	274	177
McCabe VG complexity	10	29	34	79	58	45

24.5 Conclusions

TDD focuses on, analysis, design and programming decisions. For ECS, while writing the test plan and respective test code, it is decided how to detect if the lift is present. Similarly, in case of elevator with 2 cabs, when test case is written for adding 2 floors to the elevator, the production code is developed automatically for the multiple floors making minimum changes. Thus the time taken is much less to develop the complex code and errors are easily detected and corrected. Though TDD help to get the effective production code in less time and better understanding, many problems are still faced. As understanding TDD is tough, and how to start to write test cases, is difficult. The fine granularity of test-then-code cycle gives continuous feedback to the developer. With TDD, faults and/or defects are identified early and quickly as new code is added to the system, and the source of the problem is more easily determined.

References

1. Huo M, Verner J, Zhu L, Babar MA (2004) Software quality and agile methods. In: Proceedings of the 28th annual international computer software and applications conference, vol 01. Series-COMPSAC 2004, ISBN-0-7695-2209-2-1, IEEE Computer Society, Washington, DC, pp 520–525
2. Agile Software Development (2013) http://en.wikipedia.org/wiki/Agile_software_development. Accessed 1 Dec 2013
3. Mall R (2009) Fundamentals of software engineering. Prentice Hall of India Learning Private Limited, New Delhi. ISBN: 978-81-203-3819-7
4. Boehm B (1986) A spiral model of software development and enhancement. SIGSOFT Software Engineering Notes, August 1986, vol. 11, No. 4, pp 14–24, ACM, New York. ISBN. 0163-5948
5. Kumar S, Bansal S (2013) Comparative study of test driven development with traditional techniques. Int J Soft Comput Eng (IJSCE) 3(1):2231–2307
6. Grenning JW, Carter J (2011) Test-driven development for embedded C. Series: the pragmatic programmers. Pragmatic Bookshelf, Raleigh
7. Sinaalto M (2006) The impact of test-driven development on design quality. Agile software development of embedded systems, version-1.0, information technology for European advancement (ITEA), March 2006
8. CppUnit Wiki at SourceForge (2013) http://cppunit.sourceforge.net. Accessed 01 Dec 2013
9. Hamill P (2005) Unit test frameworks—a language-independent overview. O'Reilly, Sebastopol. pp I–XII, 1–198. ISBN-978-0-596-00689-1

Chapter 25
Proto-Spiral: A Hybrid SDLC Model for Measuring Scalability Early in Development Using a Probabilistic Approach

Anirban Bhar and Sabnam Sengupta

Abstract In this paper, we propose a probabilistic model for measuring one of the many Non-functional requirements, namely, the Scalability, which is largely unexplored till now early in software development life cycle. The proposed model is a combination of Prototype and Spiral models; those are standard and accepted SDLC models. Developing software which addresses the functional requirements only, can lead to a solution, but quality attributes for that remain unanswered. It may be a less useful, slow, less reliable system. The system's 'quality characteristics' or 'quality attributes' are specified in the Non functional requirements to improve QoS which are hardly covered by functional requirements. The proposed model, named Proto-Spiral can be used for measuring many of the non-functional requirements. In this paper it has been used to evaluate scalability parameters of the software. We have used a case study of the website ebay.com, a dominant online store to illustrate our approach.

Keywords Non-functional requirements · Probabilistic approach · Spiral model · Prototyping model · Scalability

25.1 Introduction

The ability of a system or process, to overcome the problem of increasing amount of work in an efficient manner or its capability to accommodate overload caused by the growth of work is called the Scalability of the system.

A. Bhar (✉)
Narula Institute of Technology, Kolkata, India
e-mail: anirban.bhar1983@gmail.com

S. Sengupta (✉)
B. P. Poddar Institute of Management and Technology, Kolkata, India
e-mail: sabnam_sg@yahoo.com

S. Sengupta et al. (eds.), *Emerging Trends in Computing and Communication*, Lecture Notes in Electrical Engineering 298, DOI: 10.1007/978-81-322-1817-3_25,

Scalability is a non functional requirement of a system that describes the ability of the system to work efficiently when the functionality of the system has increased in size or volume without affecting much the QoS of the system.

A system or algorithm and the protocols of networking process is said to be in proper scale if it is suitable, efficient and practical when it is applied to large scales (e.g. a large amount of input data, a large number of users, or a large number of actively connected nodes in case of a distributed system). If that system fails to manage its efficiency when quantity increases, then it is not scale able. In practice, if there are a large number of things (n) for which scaling is affected, then n must grow less than n^2 [1]. An example is a search engine which should scale not only for the number of users uses the search engine, but also for the number of objects it indexes [2]. Scalability refers to the capability of a site or system to increase in size as per the demand.

The conventional approach measures scalability of a system after the system has been developed, but if the newly developed system fails to meet the desired result in terms of scalability, the system would be either a poor scalable system or the developer(s) has to re-build the system again which is a practical problem because it involves wastage of time and effort. To overcome this problem, we have tried to propose a probability based approach that measures the probable scalability of a system during development time of the spiral model in every prototype. If the probable result is as per the requirement, then it may be concluded that the partial development approach (prototype) is correct, otherwise developer have to think alternatively. It in turn will save time and effort for development.

25.2 Related Work

Work is in progress on 'scalability' and some work has been done on scalability factors on networking: integrating thousands of nodes to millions of nodes to a network system in a scalable manner [3].

Some work is done on Scalability of databases those use SQL and so called "No-SQL" data stores which was designed to run simple OLTP-style application to scale for the loads over many servers [4]. These databases are also supported by Google's BigTable as well as BigTable, memcached, and Amazon's Dynamo provided a "proof of concept" that inspired many of the data stores. A work has been done to analyze scalability of seven system applications running on Linux on a 48-core computer [5]. The key factors which influence the scalability of large complex Deeply Embedded Sensor systems has been recognized [6]. Using the concept of clustering, some research work has been done on the large scale growing e-commerce systems [7]. Scalability is an important aspect in case of networked systems also, as number of user may increase in huge amount in this modern era of computers [8, 9], even it is an important aspect for unguided communication mediums like wireless LAN [10]. In this regard a work has been done on distributed [11] networked information management system [12], and it is applicable for server systems also [13]. Finally, the Internet is facing a noticeable growth on two aspects simultaneously: (1) the amount of data getting processed, stored and passed through web, and (2) the number of users increasing day by day. An effort has been made to propose new technologies to improve Internet scalability which will support all the types of these growths [14].

These works are individually very efficient in their respective domains and provides some good solutions, but all of these works has been done with some particular already developed systems. Proto-Spiral is a hybrid model that provides an efficient solution of the development approach for a good scalable software system, which may be time saving for the developers as well as cost effective.

25.3 Scope of the Work

There are existing metrics of scalability, described by different authors [15] which are as follows:

- The Speedup (S) measures the increased rate of doing work with the number of processors k, compared to one processor, and has an "ideal" linear speedup value of $S(k) = k$.
- Efficiency E measures the rate of work per processor (that is, $E(k) = S(k)/k$), and has an "ideal" value of unity.
- Scalability $\psi(k1; k2)$ from one scale k1 to another scale k2 is the ratio of the efficiency figures for the two cases, $\psi(k1; k2) = E(k2) = E(k1)$. It also has an ideal value of unity.

All these metrics has been used to measure scalability of any existing system. In this case, at the end, if it is found that the scalability result is not satisfactory, the system needs to be re-built. However, if we can measure these metrics early in SDLC, measures can be taken to improve it. Therefore, it is required to measure scalability of the systems which are under development or which are partially developed, so that the system can provide the desired scalability after completion of the development. It may help the developer to understand the current status for scalability at any point of development and hence, it can provide cost efficiency development of the systems where scalability is one of the important issues.

In this paper Proto-Spiral, a hybrid model is proposed, which is a combination of Prototype and Spiral Model to measure scalability of software at different iterations of spiral model, at different levels. This approach has the potentiality to provide a guideline to develop a good scalable software system.

25.4 Proto-Spiral: The Proposed Model

The proposed model defines a procedure to develop a scalable software system that may be a networked system or a distributed system or a stand alone system. If the system demands high scalability, this model can provide a feasible practical solution to the approach or algorithm to develop the software system.

25.4.1 The Model

A. Constructs:

A combination of Prototyping model and Spiral model has been used in this paper to develop a software system. To overcome the limitations of waterfall model, Prototyping model was developed i.e. after developing the software system it is tedious to make any change in the system. The basic idea behind prototyping model is that instead of baggage the requirements before any design or coding can begin, a dummy prototype of simple similar functionality is built to understand the requirements. The prototype is built based on currently known requirements.

The prototype undergoes all the phases of SDLC, i.e. design, coding, testing but these phases are not formal. After developing each prototype it is delivered to client, and these prototypes are used by the clients to get the actual feel of the system. He uses and interacts with the prototype systems and understands the requirements of the desired system better; and lastly it results more stable and precised requirements from clients.

In prototyping model the focus of the development is to include those features those are not properly described or understood as prototype is anyway to be discarded.

Spiral model of software development implies the idea of iterative development (prototyping). A prototype is a preliminary sample or model built for testing a concept or process or to test as a thing to be replicated or learned from. It provides the prospective for speedy development of incremental versions of the software. Software is developed in a sequential chain of incremental releases using the spiral model. The activities in this new model are organized like a spiral. The spiral of the model consists of many cycles. The radial dimension represents the increasing cost incurred to accomplish the steps domain so far and the sharp dimension represents the progress that is made in completing each cycle of the spiral. In the spiral, each and every cycle begins with the objectives identification for that cycle and scanning for all the different alternatives those are possible to achieve those objectives considering the imposed constraints.

The risk driven nature of the spiral model allows it to accommodate any mixture of specification-oriented, prototype-oriented, simulation-oriented or some other approach. An important feature of the model is that each cycle of the spiral is completed by a review, which covers all the products developed during that cycle, including plans for the next cycle. The spiral model works for developing as well as enhancement of already developed projects.

B. Working principal of the proto spiral:
In this paper Proto-Spiral is a hybrid model where prototype-oriented spiral model is used and after development of each and every prototype (P1, P2,...Pn), the scalability factors (SF) have been analysed (Fig. 25.1). There are some pre-defined measurements of SF (Table 25.1) depending upon the project itself and obviously there is a satisfactory level for each prototype. The satisfactory level may be assigned after analysing the whole project as well as with a probability based survey report of the Scalability Factors. If a developed prototype satisfies the desired scalability factors and if it meets the customer requirements, then the process of building the next prototype gets started as per the Spiral model of software development. The Proto-Spiral model can measures the probable scalability of each prototype during development.

As per Boehm [16], each cycle of the spiral begins with the detection of:

- The objectives of the portion of the product being elaborated (performance, functionality, ability to accommodate changes etc.).
- The possible alternatives to implement that portion of the product (Design A, design B, reuse, buy etc.).
- The applied constraints on the application of the alternatives (cost, schedule, interface etc.).

After development of the final prototype 'n' if the replica is accepted as a whole by the customer and the scalability factors are still of satisfactory level, then the actual software development takes place. Then also after executing each phase, the scalability factors are analyzed for the actual system.

25.4.2 The Metrics

To measure scalability, some scalability metrics are taken in this paper [2], such as:

- Throughput—it is the rate of transactions processed by the system.
- Resource usage—it is the levels of usage for the various resources involved in the system (CPU, memory, disk, bandwidth etc.).
- Cost—it is the amount of price per transaction.

Throughput: As per Little's law, if the segment contains an average of N users, and the average user spends R seconds in that segment, then the throughput X of that segment is roughly measured by

$$X = N/R.$$

Resource Usage: It can be measured by the percentage of time that a component is actually occupied, as compared with the total time that the component is available for use. For a particular instance, if a CPU processes transactions for a total of 45 s during a single minute, its utilization during that interval is 75 %.

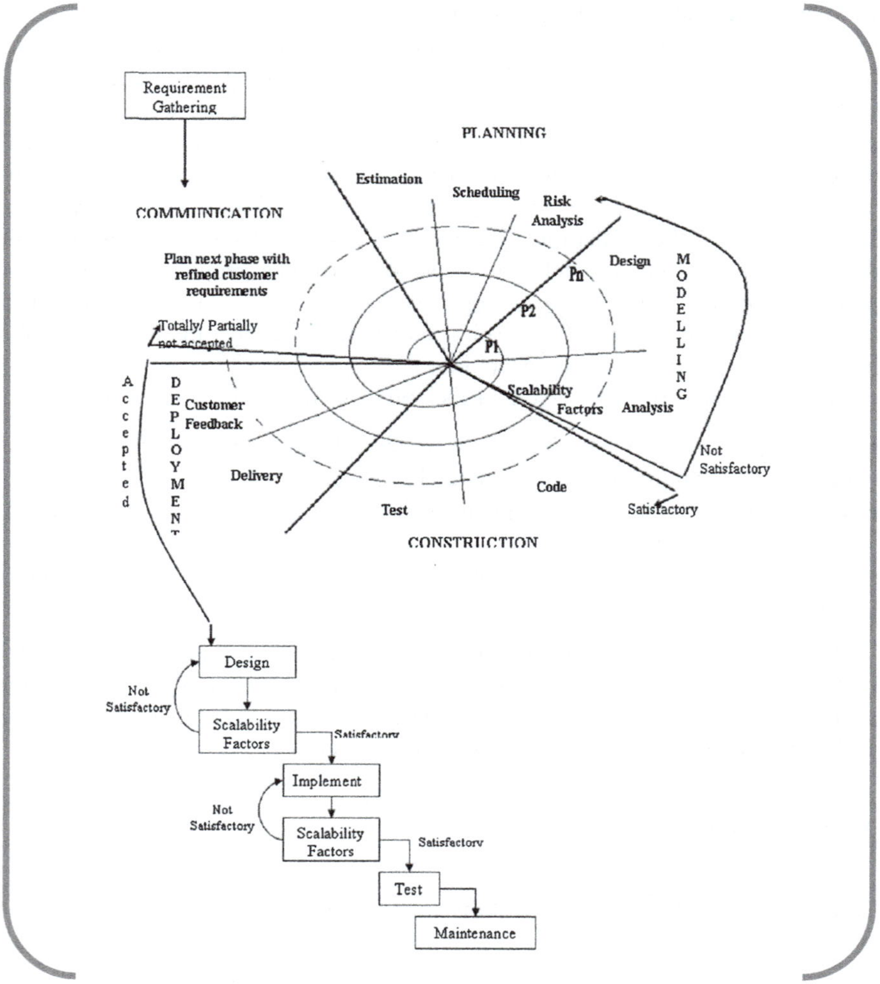

Fig. 25.1 Proto-spiral-the hybrid model

Table 25.1 Probability based scalability factors

Scalability factors		Unit	Number of users (N)			
			100	1,000	10,000	20,000
Probable throughput (X)		tps	100	93.66	89.2	83.54
Probable resource usage	CPU	%	70	70	73	75
	Memory	%	76.3	82.7	89.11	93.18
	Disk	%	0.5	5	50	100
	Bandwidth	Hertz	40	48	54	60
Probable cost per transaction		₹	83	86.5	91.5	98
			Satisfactory	Satisfactory	Satisfactory	Satisfactory

Resource usage may be monitored by measurement and utilization records of the following system resources regularly:

- CPU
- Memory
- Disk
- Bandwidth (for web-based application).

A resource is said to be critical to performance when it becomes overused or when its utilization is not optimum to that of other components. For instance, it may be considered a disk to be in critical point or overused when it has been 75 % utilized and all other disks on the system have 30 % utilization. Although utilization of 75 % does not indicate that the disk is severely overused, but it is possible to improve the performance by rearranging all the data to balance I/O requests through out the entire set of disks.

Measurement of resource utilization depends on the tools that the operating system provides for reporting system activity and resource utilization.

Cost: Probable cost of the project may be estimated by any cost estimation technique (like CoCoMo) for every prototype, for every transaction as well as the entire software system.

At this instance, as an example, it may be considered that a library management software for a growing educational institution consisting 100 students. It is also assumed that the institute wants to use the software for the next 20 years. From statistical survey the institute concludes that the number of students will be maximum 20,000 after 20 years. So the software must be efficient enough for 100 students as well as for 20,000 students. So, here arise the scalability factors. All of these scalability factors may be assumed after each prototype, using survey based probabilistic approach. If the scalability is measured after the development of the software, and if it is not satisfactory, then the whole effort as well as cost might get wasted. Our model describes, if the scalability factors may be imposed during development of the software. This model is based on prototyping model of software development.

The scalability factors (SF) may be measured with the help of Little's Law after development each prototype as well as after the design as well as implementation part using Table 25.1.

25.5 Case Study: eBay.com

EBay is an e-Commerce system where the C2B2C (Customer-to-Business-to-Customer) model is followed. Any user can browse to the website eBay.com and can search for any goods or service he want to buy, either in auction or directly from the seller, or to post some item he wants to sale to some other users, who can search for as prospective buyers. The users then arrange for online payments (using any secure electronic transaction system or eBay's payPal system, which is a distinct system designed exclusively for the purpose of secure transaction and recently it is integrated onto the eBay platform) and receive the item by postal mail (for tangible goods) or by e-mail (for services).

Like most of the internet-enabled business systems, eBay is designed using distributed object technology in a real time system. It requires a high scalability, better performance, high availability, and is a highly secured manner. It is required to have the potential to handle large volumes of requests generated by the internet community and must it be able to respond to all of these requests in a timely fashion (real-time).

The most recent statistics regarding eBay state that [source: 14]

- The registered users it manages is around 248,000,000.
- Number of photos it manages is over 1 billion.
- eBay has live applications of nearly 10,000.
- eBay currently has 30 Software Architects in its employ.
- eBay averages well over 1 billion page views per day.
- Every month around 4.4 billion API calls handles by the eBay platform.
- In every two weeks around 100,000 + lines of code are added in this system.
- There are 30,000 software builds per week.
- There are more than 44 billion SQL executions per day.

These stats give an idea of the large scale e-commerce of the eBay platform and the growth that has taken place in just a few years since the launch of this web application.

This is a case study that will summarize the core architecture with special focus on the evolution of eBay to accommodate the scalability requirement.

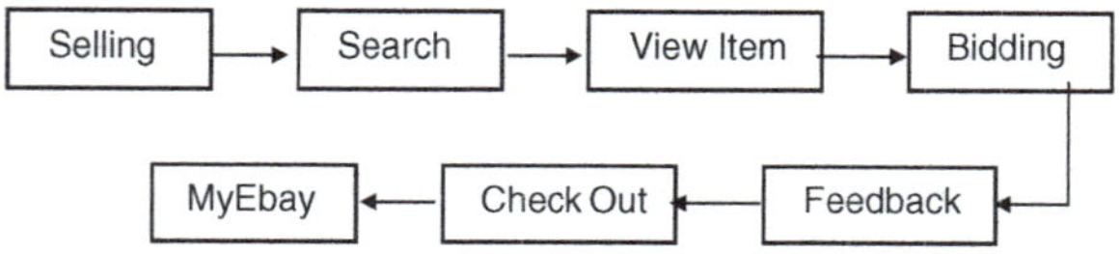

The architecture of a system of such huge size goes through several iterations. The architectural solution of this level is not only based on the architecture of the software but also on the architecture of the system, since the components those are occupied in the system are not only just web clients-servers but also includes databases, servers (security servers, application servers, proxy servers, and transaction servers). The system discussed above has a 3-tier architecture as well as a web-enabled device (a web-browser), application and transaction servers, and it also has databases at the data services layer.

The system should support increase and decrease in load without human interference. As more and more people join the internet, more and more users attempt to access the website and request information simultaneously and they must get service without fail. The system must have fault tolerance power.

By studying and analyzing the above discussion, it is clear that there was no way for the architects of the eBay platform to predict the rapid growth that took place, in the expansion and the popularity of the world wide web in particular, with millions of people gaining access to it, in just a few years, and also of the development of the frameworks and programming languages that would later be used to improve upon the architecture of the system.

25.6 Conclusion

In the situation like this, the report of probability based number of users for the coming years must be managed at the time of the development of this kind of service. The model described in this paper, may assure this point to the developer and it may provide the right development approach of a high scalable system like e-Bay.com.

References

1. Laudon & Traver (2008) E-Comerce business.technology.society, fifth edition, Prentice Hall, Upper Saddle River, New Jersey, 07458
2. Barish G (2002) Scalable high-performance web applications, java web applications using J2EE Technology, May 24
3. Metev SM, Veiko VP (1998) Laser-assisted microtechnology (Springer Series in Materials Science), 2nd edn, Berlin
4. Breckling J (ed) (1989) The analysis of directional time series: applications to wind speed and direction, ser. (Lecture Notes in Statistics), vol 61. Springer, Berlin
5. Zhang S, Zhu C, Sin JKO, Mok PKT (1999) A novel ultrathin elevated channel low-temperature poly-Si TFT. IEEE Electron Device Lett 20:569–571
6. Wegmuller M, von der Weid JP, Oberson P, Gisin N (2000) High resolution fiber distributed measurements with coherent OFDR. In: Proceedings of the ECOC'00, paper 11.3.4, p 109
7. Sarwar BM, Karypis G, Konstan J, Ried J (2002) Recommender systems for large-scale E-commerce: scalable neighborhood formation using clustering. In: Proceedings of the fifth international conference on computer and information technology, vol 1. Department of Computer Science and Engineering, University of Minnesota, Minneapolis
8. Karnik A (1999) Performance of TCP congestion control with rate feedback: TCP/ABR and rate adaptive TCP/IP, M. English thesis, Indian Institute of Science, Bangalore
9. Padhye J, Firoiu V, Towsley D (1999) A stochastic model of TCP Reno congestion avoidance and control. University of Massachusetts, Amherst, MA, CMPSCI Technical Report 99-02
10. Wireless LAN Medium Access Control (MAC) and Physical Layer (PHY) Specification, IEEE Std. 802.11, 1997
11. The IEEE website (2002) Available: http://www.ieee.org/. (Online)
12. Shell M (2002) IEEEtran homepage on CTAN. Available: http://www.ctan.org/tex-archive/macros/latex/contrib/supported/IEEEtran/. (Online)
13. FLEXChip Signal Processor (MC68175/D), Motorola, 1996
14. PDCA12-70 data sheet, Opto Speed SA, Mezzovico, Switzerland

15. Jogalekar P (2000) Evaluating the scalability of distributed systems. Murray Woodside, Natural Sciences and Engineering Research Council of Canada, CITO (Communications and Information Technology Ontario), March
16. Boehm BW (1988) TRW defense systems group. A spiral model of software development and enhancement, IEEE Computer, pp 1–4

Chapter 26
A Framework of Musical Pattern Recognition Using Petri Nets

Samarjit Roy, Sudipta Chakrabarty and Debashis De

Abstract Petri Nets are modeling tools that are used in an enormous number of real-world simulations and scientific problems. The primary objective of this paper is establishing that Petri Net is one important tool that represents quality music compositional analysis process. In this work this has been illustrated how music structures can be processed by means of a more abstract kind of representation and allow to explicitly describing the process of computational modeling of Musicology that present the attempt on music composition from the fundamental musical objects like vocal and rhythmic cycles usage using Petri Net. This attempt has been obtained to answer the query that whether Petri Net is an adequate tool for modeling such a complex process as a complete composition of music from the fundamental musical objects like vocal and rhythmic structures is. The main focus behind this work is to explore that Petri nets can be used as a good basis for retrieval of music information in World Music.

Keywords Petri nets · Musicology · Rhythmic structures · Incidence matrices · Reachability graph

S. Roy (✉) · D. De
Department of Computer Science and Engineering, West Bengal University of Technology, BF-142, Sector-1, Salt Lake City, Kolkata, 700064 West Bengal, India
e-mail: samarjit.tech89@gmail.com

D. De
e-mail: dr.debashis.de@gmail.com

S. Chakrabarty
Department of Master of Computer Application, Techno India, EM 4/1, Salt Lake City, Sector-V, Kolkata, 7000091 West Bengal, India
e-mail: chakrabarty.sudipta@gmail.com

S. Sengupta et al. (eds.), *Emerging Trends in Computing and Communication*, Lecture Notes in Electrical Engineering 298, DOI: 10.1007/978-81-322-1817-3_26,

26.1 Introduction

Computational music composition has at present time a long tradition for hardly any decades. In the context of Hindustani Classical Music, copious exigent compass reading have in researches to be projected and evaluated the defined pathways to intricate the musical blueprints. Petri Nets are a particular breed of graphs which can be directed, bipartite and weighted. The nodes used in these graphs are of two manners like places and transitions. The arcs are the connectors of these nodes by initializing either in a place and terminating in a transition, or vice versa.

A fundamental perception that Petri Nets develop is the concept of marking the location, recognized by tokens in Place. Any place holds at a given time of participant process a non-negative number of tokens. Tokens can be relocated from one place to another place enabling the system according to guidelines known as the 'Firing rules'. In this rule, a transition will be enabled when all the incoming places of the system of that transition present a number of tokens greater or equal to the weights of the consequent incoming arcs, and after fire of the transition the marking of all the output places will be less than or equal to their capacities.

26.2 Related Works

On this basis in terms of Petri nets this efforts provided innovative steps for making these labors operational in concrete computerized environments, not just at the level of formal abstraction [1]. The analysis on original orchestral score, music object recognition within a score, mapping the musical objects, testing of music modeling architecture has elaborated [2]. Paper [3] addressed the adoption of Petri Nets to describe music processes. In several efforts authors have been described the musical pattern recognition and rhythmic features exposures by object-oriented manner [4, 5]. A lot of discussions are available about the features classification using UML-oriented class diagrams and the way of implementations of musical rhythmic cycles for the percussion-based instruments [6]. In an effort by the authors have been modeled an optimized algorithm on rhythmic cycles used in Indian Classical Music using Genetic Algorithm Concept [7]. The authors have explained elaborately the notes structures as well as thhats or raga origin in music which imposed to construct the songs in Hindustani musical patterns [8].

26.3 Structural Formations of Petri Nets

In this section is given several very illustrative cases where Petri Nets are used for composition, more precisely music description and processing. Simple musical rhythmic patterns is used which are imposed to the actual vocal Simple musical

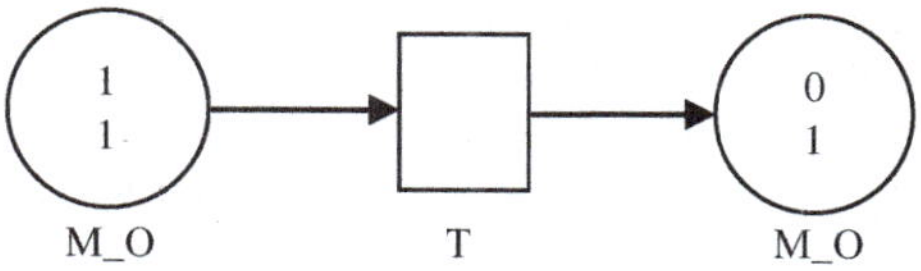

Fig. 26.1 Sequential reproduction of musical obj

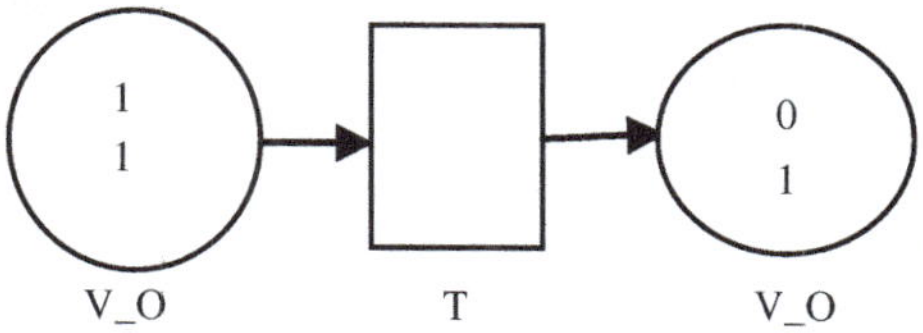

Fig. 26.2 Sequential reproduction of vocal obj

rhythmic patterns is used which are imposed to the actual vocal objects to define musical objects by using modeling perceptions of Petri Nets, denoted as "R_O" for Rhythmic Object, "V_O" for Vocal Object and "*M_O*" for Music Object. In Figs. 26.1 and 26.2 the sequential reproduction has been mentioned for musical objects and the raw vocal objects.

26.4 Petri Net Model for Complete Music Composition

The Petri Net is a meta-modeling tool by which we can model the discrete events. The nodes used of two types in Petri Nets based modeling are entitled as the places and transitions denoted by generally P and T respectively. The Places and transitions are connected via edges or arcs. Places (P) are represented by circles, transitions by bars. The Fig. 26.3 has demonstrated the entire meta-modeling for a complete music composition using Petri Net.

If the instructions have been followed closely, then only very minor alterations will be in the Table 26.1, characterizes the set of places and the set of transition functions owing to elaborate the music demonstration with the exact compositions of both the vocal and the corresponding ornaments like rhythmic cycles. The entire music composition is only possible when the experimental underdone vocal data performed solo by the vocalists primarily. Hence, at the very beginning the vocal performance is responsible for instigating the entire process of analysis and combine with accurate rhythmic cycle structures. It is enabled at P_0 when there is at least one token present.

In this context, the formal definition has been explained as of a Petri Net (PN) is described by the 5-tuple presented as, PN = $<P, T, D-, D+, M_0>$, where, P is the set of places. T is the set of transitions. $D- = P_i \times T_i$. N_1 denotes is the pre-incidence matrix that specifies the arcs which are directed from the places to the transitions (P to T). On the other hand, $D+ = T_i \times P_i$, N_2 is the post incidence matrix that specifies the arcs which are directed from the transitions to the places (T to P). M_0 is the initial marking where from the whole transaction starts, M_0: $P \rightarrow \{0, 1, 2, \ldots\}$.

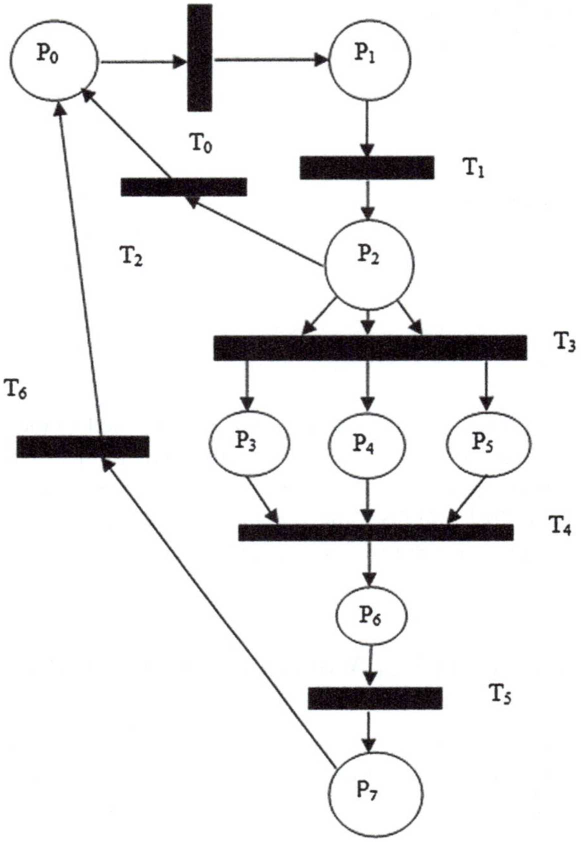

Fig. 26.3 Representation of a Petri net model for a complete music composition

The entities in this effort of modeling process are Vocal musical data, percussion-based rhythmic cycles, several variations of fundamental rhythmic cycles, entire music composition. The stochastic model of this Petri Net is intended with the vocal and the numerous variations of rhythmic cycles using eight places (from P_0 to P_7) and seven transitions (from T_0 to T_6) to check to be analyzed the reachability analysis using the reachability graph (Fig. 26.4) analysis of the complete Net.

26.5 Incidence Matrices of the Petri Net Model

The pre-incidence matrix (Table 26.2a) is representing the initial state of the system. The Post-incidence matrix (Table 26.2b) is representing operational state and the combined matrix (Table 26.2c) representing the overall circumstances with their location or places at any precise transition of the circumstances. Each location can be worn to scrutinize the Reachability of the Petri Net Model.

Table 26.1 Place and transition functions used in designing the Petri net of music composition

Places		Transitions	
P_0	Unprocessed music library	T_0	Vocal unprocessed data extracted
P_1	Identification of vocal and stored in vocal library	T_1	Vocal characteristics identifications
P_2	Instruments choice against vocal performances	T_2	Instruments chosen against vocal performances
P_3	1st multiplier (rhythm) chosen by vocal	T_3	Rhythmic variations acts upon vocal
P_4	2nd multiplier (rhythm) chosen by vocal	T_4	Musical quality identifier
P_5	4th multiplier (rhythm) chosen by vocal	T_5	Generated entire musical performances
P_6	Rhythmic variations library with vocal	T_6	Enter into library for musical fundamental characteristics
P_7	Entire music composition library		

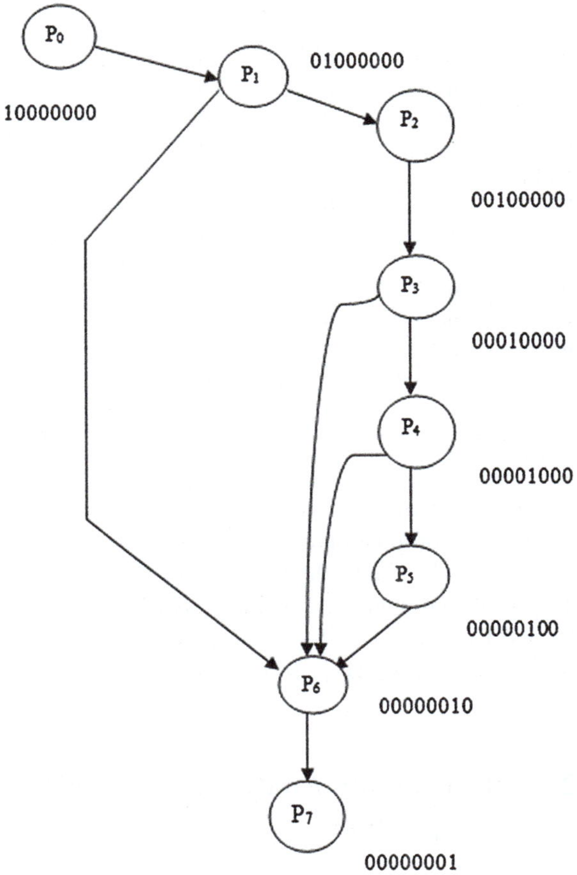

Fig. 26.4 Reachability graph for a complete music composition

In Table 26.2a, the row constituents P_0, P_1, P_2, P_3, P_4, P_5, P_6 and P_7 are representing the places or location of explanations (in Fig. 26.3). Columns T_0–T_7 are on behalf of the transitions. The pre-incidence matrix is represented by [D−] and the

Table 26.2 (a) Pre-incidence matrix. (b) Post-incidence matrix. (c) Combined-incidence matrix

	T_0	T_1	T_2	T_3	T_4	T_5	T_6
(a) Pre-incidence matrix							
P_0	1	0	0	0	0	0	0
P_1	0	1	0	0	0	0	0
P_2	0	0	0	1	0	0	0
P_3	0	0	0	0	1	0	0
P_4	0	0	0	0	1	0	0
P_5	0	0	0	0	1	0	0
P_6	0	0	0	0	0	1	0
P_7	0	0	0	0	0	0	1
(b) Post-incidence matrix							
P_0	0	0	1	0	0	0	1
P_1	1	0	0	0	0	0	0
P_2	0	1	0	0	0	0	0
P_3	0	0	0	1	0	0	0
P_4	0	0	0	1	0	0	0
P_5	0	0	0	1	0	0	0
P_6	0	0	0	0	1	0	0
P_7	0	0	0	0	0	1	0
(c) Combined incidence matrix							
P_0	−1	0	1	0	0	0	1
P_1	1	−1	0	0	0	0	0
P_2	0	1	0	−1	0	0	0
P_3	0	0	0	1	−1	0	0
P_4	0	0	0	1	−1	0	0
P_5	0	0	0	1	−1	0	0
P_6	0	0	0	0	1	−1	0
P_7	0	0	0	0	0	1	−1

token status at any [P, T]ij, where i = {0, 1, 2, 3, 4, 5, 6, 7} and j = {0, 1, 2, 3, 4, 5, 6} before origination of the process is existing. The combined incidence matrix in Table 26.2c, shows the token status at any instance after initiating the process. The combined matrix is computed as [D+ − D−]. Places are represented as: [P_0, P_1, P_2, P_3, P_4, P_5, P_6, P_7]. Transactions represents: [T_0, T_1, T_2, T_3, T_4, T_5, T_6]. Initial marking M_0 is: $[1\ 0\ 0\ 0\ 0\ 0\ 0\ 0\ 0]^T$.

26.6 Reachability Analysis Using Petri Net for Music Composition

The Reachability graph in Fig. 26.4 is a directed graph and the nodes of the graph are recognized as locations or markings of the Petri Net R (N, M_0), where the arcs or edges are represented by the transitions of N and M_0 is the initial marking.

The graph is used to define a given Petri Net N and marking M, whether M belongs to the Reachability, R(N). Each of the initial marking M has an associated Reachability set. This set consists of all the markings that can be reached from M_0 through the firing of one or more transitions. In this case the reachability graph starts with initial marking M_0 where the place is represented as P_0. The initial place P_0 is marked primarily as $[10000000]^T$ and this is reachable up to the final place $[00000001]^T$. By this Reachability graph model this can be analyzed about the tokens subsists in the initial place to the reachable places (from P_0 to P_7). The '0' indicates that there is no token and '1' indicates the existing token which will participate actively in the firing and as well as transitions. From the initial place P_0 to the eventual place P_1 all the tokens whether they exists or not are represented as the combination of '0' and '1'. This is happened a particular place engaged with an infinite number of tokens. Resulting, the Reachability Graph would be the infinite tree model.

26.7 Conclusion

Chief spotlight of this contribution is to build a robust modeling protocol for dynamic representation of quality music that comprises of vocal objects and rhythmic objects. This is related with both Sound-Oriented and Analysis-Oriented concept as Petri net is an analysis tool and that is used for music composition modeling. After completion of the tasks of the paper this can be concluded that Petri nets is a well formed, developed, high-potential, concurrent, distributed, parallel processes, modeling tool that capable of analyzing all the spheres of Musicology and useful for compositional music modeling.

References

1. Trček D, Trček G Computationally supported musical composition using Petri nets. Latest Trends Appl Comput Sci 149–152. ISBN: 978-1-61804-171-5, 2010
2. Haus G, Rodriguez A Formal music representation; a case study: the model of Ravel's Bolero by Petri nets. Intell Music Workstation. Italian National Research Council in the frame of the MUSIC Topic (LRC C4): intelligent music workstation, Goal C: Sistemi avanzati di produttivita' individuale, Subproject 7: sistemi di supporto al lavoro intellettuale, Finalized project Sistemi informatici e calcolo parallelo
3. Baratè A, Haus G, Ludovico LA Petri nets applicability to music analysis and composition. pp 97–100, 2007
4. De D, Roy S (2012) Polymorphism in Indian classical music: a pattern recognition approach. In: Proceedings of international conference on communications, devices and intelligent systems (CODIS). IEEE, pp 612–615
5. De D, Roy S (2012) Inheritance in Indian classical music: an object-oriented analysis and pattern recognition approach. In: Proceedings of international conference on radar, communication and computing (ICRCC). IEEE, pp 193–198

6. Chakraborty S, De D (2012) Pattern classification of Indian classical ragas based on object oriented concepts. Proc Int J Adv Comput Eng Archit 2:285–294
7. Chakrabarty S, De D (2012) Quality measure model of music rhythm using genetic algorithm. In: Proceedings of international conference on radar, communication and computing (ICRCC). IEEE, pp 125–130
8. Bhattacharyya M, De D (2012) An approach to identify Thhat of Indian classical music. In: Proceedings of international conference on communications, devices and intelligent systems (CODIS). IEEE, pp 592–595

Chapter 27
A Probabilistic Model for Analysis and Fault Detection in the Software System: An Empirical Approach

Gitosree Khan, Sabnam Sengupta and Kunal Das

Abstract Software reliability and estimation of defects plays an important role in software testing stage. For studying defects, one common practice is to inject faults in subject software, either manually or by using a program that generates all possible mutants based on a set of mutation constraints. Getting the optimized results for the software system while predicting defects using realistic analysis, and confirming whether that leads to valid and consistent data during software testing stage is a challenge. In this paper, we propose Process simulation Model (PSM), which is a probabilistic model-based approach that overcomes these challenges and enables prediction of software defects and its impact in the system using Bayesian estimation. Moreover, a Fault Detection Algorithm FDA is derived from PSM model that helps to predict software faults for different deterministic problems that we have taken in our experimental study to demonstrate the reliability, verification and consistency of the system. A comparative study is shown on various deterministic problems by finding set of random defects through probabilistic approach the fault may occur in the proposed software model.

Keywords Fault-tolerant · Fault injection · Mutant · Reliability · Consistency · Verification

---

G. Khan (✉) · S. Sengupta · K. Das
B P Poddar Institute of Management and Technology, Kolkata 700052, India
e-mail: khan.gitosree@gmail.com

S. Sengupta
e-mail: sabnam_sg@yahoo.com

K. Das
e-mail: kunaldasqca@gmail.com

S. Sengupta et al. (eds.), *Emerging Trends in Computing and Communication*, Lecture Notes in Electrical Engineering 298, DOI: 10.1007/978-81-322-1817-3_27,

27.1 Introduction

Fault injection has been widely used for validation of fault tolerance properties. A major problem in the development of fault-tolerant systems is the accurate determination of the dependability properties of the system. The generally accepted solution to this problem is to inject the set of faults in a simulation model or a prototype implementation, and to observe the behavior of the system under the injected faults. On the other hand, it is much more difficult to inject accurate (i.e., realistic) faults into a prototype, but the effects of faults on operational code can be readily observed. Software-based fault injection (software implemented fault injection) is the technique consists of reproducing the errors at software level. In this paper we propose a probabilistic approach for fault injection and constructing a new scheme for software fault detection model using probabilistic model called Process simulation Model (PSM). The main thing is to seed faults in subject software, either manually or by using a program that generates all possible mutants based on a set of mutation operators. We are injecting various types of faults such as logical, semantic and exception faults in our systems and thus checks how the proposed system behaves under defective conditions. Generated mutants can be used to predict the detection effectiveness of real faults. Results show that injected faults cannot be considered representative of residual software faults and they can be detected with proper methods and can be recovered.

27.2 Review of Related Work

There are various research papers that work on fault injection and it's tolerant. De Florio and Botti [1] proposes a framework to software-implemented fault tolerance for distributed applications. This approach increases the availability and reliability of the application at a justifiable cost, also thanks to the re-usability of the components in different target systems which further increasing the maintainability. Wu et al. [2], investigates an approach to incorporate the time dependencies between the fault detection, and fault correction processes, focusing on the parameter estimations of the combined model. The authors in [3] present an extensive experimental study to evaluate the representativeness of faults injected by a state-of-the-art approach (G-SWFIT). This paper analyzes the representativeness of injected faults in three complex, real-world software systems, and proposed an approach for improving fault representativeness. This aspect is important for obtaining a realistic assessment of fault tolerance. Carreira et al. [4] presents a new software implemented fault injection and monitoring environment, called Xception, which is targeted for the modern and complex processors. Xception uses the advanced debugging and performance monitoring features existing in most of the modern processors to inject quite realistic faults by software, and to monitor the activation of the faults and their impact on the target system behavior in detail. Andrews et al. [5], reports on an empirical study performed on one industrial program with known system testing faults. In paper [6], Kaustubh R. Joshi et al. develop a holistic approach to automatic recovery in distributed systems using a theoretically well-founded model-based mechanism for automated failure detection, diagnosis, and recovery that realizes benefits that could not be achieved by performing them in isolation. In [7], the authors analyzed static analysis faults and test and customer-reported failures for three large-scale industrial software systems developed at Nortel Networks. Sherlock [8] provides probabilistic diagnosis for multitier systems by using Bayesian inference on dependency graphs constructed automatically via network sniffing along with user perceived end-to-end behavior. In [9], B. Littlewood et al. define measures of fault finding effectiveness and of diversity and show how these might be used to give guidance for the optimal application of different fault finding procedures to a particular program. More closely related to our work is [10], where the authors proposes a new scheme for constructing software reliability growth models (SRGM) based on a non homogeneous Poisson process (NHPP). Rx [11] proposes techniques to recover from software bugs by restarting the application in a different environment designed to mask or remove the bug.

Our approach has the ability to detect when a problem is beyond its diagnosis and recovery capabilities, and thus, to determine when a human operator needs to be alerted. We believe that the approach is applicable to a wide variety of practical systems, and experimentally illustrate its use with several non deterministic problems like bubble sorting, knapsack problem, etc. Since, our work is an alternative probabilistic model driven approach for fault detection called Process simulation model (PSM) using Bayesian network which helps to demonstrate the fault tolerance of the system.

27.3 Scope of Work

This paper describes a probabilistic approach to software implemented fault tolerance for different software applications. This new approach can be used to enhance the reliability of the target applications of the software system. Our research work is divided into four sections. Section 27.4.1, represents a new software implemented fault injection technique and constructing a new scheme for software fault detection model using probabilistic approach called Process simulation Model (PSM). We are injecting three types of faults such as logical, semantic and exception faults in our systems and thus checking how the proposed system behaves under defective conditions.

The Sect. 27.4.2 represents the fault detection algorithm which has been derived from the Process Simulation Model directly. The proposed algorithm basically detects the all possible faults such as logical, semantic and exceptions that may be present in the proposed model using mutation analysis. In Sect. 27.4.3 shows a process flow that focuses on the use of mutation analysis in testing research which helps for systematic improvement of the test cases to assure that most bugs will be detected with the least effort. The main goal is to improve the ability of test cases to detect faults and to know the system behavior and performance. In Sect. 27.4.4, we summarize the main experimental results on various deterministic problems by injecting various mutants such as logical, semantic and exceptions. Here, we have studied the fault tolerance of the proposed system under various defective conditions. Results shows that injected faults cannot be considered representative of residual software faults and they can be detected with proper methods and can be recovered.

27.4 Process Simulation Model for Fault Detection Process

27.4.1 Definition

Process simulation model is the Bayesian network model that is done using simulation software refer in [12], which represent a system of probabilistic events as nodes in a directed acyclic graph (DAG) and can estimate the probable amount of random defects generated in the entire proposed software system. The following points represent the PSM:

Firstly, we consider a system that is denoted as function 'Main()'. It is differentiated into three portions namely before logical, before semantic and before exception i.e. P(BL), P(BS) and P(BE) which denote the probability of error before injection of logical error, semantic error and exception in the source code. Then for each of these cases further factoring is done. We have computed the probability of error before injection of logical error. We compute the conditional probability for occurrence of error after injection of logical error given that probability of error before logical error injection is known. It is denoted by P(AL/BL).

Secondly, the computation of conditional probability of occurrence of error after injection of semantic error and probability of occurrence of error after injection of exceptions with respect to the probability of error before injection of logical errors is done. It is shown as P(AS/BL) and P(AE/BL). The marginal probability for logical errors is calculated by taking the sum of all the 3 conditional probabilities i.e. P(AL/BL) + P(AS/BL) + P(AE/BL).

Further, the probability of error before injection of semantic error is calculated. We then compute the conditional probability for occurrence of error after injection of logical error given the probability of occurrence of error before injection of semantic error which is denoted as P(AL/BS). Similarly, the conditional probability of other error is done which is represented as P(AS/BS) and P(AE/BS). The marginal probability for semantic errors is calculated by taking the sum of all the 3 conditional probabilities i.e. P(AL/BS) + P(AS/BS) + P(AE/BS). Similarly, the conditional probability of exception is calculated in the similar manner. The marginal probability of exceptions is calculated by taking the sum of all the 3 conditional probabilities i. e. P(AL/BE) + P(AS/BE) + P(AE/BE).

Finally, the total marginal probability is computed which is the sum of marginal probability of logical errors, semantic errors and exceptions. Our calculation shows that the system fault tolerance is maximum in case of logical injection in compare to semantic and exception and its minimum in case of semantic. Thus, under various defective conditions system shows variations in fault tolerance. The proposed model is diagrammatically shown in Fig. 27.1.

27.4.2 Fault Detection Algorithm

In this section, we have proposed a Fault Detection Algorithm which has been derived from our proposed software model i.e. Process Simulation Model (PSM) network. The proposed algorithm basically detects the all possible faults such as logical, semantic and exceptions that may be present in the proposed system software model.

The Fault Detection algorithm is as follows:

Step 1: Add mutant to the code of our process.
Step 2: Take set of inputs from the user as per the functioning of the program.
Step 3: Set of outputs are displayed.
Step 4: Then set of inputs and outputs are inserted in the Process Simulation Model for Fault Detection.
Step 5: Features such as f(faults), p(E)(probability estimates), x(state estimates) are parameter estimator.
Step 6: Compare the features f, p(E), x with the normal behavior.
Step 7: If there is change, error (e) is detected.
Step 8: Fault diagnosis is done.
Step 9: Exit.

In this algorithm, mutants such as logical, semantic and exception were added to the process part of the model by injecting it inside the codes. User will insert the input based on the process. Based on the functioning of the process after the mutant injection, the outputs will be displayed according to the given input. Overall set of inputs and the outputs are generated in the Bayesian model-based fault detection after the mutant injection in the code of the process. Different type of process will generate different kind of feature with the output it generated after mutant injection. When there is no mutant injection the features estimated value is different, this helps in determining the normal behavior of the program when it is mutant-free. Changes in normal behavior features help us to know the presence of error in the program code. After the detection of the presence of error in the error the diagnosis of the code part is done to locate the error and debug the error. If

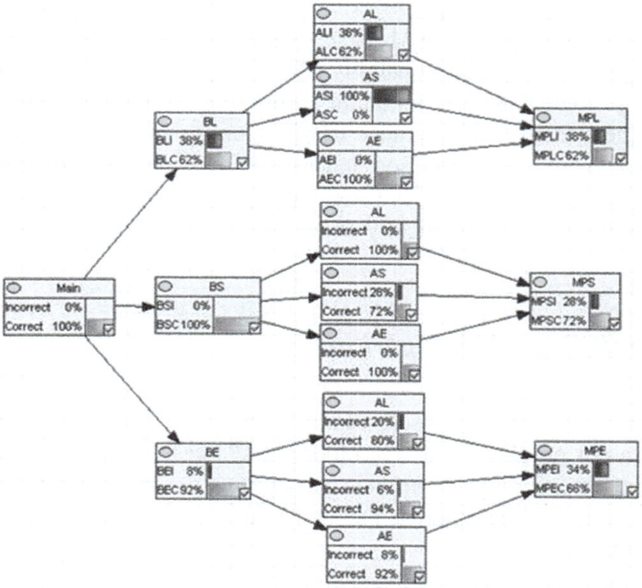

Fig. 27.1 Process simulation model network. *Main* main () function, Starting node, *BL* before logical, *BS* before semantic, *BE* before exception, *AL* after logical, *AS* after semantic, *AE* after exception, *MPL* marginal probability after logical, *MPS* marginal probability after semantic, *MPE* marginal probability after exception

there is no detection of error due to normal behavior feature of the process then we directly exit from the process model otherwise we exit after the fault diagnosis.

27.4.3 Process Flow for Process Simulation Model

In this section, we discuss the process flow of PSM in Fig. 27.2 using a Schematic block diagram based on Fault detection and diagnosis. Here in this model we have taken U as a set of inputs to be inputted in the process, where mutants are injected into the process in the form of code changes. Then the set of outputs is generated based on the inputs and mutants injected in the process. The fault is injected in the proposed system and thus it behaves according to the injected faults, from which we can study the fault tolerance under different defective conditions. The detection methods is carried out based on various parameters such as faults f, parameter estimates Θ or state estimates x, which are called features of the process model. By comparison with the normal features, if there is a change in the features newly generated then errors e is detected from our process model. Thus this feature comparison for the process model helps in fault detection and diagnosis. If there is no change in the feature then the fault is not detected and it shows the failure of the

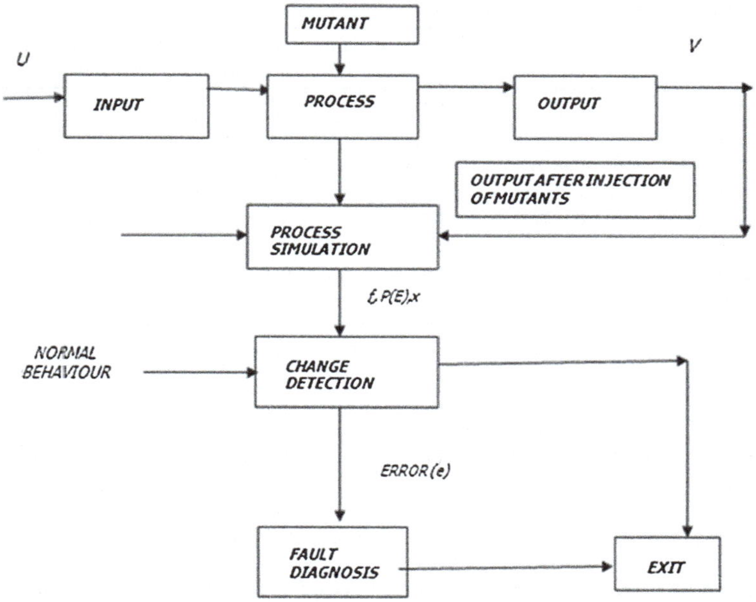

Fig. 27.2 General scheme of process model based on fault detection and diagnosis. *U* set of input data, *V* set of output data, *X* state estimates *P(E)* probability

process and we directly exit from the model scheme of fault detection. Fault tolerance can be measured under various defective conditions as per the injection of faults.

The Fig. 27.2 general scheme of process model based on fault detection and diagnosis.

27.4.4 Probabilistic Estimation Results on Various Deterministic Problems: A Comparative Study

A comparative study is a study that involves the comparison of two or more things of the same kind. We have made the comparative study table on various estimation results that we have got from different deterministic problems Firstly; we are taking the probable values before inserting logical errors and after inserting logical errors in all the cases of our programs. Then we take the average of all the values for both before and after injection of errors. Similarly we do for both semantic injection of errors and for exceptions. Now we can compare all the values of before injection and after injection of errors. These all are joint events and the probability of such joint events is determined by:

$$P(E1, E2) = P(E1)P(E2/E1)$$

Table 27.1 Probabilistic results on various deterministic problems

	After logical injection	After semantic injection	After exception injection	Marginal probability after injection
Before logical injection	0.38	0.0	0.0	0.38
Before semantic injection	0.0	0.28	0.0	0.28
Before exception injection	0.076	0.056	0.2	0.332
Marginal probability before injection	0.456	0.336	0.2	0.99

with the help of this formula we have calculated the comparative values of all. Like for example in the table we have calculated the value of P(SA/LB) by using that formula.

$$P(AS/BL) = P(SA)P(BL/AS)$$

where: AS—After semantic injection of errors, BL—Before logical injection of errors.

The values of the Table 27.1 are calculated with the help of the same formula. Then we calculated the marginal probabilistic values after and before injections by adding all the values in each row and each column individually. The total probability of fault tolerance after adding the marginal values is 0.99. In the earlier section, we have shown the Process Simulation Model Network, in which our simulation result shows the total probability as 1, by adding all the marginal probabilistic values of each row and each column individually after and before injections of logical, semantic and exceptions errors. Now, we are comparing our simulation results with the estimated results of different deterministic problems that we have shown in the comparative study table. The simulation result shows that the best case where total probability of fault tolerance is 1, which is justified and compared with Table 27.1 results shows that the marginal probability before and after fault injection is 0.99.

27.5 Experimental Study on Various Deterministic Problems

27.5.1 Definition of the Problem

In this section we have studied the probabilistic fault detection technique using mutation analysis on few deterministic problems. But, in this paper we have illustrated the Depth First Search problem. This problem is basically a systematic way to find all the vertices reachable from a source vertex. DFS is an uninformed search that progresses by expanding the first child node of the search tree that appears and thus going deeper and deeper until a goal node is found, or until it hits a node that has no children. Then the search backtracks, returning to the most recent node it hasn't finished exploring.

27.5.2 Mutant Generation

In this section we are injecting different types of mutants such as logical, semantic and exception in different section of the program code and we are getting corresponding output in different cases.

Case 1: Analysis after Injecting Logical Errors
Injection of logical errors—Logical errors occur by introducing fault through changing the logic of our program. The program gets compiled correctly but when we will run our program it will not give the proper output. It will either show infinite loop or it will not give any output. Our program will show weak fault tolerance towards the mutant (Table 27.2).

Case 2: Analysis after Injecting Exceptions
Injection of exception errors—It is a strong fault tolerance mechanism towards the mutants. Introducing exception will detect the errors in the program itself. We will not have to check our code each and every time. It will kill the mutant. In this program we are showing three cases of exception. When we will input wrong values it will throw exception. Table 27.3 shows the results of the program output after injecting the faults.

Case 3: Analysis after Injecting Semantic Errors
Injection of semantic errors—Semantic errors occurs when program statements are not constructed properly. The output results are shows in Table 27.4. Then program will not show any error during compilation but it will not run properly. Our program will show weak fault tolerance towards the mutant.

27.5.3 Experimental Analysis Using Graphical Representation

In this section, we are showing the graphical representation of different deterministic problem that are plotted based on various probabilistic values that we have got during the execution of the program after injecting logical, semantic and exceptions error.

Table 27.2 Summary of results after injecting logical errors

Case	I	O	Logical errors	I	O
Dfs()	2	DFS: 1 2	Vis[i] = = 0 Changed to Vis[i]! = 0	2	Infinite loop
Push()	2	DFS: 1 2	If(top = = 19) Changed to If(top! = 19)	2	Stack overflow

Pop()	2	DFS: 1 2	If (top = = −1)
			Changed to
			If(top! = −1)
				2	Exit

Pop()	2	DFS: 1 2	If(stack = top−−) Changed to If(stack = top++)	2	DFS: 1 2

Table 27.3 Summary of results after injecting exception

Case	I	O	Exception errors	I	O
Negative number exception	−2	Exit	If(n < 0) Print negative no. exception	−2	Negative no. exception
Array out of bounds	7	Wrong output	#define max 5 if(n * n > max * max) print(–Array out of bounds)	7	Array out of bounds
Number exception	a	Exit	if(isdigit(n));	a	Number exception

Table 27.4 Summary of results after injecting semantic errors

Case	I	O	Semantic errors	I	O
Dfs()	2	DFS: 1 2	Push(i) Changed to pop(i)	2	Dfs = 2
Dfs()	2	DFS: 1 2	Push(i) Changed to pop(i)	2	No result
Push()	2	DFS: 1 2	Top = = 19 Changed to bottom = = 19	2	Dfs = 2 1
Pop()	2	DFS: 1 2	Top = = −1 Changed to bottom = = −1	2	No result

Case 1: Depth First Search

In Fig. 27.3, we are showing the output results after injection of logical, semantic and exception in the y axis and no of mutants introduced in the x axis. The nature of the graph in case of logical error shows that in some point it is able to detect the error but after reaching a certain point it becomes stable and is unable to detect the errors. System started performing abnormally in a linear way and after reaching the peak point it linearly dropped down and where it finally stabilizes. Here, fault tolerance is high. Here our graph was stable in the beginning but the system performance dropped at some point and finally it stabilizes at the end.

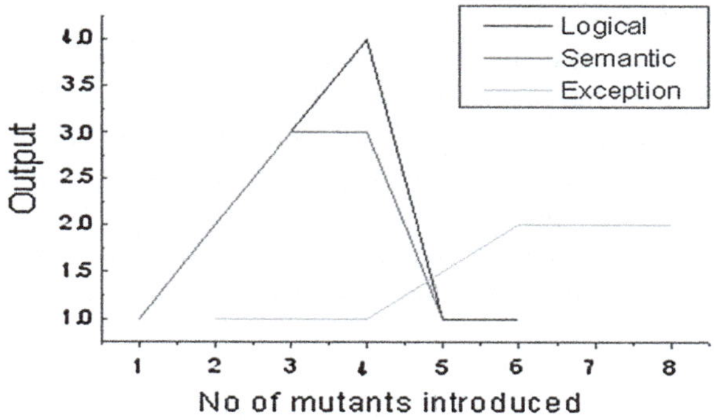

Fig. 27.3 Study of system behavior under various erroneous conditions for Depth First Search

Case 2: Knapsack Problem

Figure 27.4 is drawn on the basis of the continuous testing that was performed during the study of system behavior under various erroneous conditions. Three graphs were plotted. The first graph shows the number of mutants introduced due to change in the values of the weight of the first item after introducing logical errors, semantic errors and exceptions. The nature of graph for the three cases in the first graph is not entirely the same but after reaching a certain point the graph shows similar nature i.e., the number of mutants introduced versus output bear a constant ratio as the number of mutants are increasing continuously.

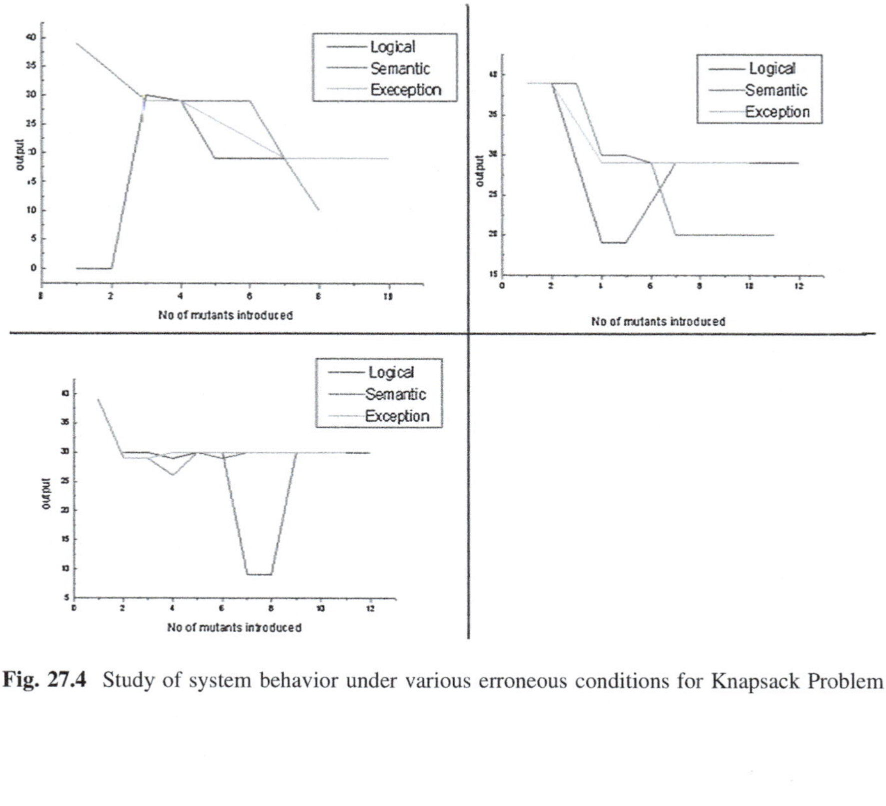

Fig. 27.4 Study of system behavior under various erroneous conditions for Knapsack Problem

Case 3: Bubble Sorting

Figure 27.5 shows the system behavior under various erroneous conditions in case of bubble sorting. We have taken two cases for logical error. We are taking options as input. Now if we take option 1 as input then we will be exited from the output window. If we take 2 or 3 or 4 then no output will be displayed and our system will hang and for rest other input default value will be displayed. Case 1 was the worst case where system performance degraded linearly till infinity and case 2 was the best case where the system performance remained stable.

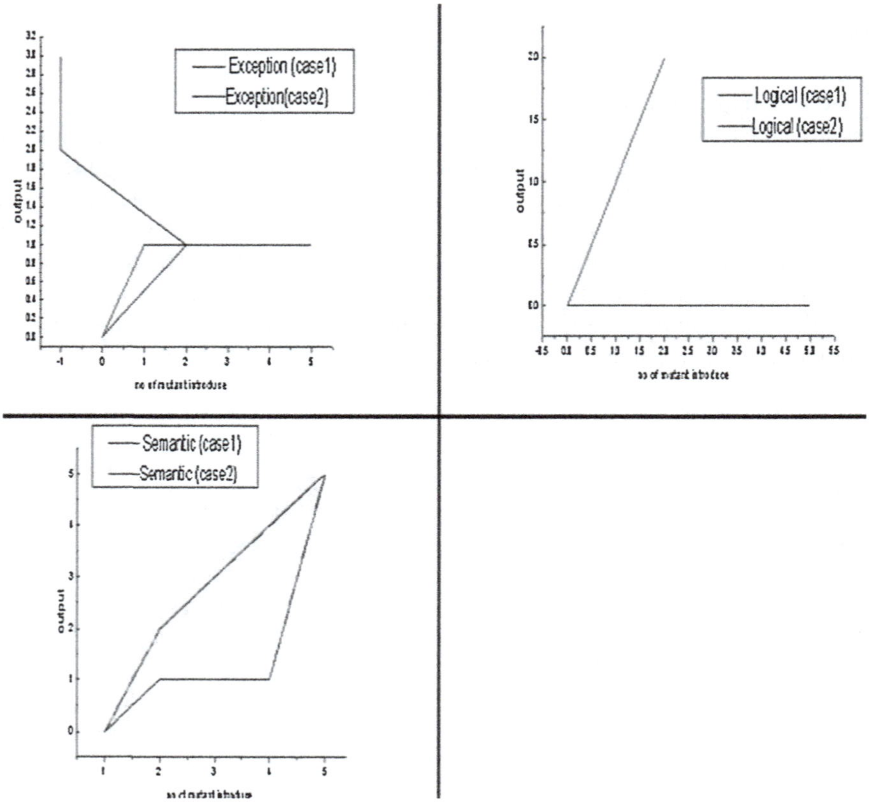

Fig. 27.5 Study of system behavior under various erroneous conditions for Bubble Sorting

27.6 Conclusion and Future Scope

Our analysis provides the idea about the fault behavior and its tolerance in deterministic problems. We have made a Process Simulation Model using Bayesian network approach. This approach is applied on various deterministic problems which significantly help in determining the fault by just looking at the proposed model without going through the coding details, but in this paper only the DFS traversal problem is defined and check how the system behaves under defective conditions. We would like to extend our work on fault behavior and its tolerance by introducing several errors at a time and can make a comparative study for the system behavior and its tolerance based on various method of injecting system faults.

References

1. De Florio V, Botti O (2002) Software-implemented fault-tolerance and separate recovery strategies enhance maintainability. IEEE Trans Reliab 51(2):158–165
2. Wu Y, Hu Q, Xie M, Ng SH (2007) Modeling and analysis of software fault detection and correction process by considering time dependency. IEEE Trans Reliab 56(4):629–642
3. Natella R, Cotroneo D, Duraes JA On fault representativeness of software fault injection. IEEE Trans Software Eng 39(1), 80–96, 03 January 2012
4. Carreira J, Madeira H, Silva JG (1998) Xception: a technique for the experimental evaluation of dependability in modern computers. IEEE Trans Software Eng 24(2):125–136
5. Andrews JH, Briand LC, Labiche Y, Namin AS (2006) Using mutation analysis for assessing and comparing testing coverage criteria. IEEE Trans Software Eng 32(8):608–624
6. Joshi KR, Schlichting RD, Sanders WH, Hiltunen MA (2011) Probabilistic model-driven recovery in distributed systems. IEEE Trans Dependable Secure Comput 8(6):913–928
7. Zheng J (2006) On the value of static analysis for fault detection in software. IEEE Trans Software Eng 32(4):240–259
8. Bahl P, Chandra R, Greenberg A, Kandula S, Maltz D, Zhang M (2007) Towards highly reliable enterprise network services via inference of multi-level dependencies. In: Proceedings of the ACM SIGCOMM, Aug 2007
9. Littlewood B, Popov P, Shryane N, Strigini L (2000) Modeling the effects of combining diverse software fault detection techniques. IEEE Trans Software Eng 26(12):1157–1167
10. Kuo SY, Huang CY, Lyu MR (2001) Framework for modeling software reliability, using various testing-efforts and fault-detection rates. IEEE Trans Reliab 50(3):310–320
11. Qin F, Tucek J, Sundaresan J, Zhou Y (2005) Rx: treating bugs as allergies—a safe method to survive software failures. In: Proceedings of the symposium on operating systems principles (SOSP), pp 235–248
12. GeNIe, version 2.0. http://genie.sis.pitt.edu/

Chapter 28
Software Coverage and Its Analysis Using ABC

Praveen Ranjan Srivastava

Abstract In software development lifecycle (SDLC), software testing holds the primary importance. Software is tested to uncover errors that were made inadvertently as it was designed; this forces us to perform software testing in a way that requires reducing the testing effort but should provide high quality software that can yield comparable results. To accomplish this, we have implemented a unique technique which takes into consideration Artificial Bee Colony (ABC) Algorithm. In Design Phase we have applied ABC on the Control Flow graph generated from the state diagram which then gives a test suite. In implementation phase newly proposed Algorithm takes CFG of the SUT and test suite from design phase as input and then generates an optimal test suite along with the software coverage. The resulting solution guarantees full path coverage keeping in view the design and implementation phase.

Keywords Software testing · ABC (Artificial Bee Colony Optimization) · Test-case · Agents · Path-coverage · Optimal path · Cyclomatic complexity · CFG (Control Flow Graph) · Test data

28.1 Introduction

Software testing is an optimization process where multiple variables are taken into account to generate the efficient number of test cases and to provide the optimal path. Infinite number of test cases will be generated by use of exhaustive testing. However, only a part of them will be more effective if implemented in testing the software. This approach leads the tester to obtain the comparable results using

P. R. Srivastava (✉)
Information Technology and System Group, Indian Institute of Management, Rohtak, India
e-mail: praveenrsrivastava@gmail.com

S. Sengupta et al. (eds.), *Emerging Trends in Computing and Communication*, Lecture Notes in Electrical Engineering 298, DOI: 10.1007/978-81-322-1817-3_28,

reduced and optimal set of test cases. Independent paths are useful in testing the path coverage and code coverage. Independent paths of SUT are identified using the control flow graph (CFG) [1, 2]. Meta-heuristic search techniques [3–6] are high-level frameworks which utilize heuristics in order to find solutions to combinatorial problems at a reasonable computational cost. From some time researchers have been trying the idea of applying artificial intelligence (AI) techniques in Software Engineering (SE) domain [6]. Recent research and development of Artificial Bee Colony Optimization (ABC) [4, 5] based systems are focusing mostly on applications such as financial decision making systems, transportation, manufacturing, and aerospace, military and so on. In our approach, we extended the functionality of the bee to do the testing and monitoring activity so that, it reduces the manual work and improves the confidence on the software by testing it with the coverage of the given software. The ABC algorithm [5], a meta-heuristic approach is used to generate the optimal number of test-cases which are sufficient to cover the paths generated by using the control flow graph (CFG). Proposed methodology is aimed at generating the optimal number of test cases and to achieve greater path coverage. To solve this problem ABC approach [5] is used where the bee agents gather the food sources (test cases) and then calibrate the fitness function which is in-turn used to identify the optimal test cases with highest path-coverage using less amount of resources. The optimal test suite generated from design phase is applied on the implementation phase to find the software coverage. We have even reworked the base Fitness function, on whose basis the solution is deemed to be fit or unfit, and has provided us with better solution in less number of test runs. To come up with the best possible solution which can be effective as well as efficient, we were required to do the basic background work of all the techniques which are being used to generate test cases. We have explained the background work that we carried out in short in the next.

28.2 Background Work

Precise demarcation of a program's input is not possible for any reasonably-sized program, thus random methods are not reliable and unlikely to exercise deeper features of software. Because of the size and complexity of the software involved, the issue of Test data generation evolves into an undesirable problem. The application of AI techniques in Software Engineering (SE) is the most sought after field of research. Researchers [3–6] use Ant Colony Optimization (ACO), Genetic Algorithm (GA), Tabu searches (TB), Bee Colony, Fuzzy Approach, Data mining Concept and many more alternate techniques. All the above mentioned approaches are being used in various areas of software engineering such as testing, quality, reliability etc. The application of meta-heuristic search techniques [1, 3–5] to test data generation is a possibility which offers much promise for the aforementioned problems. Meta-heuristic search techniques are high-level frameworks which utilize heuristics in order to find solutions to combinatorial problems at a reasonable computational cost. Such a problem may have been classified as NP-complete or NP-hard. These are the problems for which a polynomial time algorithm is known to exist but is not practical. While these heuristic techniques have achieved considerable success in their pertinent fields, they have also been marred by flaws which are inherent to their core path of approach. ABC optimizer [5] shows how to optimize the test suite using the standard ABC algorithm. In this approach the standard fitness function is used in the algorithm. In proposed approach, we have formulated a new fitness function for ABC algorithm. Analysis part of the paper shows the significance of using new fitness function. As described in [1], the test case optimization is done by taking into consideration just the implementation phase but here in proposed approach we have used both the Design Phase and Implementation Phase because generation of test suite at Design Phase and then applying this test suite in implementation phase results into most optimal test suite and guarantees maximum software coverage.

28.3 Proposed Approach

We know that Design and Implementation phases are two of the most important phases of Software Development Life Cycle (SDLC) [1]. All the work being carried out in Design Phase is used in Implementation Phase. But this is not the case in all the scenarios. We may change the code slightly during the implementation phase. We, thus propose a method that would benefit from the subtle difference between design and implementation phase. We have chosen state diagram of the design phase as the best representative for the flow of the project and to say the least about control flow graph generated using code from implementation phase.

In Phase I (Fig. 28.1) of the proposed method, the state diagram is given as input from the design phase to the ABC algorithm [5]. In ABC algorithm [5], we have slightly twisted the fitness function to our advantage. ABC algorithm generates the test suite as output. Test Suite along with Code from the Implementation phase is given as input to the new proposed algorithm which calculates the software coverage (in %) and generates the optimal test suite as well. We will be explaining the new fitness function used in the ABC algorithm [1, 5] and proposed algorithm used in the phase-II of 2-phase is explained below.

Phase-I:

We will try to explain Phase-I of the 2 phase Approach in this section. Phase-I of the proposed approach deals with the design phase of SDLC. State Diagram depicts the flow of information from one state to another as it happens in the control flow graph. This makes the state diagram, an indispensable entity in our 2 phase approach. State diagram acts as input to ABC Algorithm (with new fitness function) which generate the test suite as output. If you are interested in ABC Algorithm and its working then you can refer [1, 5].

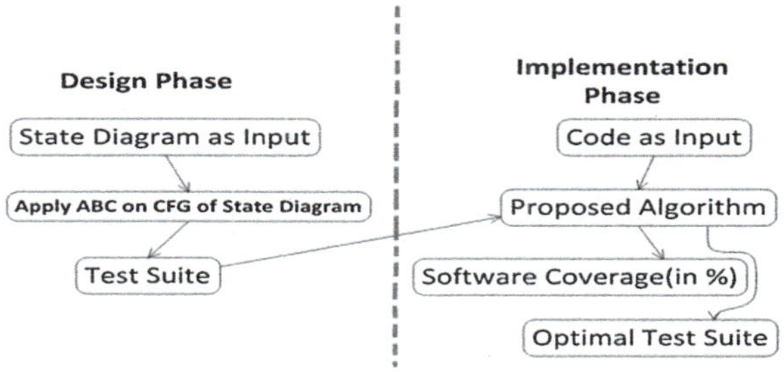

Fig. 28.1 Diagrammatic representation of the 2 Phase method to find the software coverage

The fitness function used in the ABC algorithm was doing compute extensive operation which were not required in proposed approach. To counter this, we changed the fitness function to suit our needs and which also results in faster test suite generation which can be verified by the analysis of the two fitness functions given in the analysis part of the paper. New fitness function is as given below.

$$\lambda_{ij} = \lambda_m + k(\lambda_j)$$

where i = 1 to Cyclomatic Complexity, j = 1 to no. of variables, λ_m: Mean value for all values supplied to the variables as Test Case, λ_j: Specific value present in the Test Case, K: Random number generated with respect to no. of variables. At the end of the Phase-I, Test suite is generated as output. The test suite generated here is obtained by applying ABC algorithm with new fitness function. Now this test suite along with CFG of the source code will be used in the Phase-II of our approach to generate optimal test suite and to yield maximum software coverage.

Phase-II:

As shown in Fig. 28.1, phase-II of the proposed 2-phase approach is carried out after the implementation phase of SDLC. Implementation phase gives access to the code written while implementing the software. From the code, we generate the control flow graph which acts as input along with test suite generated from the phase-I of the proposed 2-phase approach. The phase-II generates an optimal test suite as output besides calculating the software coverage of each path (in %). Algorithm used in this phase is given as below.

Algorithm used in Phase-II:

(1) Take all the test cases. Let a_{ij} contain all the test cases from the optimal Test suite generated using ABC algorithm. (2) Apply a_i. Mark all the nodes that it visits for a_i. (3) Make a note of all unmarked nodes for a_i. (4) Continue the above step for all test cases. (5) Check if all the nodes are visited. If yes then we can say we have tried to analyze the complete software with respect to Design and Implementation phase. (6) If No then product may contain errors.

`Pseudo-Code for Algorithm:`

`PO → Path with maximum number of nodes.,CO → Software coverage for Path PO,PI → Paths selected for various values of I,CI → Coverage for PI,TC → Optimized test suite,`

TSC → Software coverage for Optimized test suite, TI → No. of nodes selected, i moves from 1 to Cyclomatic complexity

```
TI ← PO
For(PI)
{
Calculate coverage for respective path
CI = (No. of nodes in PI/Total no. of nodes) * 100
Vardif = difference in number of nodes between PI& TI
If(dif > (total no. of nodes-Cyclomatic complexity))
{
Update Test Case TC
Update TI
TSC = (TI/Total no. of nodes) * 100
If(TSC == 100)
{
Break}}};
```

The 2-phase approach generates optimal test suite and total software coverage (in %) after selecting this test suite.

We applied above mentioned 2-Phase approach on various examples(C codes). Now we are presenting one particular case study for solving a quadratic equation. The state diagram of the problem with CFG of the quadratic equation problem was used to generate test suite using ABC algorithm with new fitness function. This test suite and the CFG of the source code were applied on proposed algorithm to generate optimal test suite and to find the software coverage.

28.4 Case Study

To demonstrate the efficiency of the proposed 2-Phase Approach, We tried to apply and test the approach for C programs including Quadratic Equation solving program. Here we have taken the same Quadratic Equation solving problem, to show the working of proposed approach as part of case study.

The SUT uses 3 inputs of a quadratic equation and generates its roots. Nodes are represented in the code. Applying the 2-phase approach on Quadratic Equation Code gives us the following results.

Phase-I:

We require the state diagram of the approach taken to solve the quadratic equations problem. The state diagram is as shown below in Fig. 28.2 along with the CFG of state diagram in Fig. 28.3. Now let us consider some Independent paths and the test suit, obtained after application of ABC on the CFG of Design Phase.

From the Fig. 28.3 which contains the Control Flow graph for the state diagram from the Design Phase, we deduce the Cyclomatic complexity of the graph. This in turn helps us to find the Independent Paths for the same. Now let us consider some Independent paths and the test cases to be given as input to the ABC algorithm (Table 28.1).

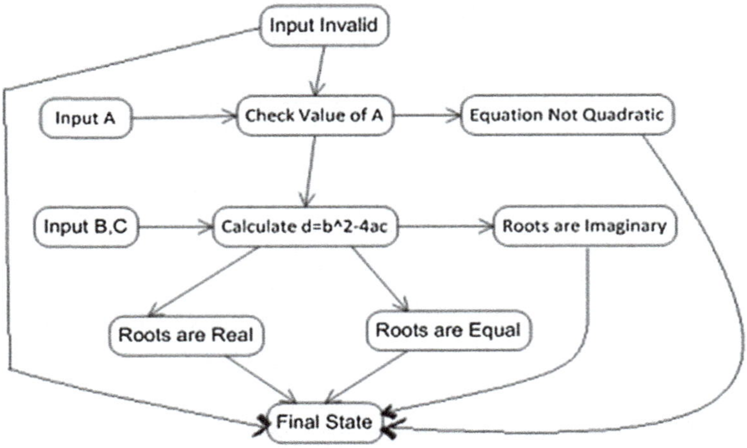

Fig. 28.2 State Chart diagram for the quadratic equations

Fig. 28.3 Control flow graph of Fig. 28.2

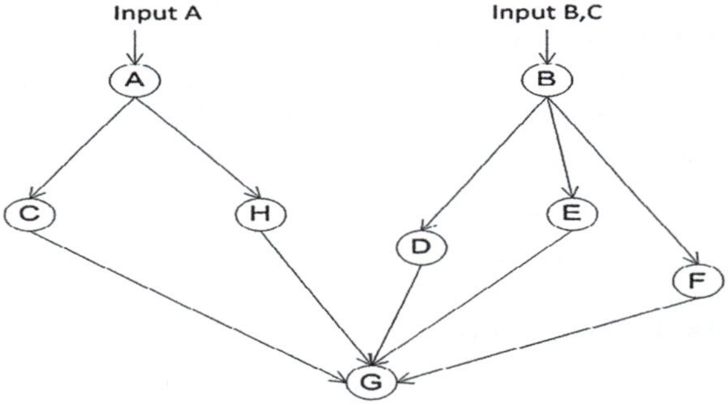

Applying ABC algorithm with a new fitness function gives us the following results as shown in Table 28.2. The step by step procedure of algorithm is as follows.

1. Let x_{ij} be the initial set of test cases as shown in the previous table.
2. $\lambda_{ij} = \lambda_m + k(\lambda_j)$ where

i = 1 to Cyclomatic Complexity, j = 1 to no. of variables, λ_m: Mean value for all values supplied to the variables s Test Case, λ_j: Specific value present in the Test Case, K: Random number generated with respect to no. of variables.

Let us consider the path A H G for the test case (50, 0, 50), $\lambda_m = -50 + 0 + 50/3 = 0$, $\lambda_{11} = 0 + (1)(-50) = -50$, $\lambda_{12} = 0 + (1)(0) = 0$, $\lambda_{13} = 0 + (0)(50) = 0$: Test case becomes $\lambda_1 = \{-50, 0, 0\}$.

Hence $v_i = \{-50, 0, 0\}$ for i = 1 and j = 1.

3. Compare v_i with x_i $\{-50,0,50\}$ for path j = 1. Follow this procedure for paths j = 1 to 3 (number of variables).

Table 28.1 Path with test cases from Fig. 28.3

Path	Test case	Value	Test decision
AHG	−50, 0, 50	N	Selected
	−50, 0, 0	Y	
AHG	50, −50, 0	N	Not selected
ACG	0, 0, 0	Y	Selected
ABDG	50, 100, 50	Y	Selected

Table 28.2 Test suit after applying AB algorithm

1	AHG	−50, 0, 50
2	ACG	0, 50, 0
3	ABDG	50, 100, 50
4	ABFG	50, 50, 1
5	ABEG	50, 50, 99

4. Get the fitness value of the node for test case at each path.
5. Select the path with highest cumulative value. Here its $x_{11} = \{-50,0,50\} + 1\{0,0,50\} - \{-50,0,50\} = \{-50,0,0\}$. This test case has the highest fitness value for path $j = 1$.
6. Memorize the test case and search for the next neighbouring nodes. Add nodes to find new test paths.
7. Continue the steps till all the nodes are covered in the CFG.

Here in we have finished the phase-I of the proposed 2 phase approach. Output of the Phase-I is given as input to Phase-II for further processing.

Phase-II:

In this phase, we generate the Control Flow Graph (CFG) from the exemplary code of Quadratic Equation problem and also perform the Cyclomatic Complexity Analysis to deduce the number of possible paths. We utilize the Test Suite generated from phase-I of the 2-phase approach to find the percentage of Software Coverage of SUT and the optimal test suite.

The Cyclomatic complexity for the above CFG is 7 and all possible Independent paths have been enumerated below in Table 28.3. Let us consider all the Independent paths for the CFG [1] shown in Fig. 28.4.

After applying the proposed algorithm discussed in the Phase-II of the 2-Phase approach.

The Algorithm takes into consideration the number of nodes which constitute the Independent path. Greater the number of nodes out of the total number of nodes present in control flow graph, greater is the probability of the respective test case to be considered in test suite. Reason being, It helps to check the software under test for software coverage. So we arrange all the Independent Paths with respect to the number of nodes covered by the same with respect to total number of nodes. Then we also take into consideration the repetition of nodes being encountered while selecting the next test case (with respect to next Independent path). As shown in Algorithm, We proceed by selecting the test cases with respect

Table 28.3 Independent path generated from Fig. 28.4

1.	A B G O Q R S	Invalid range
2.	A B C D F G O Q R S	Invalid range
3.	A B G O P R S	Not quadratic
4.	A B C D E F G O P R S	Not quadratic
5.	A B G H I J N R S	Roots are equal
6.	A B G H I K L N R S	Roots are real
7.	A B G H I K M N R S	Roots are imaginary

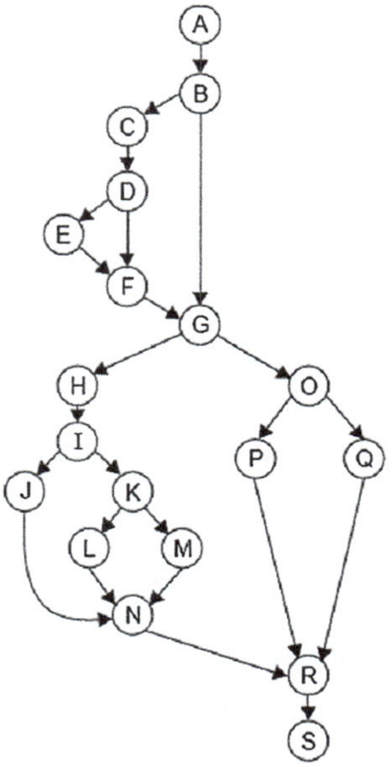

Fig. 28.4 Control flow graph

to software coverage done by respective independent path. Thus we have tried to find the optimal test suite keeping in view the software coverage using test cases for Software under Test.

Analysis Phase:

In proposed approach, we have changed the fitness function being used in ABC algorithm to suit our needs. We compared the results of these two fitness functions based on the same input given to each of them performing under the same environment. We executed the standard ABC algorithm code and the modified code containing new fitness function in Java (JDK 1.6.0.13). We got following results from the test shown in Fig. 28.5 and Table 28.4.

We have also shown a comparison between two approached i.e. Between Approach used in [5] and proposed 2-Phase approach. By taking into consideration the design phase of SDLC, we have tried to make the test suite as optimal as possible

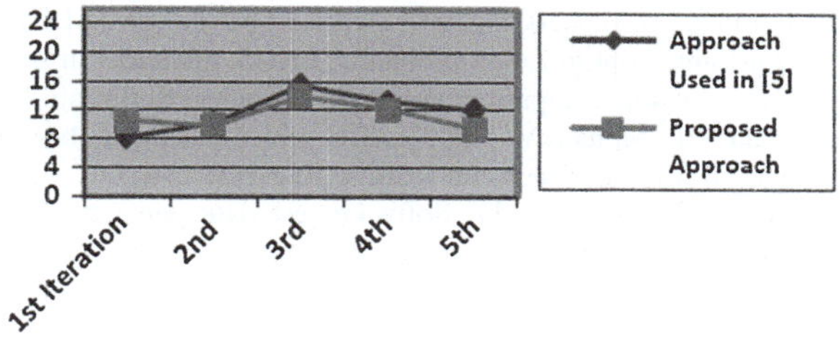

Fig. 28.5 Comparison of two fitness function

Table 28.4 Comparison of two approaches

Phase	Design	Implementation
Approach used in [5]	x	✔
Proposed approach	✔	✔

28.5 Conclusion

As proposed ABC approach is implemented on 2 different yet highly symbiotic phases of Software Development Life Cycle (SDLC) i.e. Design and Implementation, we are able to better realize what customer wants and thus deliver a product more closely to customer's expectations. Moreover as our Fitness function is inclined towards the statistical computational techniques rather than heuristics, we are able to provide a more tangible and accurate result for the problem set it is subjected to. While this open approach allows us to better analyse the Software, it also means that this testing process won't fit in our run-of-the-mill, traditional Process models of Software Engineering. This is because it requires that upon detection of a fault or error, rectification can be done to either of the both, the design of the software or the implementation code of the software. Thus it is highly usable and beneficial in the modern engineering methodologies like agile process and Scrum technique.

References

1. Srikanth A, Nandakishore JK, Naveen VK, Singh P, Srivastava PR (2011) Test case optimization using artificial bee colony algorithm. In: Advances in computing and communications. Communications in computer and information science, vol 192, Part 5. Springer, Berlin, pp 570–579
2. Briand LC (2002) Ways software engineering can benefit from knowledge engineering. In: Proceeding in 14th software engineering and knowledge engineering (SEKE), Italy, pp 3–6
3. Yu B, Qin Y, Yao G et al (2009) Tabu search and genetic algorithm to generate test data for BPEL Program. In: Proceeding in international conference on computational intelligence and software engineering (CISE), IEEE conference publication, Wuhan, China, pp 1–6

4. Karaboga D, Basturk B (2007) Artificial bee colony (ABC) optimization algorithm for solving constrained optimization problems. LNCS: advances in soft computing: foundations of fuzzy logic and soft computing, vol 4529. Springer, Berlin pp 789–798
5. Mala DJ, Mohan V (2009) ABC tester—artificial bee colony based software test suite optimization approach. Int J Softw Eng 2(2):15–43
6. Li HZ, Lam CP (2005) Software test data generation using ant colony optimization. In: Proceedings of world academy of science, engineering and technology, vol 1, pp 22–27

Chapter 29
Effective Probabilistic Model for Webpage Classification

Hammad Haleem, C. Niyas, Siddharth Verma, Akshay Kumar and Faiyaz Ahmad

Abstract World Wide Web (www) is a large repository of information which contains a plethora of information in the form of web documents. Information stored in web is increasing at a very rapid rate and people rely more and more on Internet for acquiring information. Internet World Stats reveal that world Internet usage has increased by 480 % within the period 2000–2011. This exponential growth of the web has made it a difficult task to organize data and to find it. If we categorize data on the Internet, it would be easier to find relevant piece of information quickly and conveniently. There are some popular web directories projects like yahoo directory and Mozilla directory in which web pages are organized according to their categories. According to a recent survey, it has been estimated that about 584 million websites are currently hosted on the Internet. But these Internet directories have only a tiny fraction of websites listed with them. The proper classification has made these directories popular among web users. However these web directories make use of human effort for classifying web pages and also only 2.5 % of available webpages are included in these directories. Rapid growth of web has made it increasingly difficult to classify web pages manually, mainly due to the fact that manually or semi-automatic classification of website is a tedious and costly affair. Because of this reason web page classification using machine learning algorithms has become a major research topic in these days. A number of algorithms have been proposed for the classification of web sites by

H. Haleem (✉) · C. Niyas · S. Verma · A. Kumar · F. Ahmad
Department of Computer Engineering, Faculty of Engineering and Technology, Jamia Millia Islamia, New Delhi 110025, India
e-mail: hammadhaleem@gmail.com

C. Niyas
e-mail: niyasmonc@gmail.com

S. Verma
e-mail: sidd.verma29@gmail.com

A. Kumar
e-mail: akshay061@gmail.com

S. Sengupta et al. (eds.), *Emerging Trends in Computing and Communication*, Lecture Notes in Electrical Engineering 298, DOI: 10.1007/978-81-322-1817-3_29,

analyzing its features. In this paper we will introduce a fast, effective, probabilistic classification model with a good accuracy based on machine learning and data mining techniques for the automated classification of web-pages into different categories based on their textual content.

Keywords Content classification · Machine learning · Naïve Bayesian · Web-mining · Probabilistic models · Web-page classification

29.1 Introduction

The World Wide Web (WWW) started in the year 1991 and has shown a rapid growth within last two decades. It is estimated that today more than 3.5 billion people are using internet. The number of people using the Internet is rapidly increasing. Internet World Statistics reveals that the world Internet usage growth has increased by 480.4 % during the period 2000–2011 [1]. It was also observed that more than 35 % data available in whole world is stored in Internet.

According to Netcraft's January 2012 survey, more than 584 million web sites exist on the Internet and out of which, nearly 175.2 million are active. Today there are several different tools available for an average Internet user to locate and identify relevant information on the Internet. These tools can be broadly classified as (1) Crawler based Search Engines (SE) e.g., Google, Bing, Yahoo, Duck–Duck-Go etc., (2) Meta Search engines e.g., Metacrawler, Clusty etc., and (3) Subject Directories like DMOZ (Directory Mozilla), Librarians Internet Index (LII) etc. The crawler based search engines and subject directories maintain their own data repositories, whereas meta search engines don't maintain any such data repositories, instead they depend on indices of other Search engines and subject directories to answer user queries. The database maintained by any crawler based search engine is quite large, and have considerably larger amount of data indexed in their databases as compared with subject directories. Directory Mozilla (DMOZ) [1] i.e., dmoz.com has 93,446 editors for 1.2 million categories and has indexed 5.26 million websites, which is only 2.5 % of the total active web sites available on the Internet today. Majority of the web directories are edited and maintained by human effort. Subject directories are popular due to proper classification of data in several categories. A larger website directory could be quite helpful in improving the quality of search results, filtering web content, developing knowledge bases, building vertical (domain specific) search engines. Hence need for automating the process of classification of websites based on their content arisen recently. Manually classifying data is really expensive to scale as well as quite labor intensive and in this paper we present a technique to reduce manual effort significantly and hence this paper presents a cost effective way for categorizing data. This paper presents a Naïve Bayesian (NB) probabilistic model for the automatic classification of webpages. Naive Bayes probabilistic model is one of the most effective and straightforward model for text document classification and has exhibited good results in previous studies conducted for data mining. The model is quite optimized and has quite effectively worked to classify websites based on their content in real-time. Also it can be scaled for large databases. The model presented in this paper has given accuracy around 94 % for the test data considered.

The rest of the paper is organized as follows. Section 29.2 reviews previous work on the machine learning, classification and probabilistic approaches. Sections 29.3 and 29.4 discusses the classification of web pages and Naïve Bayes Theorem respectively, Sect. 29.5 presents our approach of classifying websites based on textual content of web pages using NB technique. Section 29.6 discusses the results of our experiment. Section 29.7 summarizes the paper and gives some directions for future research.

29.2 Related Work

In this section we briefly try to review previous works in the field of text classification with a special emphasis on classification of webpages. In the starting days, task of classification of documents was generally carried out manually by experts of that domain. But very soon, ways were identified to carry out the classification in semi-automatic or automatic ways. Now we have a lot of effective ways to carry out machine assisted classification of documents. Many techniques have been researched and worked upon lately. Some of the approaches for text-categorization include statistical and machine leaning techniques like k-Nearest Neighbor approach [2–4], Bayesian probabilistic models [5, 6], inductive rule learning [7], decision trees [8, 9], neural networks [10, 11], and support vector machines [12, 13]. While most of the learning methods have been applied to pure text documents, there are numerous approaches dealing with classification of web pages. Pierre [12] discusses various practical issues in automated categorization of web sites. Machine and statistical learning algorithms have also been applied for classification of web pages [13–17]. In order to exploit the hypertext based organization of the web page several techniques like building implicit links [17], removal of noisy hyperlinks [18], fusion of heterogeneous data [19], link and context analysis [20] and web summarization [21] are used. An effort has been made to classify web content based on hierarchical structure [22, 23].

29.3 Classification of Webpages

Web page classification is different from normal text classification in some aspects. The uncontrolled nature of web content presents additional challenges to web page classification as compared to traditional text classification. The web content is semi structured and contains formatting information in form of HTML tags. A web page may contain hyperlinks to point to other pages. This interconnected nature of web pages provides features that can be of greater help in classification. In the first step of classification all HTML tags are removed from the web pages, including punctuation marks. The next step is to remove stop words as they are common to all documents and does not contribute much in classification. In most cases a stemming algorithm is applied to reduce words to their basic stem. One such frequently used stemmer is the Porter's stemming algorithm [24]. Each text document obtained by application of procedures discussed above is represented as frequency vector. Machine learning algorithms are then applied on such vectors for the purpose of training the respective classifier. The classifier is then used to test an unlabeled set of sample documents against the learnt data. In order to rank high in search engine results, site promoters pump in many relevant keywords. This additional information can also be exploited.

29.4 Bayes Algorithm

29.4.1 Why Naive Bayes

Naive Bayes classifier is preferred over other classifiers because of its simple and straight forward approach and ability to apply for a wide variety of domains. Theoretically a classifier based on Bayes theorem will have minimum error rate in comparison to all other classifiers. Also naive Bayes classifier can be easily used for any number of features without much difference in performance. It can be scaled for larger databases without rise in execution time in comparison with other classifiers.

29.4.2 Bayesian Classifiers

Bayesian classifiers are statistical classifiers which predict the class membership probabilities of tuples. It means probability of a given tuple to be in a particular class. Bayesian classifiers are based on Bayes theorem which will be explained in next section. Studies comparing classification algorithms have found that a simple Bayesian classifier known as the naive Bayesian classifier is comparable in performance with decision tree and selected neural network classifiers. Bayesian classifiers have also exhibited high accuracy and speed when applied to large databases. In theory Bayesian classifiers have the minimum error rate in comparison to all other classifiers. However, in practice this is not always the case owing to inaccuracies in the assumptions made for its use, such as class-conditional independence, and the lack of available probability data [24].

29.4.3 Bayes Theorem

This theorem was named after Thomas Bayes who did early work in probability and decision theory during the 18th century. Let X a data tuple. In Bayesian terminology. X is termed as 'evidence'. X is described by n attributes. i.e., $X = \{x_1, x_2, x_3, x_4, \ldots, x_n\}$. Let H be some hypothesis that the data tuple X belongs to a particular class C. For classification problems we want to find P(H|X), the probability that hypothesis H holds when attributes set/'evidence' X is given. That means we are looking for the probability that the tuple X belongs to class C, given that we know the attribute description of X. Bayes theorem helps us to determine the probability P(H/X) in terms of P(H), P(X/H) and P(X). Bayes Theorem is:

$$P(H/X) = \frac{P(X/H)P(H)}{P(X)} \tag{29.1}$$

where P(X/H) gives the probability for tuple to occur when it is given that the hypothesis holds. P(H) is the probability for the hypothesis H to hold. And P(X) is the probability for the attribute set X to occur.

29.4.4 Naive Bayes Classifier

Naive Bayes Classifiers are Bayesian Classifiers which assume class conditional independence. i.e., it assumes that the effect of an attribute value on a given class is independent of the values of the other attributes. This assumption simplifies the calculations a lot. Naive Bayesian classifier works as follows. Let D be a training set of tuples and their associated class labels. Each tuple is represented by n dimensional attribute vector in the form $(x_1, x_2, x_3, \ldots, x_4)$ depicting n measurements made on the tuple from n attributes, respectively, $A_1, A_2, \ldots, A_n$.

$$P(ci/X) = (P(X/ci) * P(Ci))/P(X) \tag{29.2}$$

As P(X) will be constant for all classes, we have to consider numerator term only. If we are taking equal number of training tuples for each class, then $P(C_i)$ will be same for all classes. In such cases we have to maximize $P(X/C_i)$ only. Otherwise we have to maximize $P(X/C_i)$ * $P(C_i)$. Where $P(C_i)$ can be calculated as follows

$$P(Ci) = \frac{\text{Number of training tuples belonging to Class Ci}}{\text{Total Number of training Tuples}} \tag{29.3}$$

When we have data sets with several number attributes, it is computationally expensive to calculated (X/Ci). But when we assume class conditional independence, computational cost can be reduced significantly. With this assumption P(X/Ci) can be calculated as follows:

$$P(X/Ci) = \prod_{k=1}^{n} P(xk/Ci) = P(x1/Ci)xP(x2/ci)x, \ldots, xP(xn/ci) \tag{29.4}$$

If attribute A_k is categorical, then $P(x_k/C_i)$ is given as follows:

$$P(xk/Ci) = \frac{\text{Number of tuples in class Ci Having value Xk for attribute Ak}}{\text{Total Number of Tuples in Class Ci}} \tag{29.5}$$

If attribute x_k is continuous valued, then the following function can be used for finding $P(x_k/C_i)$

$$g(x, \mu, \sigma) = \frac{1}{\sqrt{2\Pi\sigma}} e^{-\frac{(x-\mu)^2}{2\sigma^2}} \tag{29.6}$$

So that:

$$P(xk/Ci) = g(xk, \mu Ci, \sigma Ci) \tag{29.7}$$

To predict the class label of X, $P(X/C_i)$ $P(C_i)$ is evaluated for each class C_i. The classifier predicts that the class label of tuple X is the class C_i if and only if $P(C_i/X) > P(C_j/X)$ for $1 <= j <= m$ and $i! = j$

$$P(H/X) = \frac{P(X/H)P(H)}{P(X)} \tag{29.8}$$

Modification on Naive Bayes for document classification: For document classification problems calculation of $P(x_k/Ci)$ will be modified. It is required since attributes are not either categorical or continuous valued in case of document classification problems. In document classification problem we have the relation,

$$P(xk/Ci) \propto P(Ci) \prod_{1 \le K \le nd} * (Ptk/Ci) \tag{29.9}$$

where n_d is number of tuples in class C_i and total number of occurrences of word x_k in category C_i. P (t_k/C_i) can be calculated as follows:

$$P\left({}^{Tk}/_{Ci}\right) = \frac{Tci}{\sum_{tr \in V} Tct'} \tag{29.10}$$

where T_{ci}, total number of occurrences of word T_k in category C_i. V is the vocabulary, all words in training set.

29.4.5 Laplacian Correction

To avoid zero computational value in order to avoid the zero probability value, when a word in testing tuple do not come in training set, we use Laplacian correction. Laplacian corrected formula is where V is number of terms in vocabulary.

$$P(tk/Ci) = \frac{Tci + 1}{\sum_{tr \in V} (Tct' + 1)} = \frac{Tci + 1}{\sum_{tr \in V} Tct' + |V|} \quad (29.11)$$

29.5 Experimental Setup

This section explains the setup of the entire experiment. We collected different web pages belonging to predefined categories. Then popular data cleaning and data extraction techniques were applied to extract useful data from collection of webpages. Once the data extraction process is finished we trained our classifier according to k-fold strategy for various values of k and used to predict category of webpages' in test set. All the processes are briefly discussed below.

29.5.1 Creation of Data Set

The data set is collection of webpages within predefined categories. There were 6 predefined categories. Number of document in each category, number documents used for train set and number of document used for test set is given in Table 29.1. Most of the documents were taken from popular news networks like BBC, CNN, and Reuters since they are already classified into various categories. We designed a crawler to automatically scan these web documents and to store these documents into different local directories corresponding to their category. In this experiment we have only considered the data in English language. Since we were only interested in the text content we simply removed all multimedia contents on the web page like Videos, Images, Flash plugin content and JAVA applets. Other than these, pages with relatively less content were also removed from the collection. The dataset consisted of total of 3,183 documents in six categories.

29.5.2 Cleaning HTML Documents

We used the Python3 regular expression module and BeautifulSoup module to extract the information from web pages. Beautiful Soup is a Python library designed for easy and effective scraping of webpages and provides a few simple methods and Pythonic idioms for navigating, searching, and modifying a parse tree. Rather than extracting information from specific tags we removed all the HTML tags present in the document with the exception of title tag because it offers considerable amount of information regarding the content of webpage.

Table 29.1 Composition of testing data

Category	Total	Train	Test
Education	529	432	97
Entertainment	537	429	108
Politics	508	406	106
Religion	536	428	108
Sports	530	424	106
Technology	543	434	109
Total	**3,183**	**2,553**	**630**

All non ASCII characters and special symbols were removed from the data. CSS and JavaScript content present in the HTML documents were also cleaned during this phase. In many webpages images were used as buttons to be clicked in place of hyperlinks or images were used to display name of the organization. Such information is a very important feature for classification purpose, however our experiment concentrates on text based retrieval so such graphical text was ignored. The standard Stop-Word list used in Bow [25] was used along with a comprehensive Stop-word list provided In the NLTK [26, 27] (Natural language processing toolkit) library in python. Along this we used the stemming and lemmatization techniques to reduce words to their least forms. We used the post tagger present in the NLTK library to individually analyze each word. Then based on the word form i.e., Noun, Adjective or Verb the wordnet-lemmatizer was used to strip down all the words in their least forms. Those words which were not recognized by the wordnet-lemmatizer were processed further with the Porter Stemming [24] algorithm to generate strip down version of the word. After all the above processing the HTML document was converted into a list of words in their most basic forms.

29.5.3 Vocabulary Generation

The words that occurred commonly in most of the webpages were considered as stop words. A list of such words is present in Table 29.2. These common words were considered as Stop-Words and removed directly from the document. It was also observed that webmasters inflated the title, Meta description and keyword tags with multiple keywords. We normalized such repeating keywords to reduce the impact of site promotion techniques applied by webmasters. This step was performed during the cleaning phase. All the words with the occurrence less than two were removed from the Documents and also the words with length less than three were also discarded from the Data. The frequency of each of the words was stored in a Hash table along with their category, for faster and efficient access in the later stages of experiment. The keywords that appear in two sample categories are given in Tables 29.3 and 29.4.

Table 29.2 Web-page specific stop words

Login, view, browser, website, web, online, search, keyword, designed, copyright, rights, reserved, click, search, welcome, email, click, contact, developed, mail, home, page, feedback, webmaster

Table 29.3 Keywords related to educational category

Student western university offer alumni news degree view program menu read bull calendar study experience graduate visit state academics admission major meet amp business college mountain campus life sport education service inform library common event institute watch music aid recreate history faculty graham athlete finance school center Wheaton art census Wichita request depart William online office form student day scholarship

Table 29.4 Related to religion category

God worship religion Hinduism prayer will Muslim time amp before symbol john people good quote practice order oxford year king source church refer idea pope power influence relate belief include life number faith doctrine religion follow century origin word work great call accord scripture book Christian article ecclesiast reform history state universe nation yantra India Iranian Buddhism Zoroastrian group adhere culture tradition pupil jum Indian Islam Abraham create retrieve Zoroaster Mazda text evil

29.5.4 Testing and Training the Classifier

We followed the k-fold cross validation strategy (with $k = 5$) to decide the number of training and testing examples. The documents in each category are divided into 5 equal partitions say D_1, D_2, D_3, ..., D_5. Training and testing will be performed k times. In iteration I, partition Di in each category will be reserved as a training set, and the remaining partitions will be collectively used to train the model. That is, in the first iteration, subsets D_2, ..., D_5 collectively serve as the training set to obtain a first model, which will be tested on D_1 of each category; the second iteration will be trained on sets D_1, D_2, D_3, ... D_5 and will be tested on D_2. And so on. Prior probability of each category will be same since we have taken equal number of documents in each category. Training set was stored in the form of hash tables of hash tables. Each hash table in first level stands for category. And hash tables in second level will be containing the frequency of words in that particular category. The posterior probability with Laplace's correction was calculated using the formula [28, 29]:

$$P(wk|c) = (n_k + 1)/(n + |Vocabulary|). \tag{29.12}$$

where N_k stands for number of occurrences of word W_k in category C, N is total number of words in given category and |vocabulary| stands for number of words in training set. In order to classify a document say X, the probabilities of each word of document in a given category were calculated from hash table, then they were multiplied together. The probability values obtained for individual categories were sometimes going below the floating point limit of Python (in order of x $\times$ 10^{-300}). We were able to find a solution for this problem by taking the $Log_2(P_x)$, of the obtained probability and summing them up. The obtained number was then further multiplied by -1, to get overall positive value.

$$C = \arg \text{Max}(-\log_2(P_c) - \sum \log_2(P_{wk}|C) \quad (29.13)$$

The calculated values were compared to find the minimum of all. The category with minimum log of probability was selected as the classified class for a document. Equation 29.12 helps to clearly understand the solution of the mentioned problem. The Naïve Bayes algorithm to train and test the classifier is given below:

Algorithm 1 Algorithm for vocabulary generation

```
INPUT: train_set->set of training documents in the format {name,category}
categories->A list containing available categories
OUTPUT: database-> a 2d hash array listing frequency of each word in each category

Function vocabulary_generation(train_set)
database={} // define database as a hash function
 for each category in categories:
     database[category]={} //define each item in database    //as hash function
   for each document in train_set:
     freq=preprocess(document.name)
     for each word in freq:
       if word in database[document.category]:
         database[document.category][word]+=freq[word]
       else:
          database[document.category][word]=freq[word]
   return database
```

Algorithm 2 Algorithm for webpage classification

```
INPUT: freq-> word frequency list of test web page.
Database-> 2d hash table containing list of word frequencies in each category
categories->List of available categories for classification
OUTPUT :category->category of given web page
probability_model(freq_list,database):
v=total_number_words_in_database ; pc={}
for category in categories: {
    attributes=database[category]
     n=total_number_of_words_in_category
     pc[category]=MAX_VAL
     for word in freq_list: {
         if word not in attributes:
            pc[category]+=-log(1/n+|v|)
          else:
             pc[category]+=-log((1+attributes[word])/(/(n+|v|))
    Category=category for which pc[category] is maximum
}}
   return Category
```

Graph 29.1 Average accuracy versus number of folds

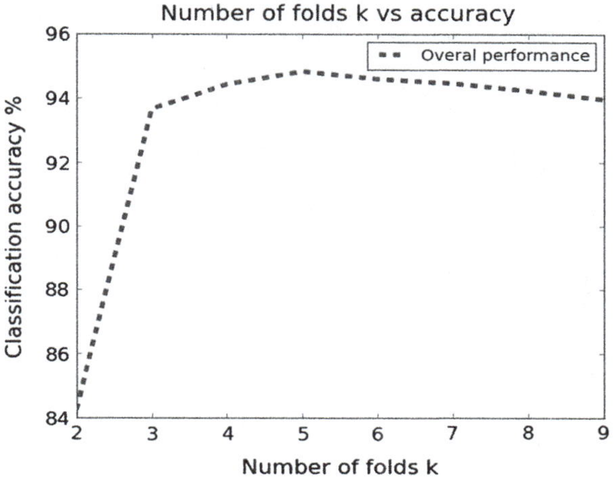

29.6 Experimental Results

The classifier was executed for the given dataset with 3,183 examples in six categories for various values of k. With each value of K, We were able to record corresponding accuracy. Graph 29.1 depicts the relation between the increasing value of K and its effect on the accuracy of classification of documents for the range $2 \leq k < 10$.

Initially there is a rise in classification accuracy with increase in number of folds which can be explained in terms of better training. Then there is a slight fall in accuracy which can be explained in terms of over training of classifier. i.e., classifier is trained too much and is not able retrieve a result.

It is observed that the classifier was able to achieve maximum accuracy which is above 94 % for K = 5, i.e., when 4/5th of the documents were used as training set and rest for test set. Using our approach we were able to get a good accuracy. Table 29.5 depicts confusion matrix. A classifier is good if majority of the documents lie along the diagonal of a confusion matrix. This trend is observed in Table 29.5. This confusion matrix can be further used to find most common classification measures like accuracy, recall, error rate, specificity and f-measures etc. Based on each of the K value we generated confusion matrix for each of the category. We calculated the accuracy for each value of K.

Graph 29.2 show the relation between the K value and accuracy achieved for specific category. We can see initially when the K value was lower the accuracy was lower. But as the K value increase efficiency also increased. Similar trend is also shown in the Graph 29.1. Thus we can say the classification model presented in this paper can be used effectively for real-time classification of webpage with high accuracy.

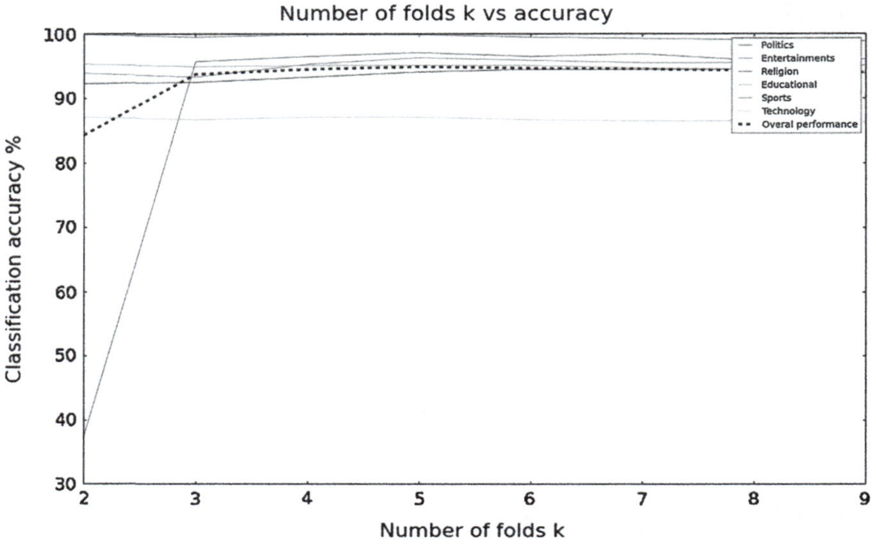

Graph 29.2 Relation between increasing K and accuracy of classifier

Table 29.5 Classification of data in various categories'

Category	Pol	Entr	Sport	Educ	Rel	Tech
Pol	470	5	10	0	14	1
Entr	7	481	6	0	6	0
Sport	11	3	485	0	1	0
Educ	17	0	0	475	6	2
Rel	1	0	0	0	499	0
Tech	64	0	0	1	0	235

Pol Politics, *Entr* Entertainment, *Sport* sports
Educ Educational, *Rel* Religion, *Tech* Technology

29.7 Conclusion and Future Works

The naive Bayes document classification algorithm discussed in this paper intelligently exploits the richness of content of webpage for effective classification into industry type category.

The algorithm here classifies the webpages into a very broad set of categories. Naive Bayes approach for of web pages based on their textual content, for six categories considered in this paper yielded above 94 % accuracy. It has been observed that the classification accuracy of the classifier is proportional to number of training documents up to a limit. The results are quite encouraging. This approach could be utilized by the search engines and other online directory projects like DMOZ [30] for effective categorization of web pages to build an automated web directory based on the content available on the web page and type of organization. Although in this experiment, only nonhierarchical and distinct categories are considered the above algorithm can also be used to classify the pages into more specific categories (hierarchical classification) by changing the feature set e.g., a web site that is ecommerce may be further classified into electronics, clothes or a book selling website.

References

1. Netcraft's internet survey for August 2012. http://news.netcraft.com/archives/2013/08/09/august-2013-web-server-survey.html
2. Simple Knn approach for text classification. http://www.cis.uab.edu/zhang/Spam-mining-papers/A.Simple.KNN.Algorithm.for.Text.Categorization.pdf
3. Optimized approach for text classification. http://www.cis.uab.edu/zhang/Spam-mining-papers/An.Optimized.Approach.for.KNN.Text.Categorization.using.P.Trees.pdf
4. Guo G, Wang H, Greer K (2004) A kNN model-based approach and its application in text categorization. In: 5th international conference, CICLing Springer, Seoul, Korea, pp 559–570
5. McCallum A, K. Nigam (1988) A comparison of event models for Naïve Bayes text classification. In: AAAI/ICML-98 workshop on learning for text categorization, pp 41–48
6. Lewis DD, Ringuette M (1994) A classification of two learning algorithms for text categorization. In: Proceedings of 3rd annual symposium on document analysis and information retrieval (SDAIR'94), pp 81–93
7. Apte C, Damerau F, Weiss SM (1994) Automated learning of decision rules for text categorization. ACM Trans Inf Syst 12(3):233–251
8. Wermter S (2000) Neural network agents for learning semantic text classification. Inf Retrieval 3(2):87–103
9. Weigend AS, Weiner ED, Peterson JO (1999) Exploiting hierarchy in text categorization. Inf Retrieval 1(3):193–216
10. Leopold E, Kindermann J (2002) Text categorization with support vector machines. How to represent texts in input space? Mach Learn 46(1–3):423–444
11. Bennett D, Demiritz A (1998) Semi-supervised support vector machines. Adv Neural Inf Process Syst 11:368–374
12. Pierre JM (2000) Practical issues for automated categorization of web sites. In: Electronic proceedings of ECDL 2000 workshop on the semantic web, Lisbon, Portugal
13. Sun A, Lim E, Ng W (2002) Web classification using support vector machine. In: Proceedings of the 4th international workshop on web information and data management, McLean, Virginia, USA, pp 96–99
14. Zhang Y, Xiao BFL (2008) Web page classification based on a least square support vector machine with latent semantic analysis. In: Proceedings of the 5th international conference on fuzzy systems and knowledge discovery, vol 2. pp 528–532
15. Kwon O, Lee J (2000) Web page classification based on k-nearest neighbor approach. In: Proceedings of the 5th international workshop on information retrieval with Asian languages, Hong Kong, China, pp 9–15
16. Dehghan S, Rahmani AM (2008) A classifier-CMAC neural network model for web mining. In: Proceedings of the IEEE/WIC/ACM international conference on web intelligence and intelligent agent technology, vol 1. pp 427–431
17. Dou S, Jian-Tao S, Qiang Y, Zheng C (2006) A comparison of implicit and explicit links for web page classification. In: Proceedings of the 15th international conference on World Wide Web, Edinburgh, Scotland, pp 643–650

18. Zhongzhi S, Xiaoli L (2002) Innovating web page classification through reducing noise. J Comput Sci Technol 17(1):9–17
19. Xu Z, King I, Lyu MR (2007) Web page classification with heterogeneous data fusion. In Proceedings of the 16th international conference on World Wide Web, Banff, Alberta, Canada, pp 1171–1172
20. Attardi G, Gulli A, Sebastiani F (1999) Automatic web page categorization by link and context analysis. In: Hutchison C, Lanzarone G (eds) Proceedings of THAI'99, pp 105–119
21. Dou S, Zheng C, Qiang Y, Hua-Jun Z, Benyu Z, Yuchang L, Wei-Ying M (2004) Web-page classification through summarization. In: Proceedings of the 27th annual International ACM SIGIR conference on research and development in information retrieval, Sheffield, United Kingdom, pp 242–249
22. Dumais S, Chen H (2000) Hierarchical classification of web content. In: Proceedings of the 23rd annual international ACM SIGIR conference on research and development in information retrieval, Athens, Greece, pp 256–263
23. Porter MF (1980) An algorithm for suffix stripping. Program 14(3):130–137
24. Porter Stemming algorithm, with various implementations. http://tartarus.org/martin/PorterStemmer/index.html
25. The BOW or libbow C Library. Available http://www.cs.cmu.edu/~mccallum/bow/
26. Bird S, Loper E, Klein E (2009) Natural language processing with python. O'Reilly Media Inc. Python NLTK, Sebastopol
27. Python based open source mathematical analysis toolkit. www.matplotlib.org
28. Mitchell TM (1997) Machine learning. McGraw-Hill Companies, Inc, New York
29. Data Mining concepts and techniques Han Kamber Lee Morgan Kaufman publications, 3rd edn. 2012
30. DMOZ open directory project. Available http://dmoz.org/

Chapter 30
Clustering Web Search Results to Identify Information Domain

Santa Maiti and Debasis Samanta

Abstract Of late, people are using Internet to retrieve information from a vast web repository. As a query, a word or a group of words can imply multiple meanings in different contexts. Web search engine, however, cannot distinguish the context and hence retrieves huge information. It becomes a tedious job for users to browse all web pages until they reach to the target one. To overcome the problem, researchers proposed to present search result in cluster form where web pages of a search result are grouped based on their similarity measures. So, the user gets proper guidance to find out their target web pages with fewer trials. After analyzing the basic clustering approaches we propose a clustering algorithm combining *k*-Means and hierarchical clustering to obtain better cluster quality with affordable time delay. We compare the proposed method with the existing methods. The experimental result substantiates the efficacy of the proposed method.

Keywords Web searching · Information retrieval · Document clustering · Clustering algorithm

This work has been carried out as a part of my MS work at Indian Institute of Technology Kharagpur.

S. Maiti (✉)
Innovation Labs, Tata Consultancy Services, Kolkata, India
e-mail: santa.maiti@tcs.com

D. Samanta
School of Information Technology, Indian Institute of Technology Kharagpur, Kharagpur, India
e-mail: dsamanta@iitkgp.ac.in

S. Sengupta et al. (eds.), *Emerging Trends in Computing and Communication*, Lecture Notes in Electrical Engineering 298, DOI: 10.1007/978-81-322-1817-3_30,

30.1 Introduction

At present, web repository works as mine of information. In recent years, World Wide Web has expanded by about 2,000 % and is doubling in size every six to ten months [1]. According to a recent survey,[1] it is reported that the indexed web contains at least 3.6 billion pages. However, only 34.3 % (as on December 31, 2011) of world population are familiar to Internet search.[2] One of the reasons behind this poor participation is that, the ranked representation of web search result is incomprehensible and not fully acceptable by all types of computer users. A search engine returns probable web pages related to the query which are huge in size and in general, from different domains. An expert computer user can use several advanced options (e.g. pages containing particular phase, page language, file type, time of upload etc.) to get accurate information in least response time. Relevancy of retrieved web pages for a query mainly depends on the accuracy of query formation.

Retrieved web page snippets are ranked depending on the PageRank [2], relevancy measure etc. A snippet contains the title of the web page, content summary and the link to that web page. The user has to predict the target snippet(s) and the desired information can be obtained by navigating the predicted web pages by trial and error method. Accuracy of query formation by user increases with time and with the familiarity of searching mechanism. If the query is imprecise, the target snippet(s) may present in the search result but with a higher rank. As a consequence, searching and extracting the required information become a tedious and time-consuming job. Again, it has been observed that more than 50 % of the users consult no more than first two screens of results [1] which results failure of obtaining required information though it is present in the web repository. As a way out, clustering mechanism has been advocated where similar web pages are grouped together.

In general, clustering is the assignment of a set of observations into subsets (called clusters) so that observations in the same cluster are similar in some sense. It is a method of unsupervised learning [3]. The systems that perform clustering of web search results, also known as clustering engines, have become popular in recent years. The first commercial clustering engine was probably Northern Light, in 1996. It was based on a predefined set of categories, to which the search results were assigned. A major breakthrough was then made by Vivisimo [4], whose clusters and cluster labels were dynamically generated from the search results. In recent times, several commercial clustering engines have been launched in the market [4, 5] namely Grouper (1997) [6], WISE (2002) [7], Carrot (2003) [8], WebCat (2003) [9], AISearch (2004),[3] SnakeT (2005),[4] Quintura (2005),[5]

[1] The size of the World Wide Web, http://www.worldwidewebsize.com.

[2] World Internet World Stats, http://www.internetworldstats.com/stats.htm.

[3] AI Search Engine, http://www.netpaths.net/blog/ai-search-engine-from-mit.

[4] SnakeT, http://snaket.di.unipi.it.

[5] Quintura—visual search engine, http://www.quintura.com.

WebClust (2006),[6] YIPPY (2009)[7] etc. These engines [4] consider search result snippets as an input to achieve faster response time. As a consequence cluster quality degrades because snippet is not always a good representative of a whole document [10].

There exist two basic clustering mechanisms: partitional and hierarchical clustering. *k*-Means clustering is the most common type of partitional clustering which produces flat clusters. Hierarchical clustering, on the other hand, creates a hierarchy of clusters which may be represented in a tree structure called dendrogram [11]. Hierarchical clustering is usually either agglomerative ("bottomup") or divisive ("top-down") [2]. Both of these clustering techniques have some limitations to apply directly in clustering of web search results. Hierarchical clustering technique results better quality of clusters, though the computational complexity of *k*-Means is less. But hierarchical clustering is trapped in past mistakes whereas *k*-Means offers iterative improvement. So, noticing the limitations of these two algorithms, we propose a combination of both the hierarchical and *k*-Means algorithms to cluster web documents. Our main objective is to obtain better clustered search results with reasonable time delay. In this work, we have proposed a hybrid clustering method. The entire web page content instead of snippet is considered as an input to improve the cluster quality. Finally, we compare the performance of the proposed technique with the existing benchmarked approaches.

30.2 Proposed Methodology

In this work, we have proposed a new clustering algorithm to group the web search results into a number of clusters depending on the similarity between the documents. Our proposed clustering algorithm follows four tasks: web page content extraction, preprocessing of web documents, document feature extraction and inter document similarity measure. An overview of our entire work is shown in Fig. 30.1.

30.2.1 Web Page Content Extraction

Search engine returns a ranked list of snippets containing web page links. Since a snippet not necessarily represents actual web content it links, we consider the whole web page content instead of snippets while clustering. Obviously, it takes a reasonable amount of time compared to the snippet downloading time. To speed up the process we download the web page contents concurrently using threading.

---

[6] WebClust—Clustering Search Engine, http://www.webclust.com.

[7] Yippy Clustering Search Engine—iTools, http://itools.com/tool/yippy-web-search.

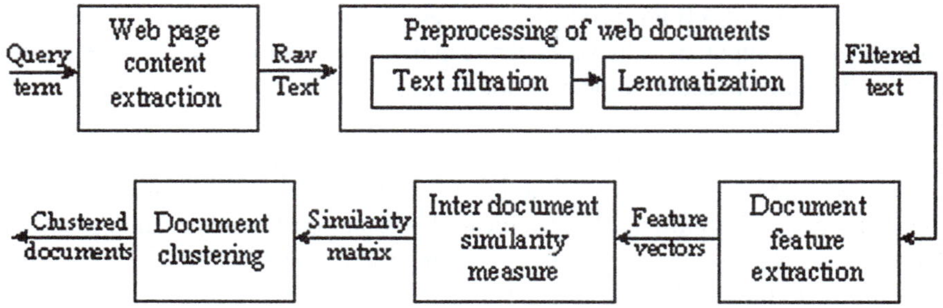

Fig. 30.1 Overview of the proposed approach

So, at first we identify the web page links from the snippets. These links are used to extract the content, that is, the source code of the web pages. The web page contents are then saved for filtration process. In our work, we consider first 100 web pages. Nevertheless, the number is not limited to this number according to our approach. Five different tourism related queries are used to test the effectiveness of threading. As download speed varies, each query is run thrice and average download time is calculated. The comparison of time requirement for downloading documents with threading and without threading is given in Table 30.1. It shows that threading reduces the downloading time by a factor of 13 for 100 documents.

30.2.2 Preprocessing of Web Documents

The web pages contain a lot of extra information, noises (e.g. advertisement) along with the desirable information. Presence of these elements is often troublesome to identify the important document features. It may also mislead the clustering process. Hence, before the feature extraction and proper clustering, document preprocessing is necessary. In this stage, we first identify extra information and noises and filter out those. Finally, we do lemmatization [2] to map inflected words into their root forms.

Text Filtration: Source code of a web page is likewise raw data which cannot be used directly for clustering. It contains a lot of extra information and noises such as HTML (Hyper Text Markup Language) tags, special characters, scripts, non-letter characters, non-printable ASCII characters etc. After downloading the source code these elements are eliminated using regular expression to extract only text (such as sentences, phases and words) from web documents.

Lemmatization: In order to perform clustering, web documents are needed to represent in terms of term-vector. Important terms are identified by using term-weighting scheme as discussed in Sect. 30.2.3. Generally, a word can occur in a document in different morphological forms. Lemmatizer can map an inflected word to its base form with the use of a vocabulary and morphological analysis [2].

Table 30.1 Web document download time with threading and without threading

# web pages	Download time (m:ss)									
	Q: Culture Assam (3.64 MB)		Q: Delhi guide (4.67 MB)		Q: Market Kolkata (4.83 MB)		Q: Royal Rajasthan (3.56 MB)		Q: Wildlife India (3.44 MB)	
	WT	WOT	WT	WOT	WT	WOT	WT	WOT	WT	WOT
10	0:08.85	0:21.99	0:09.10	0:39.21	0:08.92	0:18.94	0:08.87	0:29.76	0:08.85	0:31.78
20	0:08.97	1:33.48	0:09.91	1:24.51	0:09.87	0:43.63	0:09.02	1:11.86	0:09.18	1:36.18
30	0:11.58	2:13.70	0:11.91	2:08.51	0:13.26	1:06.27	0:09.25	1:58.62	0:09.52	2:14.18
40	0:13.51	2:44.42	0:18.36	2:49.64	0:17.84	1:27.62	0:10.69	2:34.39	0:10.21	2:44.56
50	0:14.32	3:32.07	0:24.85	3:28.76	0:24.53	3:11.93	0:14.18	3:14.52	0:14.60	3:14.04
60	0:23.67	4:16.47	0:30.78	5:15.68	0:33.10	3:35.56	0:22.16	4:24.06	0:21.66	3:57.92
70	0:25.47	4:59.79	0:31.64	6:04.89	0:33.98	4:24.28	0:22.35	5:15.69	0:25.97	4:29.43
80	0:27.19	5:39.02	0:32.28	6:49.74	0:35.01	4:55.78	0:23.42	5:52.83	0:26.50	5:14.39
90	0:28.42	6:16.56	0:33.56	7:22.20	0:35.69	5:18.35	0:24.07	6:35.51	0:26.83	5:52.23
100	0:28.97	6:58.62	0:34.71	8:49.76	0:39.37	5:36.21	0:26.75	7:33.74	0:29.65	6:17.58

Notation: Q → Query, WT → With Threading, WOT → Without Threading
Average download time for 100 documents without threading = 07:03.18 m:ss
Average download time for 100 documents with threading = 00:31.89 m:ss
Speed up = 13.27

We use LemmaGen[8] to convert all the words of web documents in their base form without changing their order. This guarantees that all the inflected forms of a term are mapped to a single term, which helps to determine each term's importance.

30.2.3 Document Feature Extraction

The aim of this phase is to identify the important terms from extracted text which are potentially capable to represent the documents. In our work, vector space model [2] is used to represent a document and term frequency-inverse document frequency is considered as dimension. At first we find out the unique terms among the bag of words of all documents. Next, we have to identify the document related important terms. In order to do this, we use the term frequency-inverse document frequency (*tf-idf*) [12] metric to measure terms' importance. Note that, stopwords (e.g. about, the, in etc.) are frequent in any document but not important. We identify 571 stopwords[9] which are filtered out from every document. After calculating the *tf-idf* value of each unique term in each document, we calculate the average *tf-idf* value of each term. Then the terms are sorted on the average value in descending order. Top *m* terms are considered as important terms. Any document

[8] LemmaGen, http://lemmatise.ijs.si/.

[9] http://jmlr.csail.mit.edu/papers/volume5/lewis04a/a11-smart-stoplist/english.stop

can be represented in terms of these terms. The value of m would be decided experimentally (discussed in Sect. 30.3.5).

Term-Document Matrix (TDM): Our next task is to represent all documents in terms of m terms identified in the previous step. We consider term-document matrix TDM to represent all documents using *tf-idf* value of m terms. Generally, column length normalization [8] is used to represent term-document matrix to avoid the biasness for very short or very long document.

30.2.4 Inter Document Similarity Measure

The main objective of the document clustering phase is to group similar documents. For this, we need to compute similarity values between every pair of documents. In this work, cosine similarity is used to measure the similarity between two documents. We have used similarity matrix to represent the pairwise similarity between two documents. A similarity value varies between 0 and 1, where 0 indicates no similarity and 1 indicates maximum similarity or identical.

30.2.5 Our Proposed Clustering Algorithm

We propose an algorithm to cluster documents. It combines divisive hierarchical approach and k-Means. Hence, we named it *HK*-Clustering. This approach resembles with bi-secting-k-Means approach. But, unlike bi-secting-k-Means the seed documents are always the most dissimilar documents in a cluster. It also overcomes the problem of local assignment of non-seed documents. Our objective is to produce coherent clusters which means a document is not strictly belongs to a single cluster. Depending on some conditions it may belong to a number of clusters as often a document covers multiple topics. Let us consider a set of n documents $\{d_0, d_1, d_2, \ldots d_{n-1}\}$ as clustering elements. The documents based on which the clusters are formed we term them as seed documents. Suppose, *S* and *Old-S* denote the sets of seed documents at current and previous level, respectively. We store all clusters generated at a level in a pool of clusters called *CP*.

- Step 1. *Cluster initialization*: Initially, all documents are in a single cluster C_0, that is, $C_0 = \{d_0, d_1, d_2, \ldots d_{n-1}\}$ and cluster-pool $CP = C_0$. *S* and *Old-S* are initialized to null as in the root level there is no seed documents.
- Step 2. *Seed selection*: Suppose, $C_{i,j}$ is the ith cluster at level j. We select seeds in $C_{i,j}$ for $(j + 1)$th level as follows. We find out a document pair say d_i, d_j with minimum similarity value among all the document pairs in the cluster $C_{i,j}$. If the similarity value of the document pair d_i, d_j is less than a limit called dissimilarity threshold denoted as α, we consider d_i and d_j as seeds for $(j + 1)$th level. Otherwise, the current seed documents of $C_{i,j}$ at level j remain the seeds for

$(j + 1)$th level. To find out all the probable seeds for $(j + 1)$th level, we consider all the clusters of jth level one by one.

Next, we find if any two or more seeds are mergeble or not. We check the similarity value of each pair of seeds. If the similarity value of any seed pair exceeds a threshold value called merging similarity threshold denoted as β, we merge those two seeds into one. It may happen that two or more seed pair satisfy the condition and have a common seed document. As an example, the similarity value of d_i, d_j and d_j, d_k is greater than β where d_i, d_j, $d_k \in S$. In this case, we merge all three of them into a single representative seed. So, a seed may contain more than one document. We term such a seed as composite seed. We store all seeds selected for the next level in S.

- Step 3. *Check terminating condition*: The process terminates if $CP =$ null and $S = Old\_S$. If the terminating condition does not satisfy, go to Step 4 else, go to Step 5.
- Step 4. *Assign documents to their nearest seeds*: For each seed in S, we create a cluster initially containing the seed document only. Thus, we have $|S|$ number of newly created clusters, where $|S|$ denotes the number of seeds in S. Next, we assign the non-seed documents to those clusters as follows.

Suppose, d_i is a non-seed document which we want to assign to a seed document. Among all seeds in S let, $d_j \in S$ has the maximum similarity with d_i. Therefore, we assign d_i to the cluster corresponding to the seed d_j. In case of composite seed where a single seed is represented by two or more documents, we compare the similarity value of a non-seed document with each of the document of the composite seed as well as with other seed documents. All newly obtained clusters are kept in CP. Then we set $Old\text{-}S = S$ and go to Step 2.

- Step 5. *Produce final coherent cluster*: Finally, we produce the coherent clusters as stated below. Let, $C'_{i,j}$ be the centroid of the cluster $C_{i,j}$. We calculate $C'_{i,j}$ by taking the arithmetic mean of all the document vectors (discussed in Sect. 30.2.3) corresponding to the cluster $C_{i,j}$. Next, we check the similarity value of each document with other cluster centroids apart from which it belongs to. Suppose, a document d_i belongs to cluster $C_{i,j}$. So, we check the similarity values of document d_i with the cluster centroids other than the centroid $C'_{i,j}$. If any of the similarity value crosses a limit called belonging similarity threshold denoted as γ, then we assign d_i to that cluster also. It is possible that a document satisfies the condition for more than one cluster centroids. In that case, that document will be assigned to all those clusters.

The Complexity analysis of *HK*-Clustering algorithm: There are three main tasks in the algorithm: seed selection for next level, assigning documents to their nearest seeds and producing final coherent clusters. Let, n is the number of documents to be clustered and h is any level. The time complexity of first task is $O(n^2/2^h + 2^h(2^h - 1 - 1))$. To assigning documents to their nearest seeds $O((n - 2^h)2^h)$ time is required. Finally, we can produce coherent clusters in $O(n^2/2)$ time. Now, the first two tasks occur repeatedly in each level. Total time required for these two tasks including level 0 to $log_2 n$ is $O(n^2)$. So, the time complexity of the proposed algorithm is $T = O(n^2)$. The space complexity of our proposed algorithm is $O(n^2 + k \cdot n)$, that is, $O(n^2)$ where k is the number of clusters generated in final state.

30.3 Experimental Results

This section presents a detail description of experiments and the results observed.

30.3.1 Experimental Setup

All experiments are carried out in Windows environment (Windows 7) with Intel Core2Duo (2.0 GHz) processor and 2.0 GB memory. The proposed approach is implemented in C# language in .Net 3.5 platform using Microsoft Visual Web Developer 2008 Express Edition. Internet explorer is used as default web browser to access Internet with speed around 3.45 Mbps. Google search engine is used for the Internet searching.

30.3.2 Experimental Data

In order to substantiate the efficacy of the proposed algorithm and compare different clustering algorithms we need a tagged or well classified data set. We use the standard document collections of Reuters 21578.[10] We collect only the tagged documents from Reuters 21578. Thirteen sample data sets are created, from reut2-000 to reut2-012 each having different number of classes. We also prepare a tagged document collection using the web search results for some specific query. To create our own document collection, ten different queries related to tourism domain are feed to Google search engine. We consider first 100 returned results for each query to build up a collection. 10 expert users are asked to tag the web pages to defined classes. Each web page is tagged to one or more topic(s) by the users. So our own created database contains 1,000 tagged documents. These dataset are used for training as well as testing purpose.

[10] Reuters-21578, http://www.daviddlewis.com/resources/testcollections/reuters21578.

30.3.3 Performance Metrics

To evaluate the performance of the proposed algorithm, cluster quality is analyzed. To evaluate the cluster quality two different measures: internal measure and external measure are used [13, 14]. Internal measure allows comparing different sets of clusters without reference of any external knowledge like already classified dataset. Whereas external measure quantifies how well a clustering result matches with a known classified dataset. In our work, the known classified dataset is the previously categorized document collection provided by human editor.

Internal Measure: We use Dunn index [13] as an internal measure to quantify cluster quality. It aims to identify dense and well-separated clusters. It is defined as the ratio of the minimal inter-cluster distance to maximal intra-cluster distance. Equation 30.1 shows the Dunn index, where $\delta(C_i,C_j)$ represents the distance between clusters C_i and C_j and $\Delta(C_l)$ measures the intra-cluster distance of a cluster C_l.

$$DI(C) = \frac{min_{i\neq j}\{\delta(C_i, C_j)\}}{max_{1\leq l\leq k}\{\delta\Delta(C_l)\}} \tag{30.1}$$

$$\delta(C_i, C_j) = \frac{1}{|C_i||C_j|}\sum\nolimits_{d_i\in C_i, d_j\in C_j} \varphi(d_i, d_j) \tag{30.2}$$

$$\Delta(C_i) = 2\left(\frac{\sum_{d_i\in C_i, d_j\in C_j} \varphi(d_i, C_i')}{|C_i|}\right) \tag{30.3}$$

Here, $\varphi(d_i, d_j)$ is distance between two documents d_i and d_j, where d_i and d_j are any two documents belong to C_i and C_j, respectively. $|C_i|$, $|C_j|$ are the size of clusters C_i,C_j. $C'_{i,j}$ is the centroid of cluster C_i.

External Measure: F measure is widely used to measure the external quality which combines the precision (P) and recall (R) ideas from information retrieval [2, 13]. Let for the document set D, C_j is one of the output clusters and C_i^* is corresponding to human edited class. Then the precision and recall is,

$$P = \frac{C_i^* \cap C_j}{C_j} \tag{30.4}$$

$$R = \frac{C_i^* \cap C_j}{C_i^*} \tag{30.5}$$

F measure is the harmonic mean of precision and recall. The F measure ($F_{i,j}$) and overall F measure (F) is computed as shown in Eqs. 30.6 and 30.7 where l is the total number of human edited class, k is the number of output cluster and $|V|$ is the total number of documents present in l number of human edited class.

$$F_{i,j} = \frac{2P \cdot R}{P + R} \tag{30.6}$$

$$F = \sum_{i=1}^{l} \frac{|C_i^*|}{|V|} max_{j=1...k}\{F_{i,j}\} \tag{30.7}$$

The value of Dunn index is high for the clusters with high intra-cluster similarity and low inter-cluster similarity. Again high overall *F*-measure indicates the higher accuracy of the clusters mapping to the original classes. So, algorithms that produce clusters with high Dunn index and high overall *F*-measure are more desirable.

30.3.4 Evaluations

We compare the *HK*-Clustering algorithm with basic clustering methods as well as the benchmarking clustering algorithms specially used for clustering web search results. Seven datasets of Reuters from reut2-005 to reut2-012 and five dataset of our data collection are used for the comparison. Some data sets containing large number documents of Reuters are used for testing to show the efficiency of algorithm for higher number of document collection. We apply *HK*-Clustering algorithm on the document collections to obtain clusters. Basic clustering methods: *k*-Means and agglomerative hierarchical clustering (group average) are also used to cluster the document collections. Similarly, the latest benchmarking algorithm Lingo using Singular Value Decomposition (SDV) [8] is used for clustering. Then we check the quality of clusters obtained from different algorithms with respect to both internal and external measures. Table 30.2 presents the comparative study of *k*-Means, agglomerative hierarchical clustering, Lingo and *HK*-Clustering algorithm with respect to cluster quality. From Table 30.2 it is clear that proposed *HK*-Clustering algorithm produces better clusters compared to *k*-Means and Lingo algorithm. Again Lingo produces better clusters compared to another benchmarking algorithm Suffix Tree Clustering (STC) used for web search results clustering [8]. Time complexity of Lingo is in $O(n^3)$ and a high number of matrix transformations leading to more memory requirements [15]. In spite of high computational complexity and space requirement Lingo is well accepted for clustering search results as it produces better quality of clusters by considering semantic approach. The time complexity of the above mentioned four algorithms are shown in Table 30.3. From the time complexity point of view, *HK*-Clustering performs better than hierarchical agglomerative clustering algorithm as well as Lingo algorithm. Hence, *HK*-Clustering offers an optimal solution by balancing both cluster quality and time complexity.

Table 30.2 Comparison of cluster quality

Document name	No. of doc	No. of class	No. of clusters	Hierarchical agglomerative		*k*-means		Lingo		*HK*-clustering	
				DI	F	DI	F	DI	F	DI	F
reut2-005	100	6	8	2.71	0.70	2.41	0.49	2.55	0.56	2.64	0.68
reut2-006	100	10	9	2.40	0.78	2.33	0.42	2.3	0.62	2.36	0.71
reut2-007	100	4	5	1.42	0.58	1.27	0.34	2.29	0.41	1.31	0.58
reut2-008	100	7	10	2.67	0.55	2.43	0.32	2.46	0.38	2.49	0.52
reut2-009	100	6	7	1.57	0.67	1.42	0.51	1.42	0.55	1.46	0.64
Adventure India	100	9	8	2.71	0.78	2.49	0.51	2.52	0.58	2.53	0.69
Hotel Hyderabad	100	14	16	4.61	0.77	4.35	0.53	4.39	0.73	4.66	0.81
India pilgrim	100	10	13	3.89	0.38	3.71	0.24	3.73	0.26	3.83	0.35
Kolkata travel	100	14	16	4.79	0.67	4.62	0.52	4.65	0.49	4.77	0.59
Weather Kolkata	100	8	7	1.59	0.68	1.48	0.41	1.52	0.53	1.57	0.67
reut2-010	400	14	17	4.51	0.56	4.14	0.50	4.37	0.51	4.48	0.53
reut2-011	800	12	15	3.45	0.49	3.19	0.24	3.52	0.33	3.36	0.44
reut2-012	1,000	15	16	4.26	0.46	4.11	0.31	4.19	0.34	4.24	0.37

Table 30.3 Comparison of time complexity

Clustering algorithm	Time complexity
k-means	O(n)
HK-clustering	O(n^2)
Hierarchical agglomerative	O(n^2logn)
Lingo	O(n^3)

*Considering group-average linkage criterion

30.3.5 Discussion

Apart from the proposed algorithm's efficiency another key issue in our proposed methodology is setting the threshold values and number of feature vectors used in the algorithm. We have done an experiment to set the threshold values and number of feature vectors used in the algorithm.

Determination of Optimal Threshold Values and Number of Feature Vectors: We have conducted another experiment to set the threshold values and number of feature vectors used in our proposed algorithm. This experiment basically trains our algorithm using a classified set of data. From the total collection of five data set of Reuters from reut2-000 to reut2-004 and five data sets of our own data collection are used to set the threshold values. First, we calculate the Dunn index for the samples with given class. The threshold values α, β and γ can vary from 0 to 1. As the number of classes is known to us, we try to maximize Dunn index for a given number of classes. We consider the set of threshold values for

Table 30.4 Determination of threshold values and number of feature vectors

Document name	No. of doc	No. of Classes	DI (given classification)	Threshold values α	β	γ	No. of features	DI (proposed classification)
reut2-000	100	09	2.42	0.025	0.335	0.8	200	2.57
reut2-001	100	13	3.37	0.03	0.365	0.6	200	3.49
reut2-002	120	12	3.51	0.025	0.44	0.6	300	3.55
reut2-003	120	18	4.63	0.02	0.345	0.8	300	4.71
reut2-004	100	13	3.48	0.025	0.54	0.8	200	3.43
Culture Assam	100	15	2.89	0.045	0.46	0.7	300	3.65
Delhi guide	100	24	4.66	0.035	0.665	0.7	400	4.72
Market Kolkata	100	14	4.49	0.025	0.55	0.5	300	4.68
Royal Rajasthan	100	12	3.41	0.03	0.755	0.6	300	3.59
Wildlife India	100	14	4.57	0.035	0.765	0.8	300	4.61

Arithmetic mean $\alpha = 0.029$, $\beta = 0.522$, $\gamma = 0.690$, number of features = 280
Harmonic mean $\alpha = 0.028$, $\beta = 0.479$, $\gamma = 0.673$, number of features = 267

which the cluster number is closest with given class number and Dunn index is maximized. After considering all samples, the arithmetic mean and harmonic mean of individual thresholds and number of features are calculated. Using these average values we again find the Dunn index of the training sample. We have plotted the Dunn index for every sample considering optimal values as well as average values. The deviation of Dunn index for average threshold values compared to optimal threshold value. We find that the deviation of Dunn index for optimal values and average values is moderate. The difference is very less for arithmetic mean and harmonic mean values. We consider the harmonic mean values for testing purpose. The experimental result is summarized in Table 30.4.

30.4 Conclusion

With the advancement of Information Technology, the amount of web repository is increasing rapidly and this trend of expanding will continue in coming years. As a consequence, for a given user query, search engine jumble with a huge retrieval and it then becomes user's prudence to extract the right information. To tackle with this, web engineer advocates clustering of the results. A number of clustering techniques can cluster web documents. But the existing techniques either need user intervention to decide the number of clusters or computationally expensive. The present work addresses this limitation and proposes a new clustering approach. The proposed approach takes the advantage of both *k*-Means and hierarchical approaches. Our clustering technique is able to produce coherent clusters of good quality without compromising the computational overhead. The proposed clustering technique is useful to cluster a large number of web search results into a group of similarity in real-time. This grouping would then help users to direct their focus into their search of interest. The technique presented in this work follows some sub-tasks. We consider naive approaches for these sub-tasks which results the time complexity of the technique in $O(n^2)$. There is enough scope of improvement to reduce this time complexity considering more efficient ways of solving the tasks. We take up this investigation as the future work.

References

1. Duhan N, Sharma AK (2010) A novel approach for organizing web search results using ranking and clustering. Int J Comput Appl 5(10)
2. Christopher D, Manning PRHS (2008) Introduction to information retrieval. Cambridge University Press, Cambridge
3. Dipa D, Kiruthika M (2010) Mining access patterns using clustering. Int J Comput Appl 4(11):22–26
4. Carpineto C, Osinski S, Romano G, Weiss D (2009) A survey of web clustering engines. ACM Comput Surv 41(3)
5. Kanda Y, Kudo M, Tenmoto H (2009) Hierarchical and overlapping clustering of retrieved web pages. In: Advances in intelligent information systems, pp 345–358
6. Zamir O, Etzioni O (1999) Grouper: a dynamic clustering interface to web search results. Comput Netw: Int J Comput Telecommun Networking 31(11–16):1361–1374
7. Campos R, Dias G, Nunes C (2006) Wise: hierarchical soft clustering of web page search results based on web content mining techniques. In: Web intelligence'06. IEEE Computer Society Washington, DC, USA, pp 301–304
8. Osinski S (2003) An algorithm for clustering of web search result. Master's thesis, Poznan University of Technology, Poland
9. Giannotti F, Nanni M, Pedreschi D, Samaritani F (2003) Webcat: automatic categorization of web search results. In: Sistemi Evoluti per Basi di Dati, pp 507–518
10. Varlamis I, Stamou S (2009) Semantically driven snippet selection for supporting focused web searches. Data Knowl Eng 68(2):261–277
11. Zhang Z, Cui X, Jeske DR, Li X, Braun J, Borneman J (2010) Clustering scatter plots using data depth measures. In: DMIN'10, pp 327–333
12. Whissell JS, Clarke CLA (2011) Improving document clustering using Okapi BM25 feature weighting. Inf Retrieval 14:466–487
13. Eissen SMZ, Stein B (2002) Analysis of clustering algorithms for web-based search. In: Proceedings of the 4th international conference on practical aspects of knowledge management. Springer, London
14. Steinbach M, Karypis G, Kumar V (2000) A comparison of document clustering techniques. In: KDD workshop on text mining
15. Osinski S, Weiss D (2005) A concept-driven algorithm for clustering search results. IEEE Intell Syst 20(3):48–54

Chapter 31
Analysis of Multithreading in Java for Symbolic Computation on Multicore Processors

Pawan Raj Murarka, Motahar Reza and Rama Ranjan Panda

Abstract In this paper we have described the impact on efficiency of algebraic computation due to multi core systems using java as the programming language. Hence we had taken two machines with different specification having variants of Windows in them and made a comparative analysis taking five different input samples. During this process we came across several aspects on which the computation performance depends upon. In succeeding discussion we have given a vivid description of how these factors show variations when they are blended in different quantities thereby justifying the need of a robust algorithm and a high performance system for efficient computation of mathematical expressions with varying complexities.

Keywords Mathematical pseudo language (MPL) · JVM · Multi-threading

31.1 Introduction

Mathematical computation and performance have been two major issues of concern for scientists, scholars, students, teacher and developers. So we chose Java which was designed to meet real world requirements of creating an interactive interface and writing concurrent programs. Java's support for multithreaded programming is a sophisticated solution for making maximum utilization of multi-

P. R. Murarka (✉) · M. Reza (✉) · R. R. Panda
High Performance Computing Lab, School of Computer Science and Engineering, National Institute of Science and Technology, Berhampur, Odisha, India
e-mail: pawanrajmurarka@gmail.com

M. Reza
e-mail: nist_reza@yahoo.com

R. R. Panda
e-mail: pandaramaranjan@gmail.com

S. Sengupta et al. (eds.), *Emerging Trends in Computing and Communication*, Lecture Notes in Electrical Engineering 298, DOI: 10.1007/978-81-322-1817-3_31,

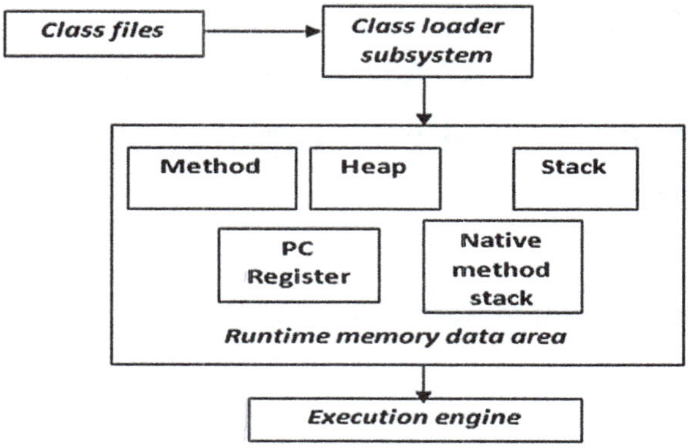

Fig. 31.1 Java virtual machine internal architecture

core processors by achieving usable parallelism with to have an efficient and interactive system. This is reduce wastage of CPU cycles [1] and make more utilization of multi-level caches. Multi-threading mechanism used by different languages on a multi-core Intel processor are different. It depends on the capability of programmer and programming language constructs to utilize the technology. Here we have chosen Java as our preferred language. It is a virtual machine which has stacked based architecture that provides the instruction level parallelism and is an add-on for a platform independent language.

Figure 31.1 shows a block diagram of the Java virtual machine that includes the major subsystems and memory areas. Each Java virtual machine has a class loader subsystem: a mechanism for loading classes and interfaces. It also has an execution engine: a mechanism responsible for executing the instructions contained in the methods of loaded classes. When a Java virtual machine runs a program, it needs memory to store many things, including byte code and other information it extracts from loaded class files, objects the program instantiates, parameters to methods, return values, local variables, and intermediate results of computations. Hence it organizes the memory it needs to execute a program into several runtime data areas as depicted in Fig. 31.2.

Since architecture determines resources utilization. Hence we can infer that performance of a multithreaded system is much dependent on its resource usage mechanism. Once such approach used is polling or event loop. In this model a thread of control runs in an infinite loop, polling a single event queue to decide what to do next. Once this polling mechanism return with a signal that a file is ready to read, then only the event loop dispatches the control to the appropriate event handler. Until this event handler returns nothing else can happen. This not only wastes a CPU cycles but also results in a part of program dominating the system. As a result the needed efficiency is not achieved. Now if the program is designed to run on multi-core CPU then we observe that application yields quick results. The reason is proper resource utilization with synchronized access. Other contributors to them are thread priority, scheduling algorithm, resource allocation, synchronization among processors and the probability of scheduling each thread on a different processor.

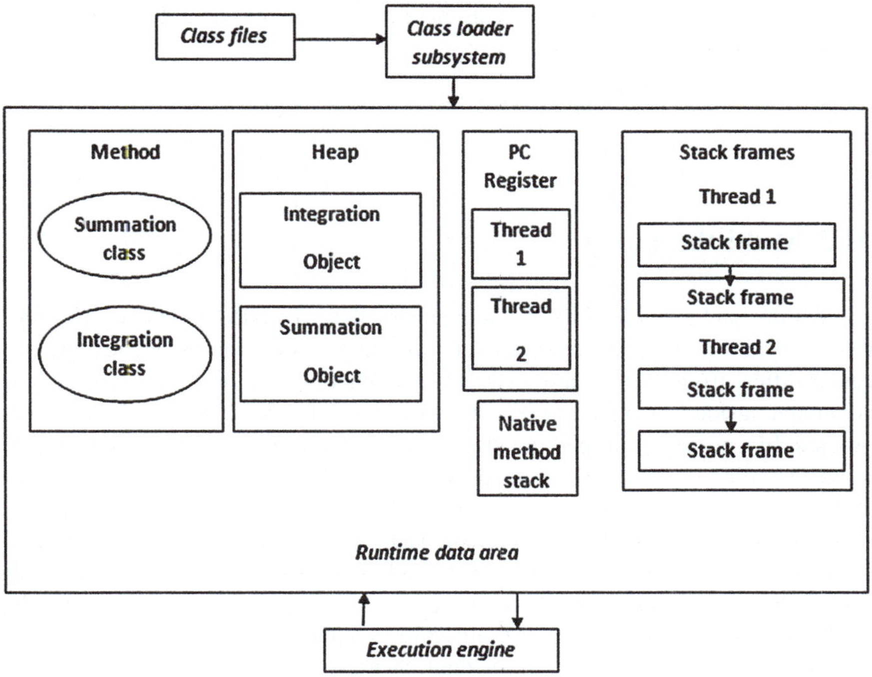

Fig. 31.2 Runtime data area usage by threads

31.2 Symbolic Computation

Symbolic computation is a Computer algebra system which takes the mathematical expression consisting of alphabets, numbers and symbolic notations as input. Further the expression is parsed using a recursive parser thereby forming lexemes having a key associated with it. Then these tokens are analyzed and prepared for different modules. The array of prepared strings is then concurrently executed using the java thread approach. Further the results calculated by different modules are collected. Finally the gathered results are scanned and accordingly displayed as output on the display area [2] in the user required format.

The above described mechanism has been successfully executed using Mathematical pseudo language (MPL) algorithms. MPL is an algorithmic language that describes the concepts and algorithms of computer algebra.

Expressions are constructed using the following set of symbols.

- Integers and Fractions that utilize rational number arithmetic.
- Identifiers that are used in expressions like summation, integration etc.
- The algebraic operators +, −, *, /, ^ (Power), and ! (Factorial) Some of the sub-expression constructed are as follows. sin(x), indefinite(x2 + sin(x),x),

log[b](expr), permu[n, r] etc. Using these a large expression is formed which is of follow kind.

$$f(x) := \sin(x) + \log_b\left(x^3 + x^2 + 1\right) + x^2 + {}^nP_2 \tag{31.1}$$

31.3 Multi-threading

A program can be divided into executable parts that are called as Threads [3]. Each thread defines its separate path of execution. Implementation of threads and processes differs from one operating system to another but one or more threads can be contained inside a single process. So thread is also known as a light weight process. In a process multiple threads can be created which run simultaneously. So multi-threading is known as a special form of multi-tasking. It also provides the abstract model for concurrent execution. With the help of multi-threading multiple parts of a program can be run simultaneously in each processor in order to optimize the execution time and efficiency by making maximum utilization of the CPU cycles. So it is always beneficial to use multi-threading concept to develop application for multiprocessor or multi-core system.

These concurrent programs have their own priority based on which they are dispatched for execution to the individual cores [4, 5]. This improves the efficiency of the process. Hence for this reason there will be creation of multiple threads working on a shared memory environment thereby making an optimum use of multi-core processors. An illustration of multithreading is as shown in Fig. 31.2.

31.4 Computation Algorithm

Extract Input

In this phase first the mathematical expression entered is converted to Mathematical Pseudo Language format [6, 7]. This is done using the following approach. Algorithm: To transform the mathematical expression in a suitable format following conversions take place.

- If input is Σ
- Add sum[delim, upper_limit, lower_limit]
- If it contains ${}_1\int^u$
- Add integrate[upper_limit, lower_limit]

The above can be explained by the following example.

$$\text{Phase 1}: \mathbf{F(x)} = \mathbf{log[2](log1(x))} \tag{31.2}$$

$$\text{Phase 2}: \mathbf{log_2(x^2 + x) \rightarrow log[2](log1(x))} \tag{31.3}$$

Hence we get,

$$\mathbf{f(x) := log[2](log(x))} \tag{31.4}$$

31.4.1 Parser

This class checks the input string for the correct programming language constructs. Example:

$$\mathbf{Expr \rightarrow (subexpression)} \tag{31.5}$$

31.4.2 String To Error

The output returned from the parser may contain errors. The errors are represented by error codes. String To Error class is used to extract the errors from the output string and shows the top five errors in the error stream in the output panel. Algorithm

- Tokenize the input string and store in a array named input
- Scan each token
- For i → 1 to length of tokens
- If it starts with ERC
- take the complete ERC code
- Check the code in database
- Retrieve the complete information and stores in a string
- Display the errors in error stream

Some of errors displayed are as follows sin(tan(90)) → Error: Parenthesis missing $^{n}p_{n+r}$ → Error: N value should be greater than r.

31.4.3 Evaluator

This class uses a divide and conquer approach [8, 9] towards multithreading. The main process is divided into smaller sub tasks. Now each task behaves as separate thread which is scheduled to run concurrently on cores. The threads share its heap and method area but the call stack, native method stack and pc registers are private to each thread. This can be demonstrated by the Figs. 31.3, 31.4, 31.5, 31.6 and 31.7.

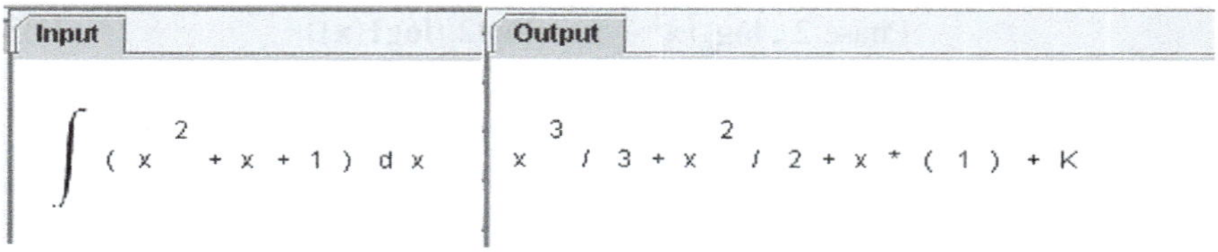

Fig. 31.3 Sample 1

Fig. 31.4 Sample 2

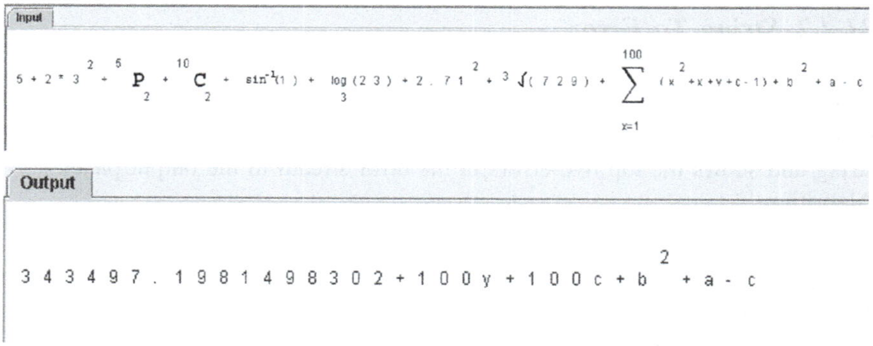

Fig. 31.5 Sample 3

Figure 31.2 depicts that the instance of the Java virtual machine has one method area and one heap. These areas are shared by all threads running inside the virtual machine. When the virtual machine loads a class file, it places this information into the method area. As the program runs, the virtual machine places all objects the program instantiates onto the heap. As each new thread comes into existence, it gets its own pc register (program counter) and Java stack. If the thread is executing a Java method (not a native method), the value of the pc register indicates the next instruction to execute. The state of a Java method invocation includes its local variables, the parameters with which it was invoked, its return value (if any), and intermediate calculations.

The state of native method invocations is stored in an implementation-dependent way in native method stacks, as well as possibly in registers or other implementation-dependent memory areas. The Java stack is composed of stack frames (or frames). A stack frame contains the state of one Java method

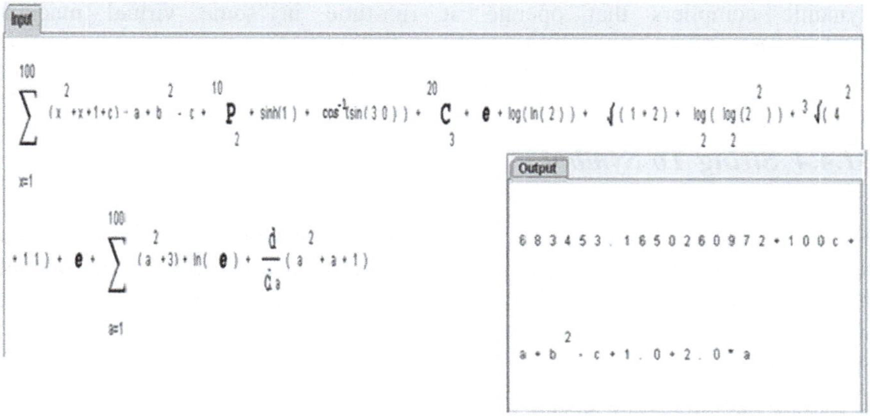

Fig. 31.6 Sample 4

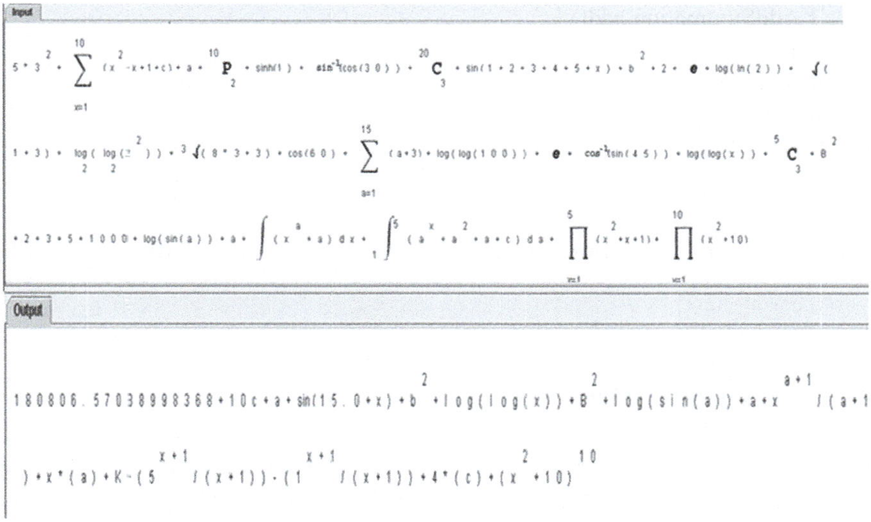

Fig. 31.7 Sample 5

invocation. When a thread invokes a method, the Java virtual machine pushes a new frame onto that thread's Java stack. When the method completes, the virtual machine pops and discards the frame for that method. The Java virtual machine has no registers to hold intermediate data values. The instruction set uses the Java stack for storage of intermediate data values. This approach was taken by Java's designers to keep the Java virtual machine's instruction set compact and to facilitate implementation on architectures with few or irregular general purpose registers. In addition, the stack-based architecture of the Java virtual machine's instruction set facilitates the code optimization work done by just-in-time and dynamic compilers that operate at run-time in some virtual machine implementations.

31.4.4 String To Symbol

String To Symbol class is used to convert a string which is in MPL (pseudo language) to Symbolic Expression. It takes two objects as input to the method. First one is the string and second one is the panel where the symbolic expression will be added.
Algorithm

- Extract the input from String
- Tokenize the input string and store in a array named input
- For i → 1 to length of component array
- If name is sum
- call AddSummation.add()
- If name is prod
- call AddProd.add()
- If name is ln
- call AddLn.add()

The above algorithm can be explained using the following example

$$\mathbf{combi[n,r] + defint[u,v](x\hat{}2+1) \rightarrow {}^{n}C_{r} +_{u} \int^{v}(x^{2}+1)dx} \tag{31.6}$$

31.5 Result and Analysis

A comparative analysis of a set of programs without multithreading and one involving was designed and tested on machines with different specifications using multiple samples. It was observed that when expressions length and complexity level were low as seen in samples depicted in Figs. 31.3, 31.4 and 31.5 then there was no remarkable difference in the execution time of two structurally different programs. Further when the complexity of the expression was increased as seen in Figs. 31.6 and 31.7, a distinguishing performance of structurally different programs (designed with serial and multithreaded code) were observed. Following were the results obtained (Table 31.1).

Table 31.1 Comparison between with and without multithreading on Machine 1 and Machine 2

Machine 1 specification			Machine 2 specification	
RAM: 4 GB(2.58 GB usable) PROCESSOR: Intel^{R} Core^{TM} i5-2400 CPU @ 3.10 GHZ 3.10 GHZ(Quad Core Processor)			RAM: 2 GB; PROCESSOR: Intel^{R} Core^{TM} 2 Duo CPU T6570 @ 2.10 GHZ 2.10 GHZ(2 Core Processor)	
System type: 32 bit operating system			System type: 32 bit operating system	
Platform: Windows 7 Professional			Platform: Windows 7 Ultimate	
Total virtual memory :5.71 GB			Total virtual memory: 3.93 GB	
	Parallel execution (ms)	Serial execution (ms)	Parallel execution (ms)	Serial execution (ms)
Sample 1	15	15	15	16
Sample 2	15	16	32	32
Sample 3	34	32	32	32
Sample 4	32	94	16	16
Sample 5	63	79	63	83

References

1. Perez JM, Sanchez LM, Garcia F, Calderon A, Carretero J (2005) High performance Java input/output for heterogeneous distributed computing. In: Proceedings of 10th IEEE symposium on computers and communications, 2005. ISCC 2005, pp 969–974, 27–30
2. Bull JM, Kambites ME (2000) JOMP—an OpenMP like Interface for java. In: Proceedings of the ACM 2000 conference on Java Grande, pp 44–53
3. Zhong H, Mehrara M, Lieberman S, Mahlkes S (2008) Uncovering hidden loop level parallelism in sequential applications. In: IEEE 14th international symposium on high performance computer architecture, 2008, pp 290–301
4. Shafi A, Carpenter B, Baker M (2009) Nested parallelism for multi-core HPC systems using Java. J Parallel Distrib Comput 69:532–545
5. Broquedis F, Diakhaté F, Thibault S (2008) Scheduling dynamic OpenMP applications over multicore architectures. vol 5004. pp 170–180
6. Cohen JS (2003) Computer algebra and symbolic computation: mathematical methods. AK Peters, Ltd, Library of Congress Cataloging in Publication Data, ISBN-156881-159-4
7. Mathematica 6.0, Wolfarm research. http://www.wolfram.com/mathematica/
8. Taboada GL, Tourino J, Doallo R (2007) High performance Java sockets for parallel computing on clusters. In: IEEE international symposium on parallel and distributed processing, 2007. IPDPS 2007, pp 1–8, 26–30
9. Launay P, Pazat JL (2001) Easing parallel programming for clusters with Java. Future Gener Comput Syst 18(2):253–263

Chapter 32
Hyper Object Data Model: A Simple Data Model for Handling Semi-Structured Data

Diptangshu Pandit, Nabendu Chaki and Samiran Chattopadhyay

Abstract This paper introduces a new data model based for handling semi-structured data extending the existing HyperFile data model that uses the hypertext notion of free-form objects connected by links. The HyperFile model provides flexibility to store semi-structured data which cannot be stored by relational database efficiently. However, the model suffers from some incompleteness in its definition towards proper representation and retrieval of stored objects. This makes it difficult to implement HyperFile data model towards storage and manipulation of data. The proposed data model, namely, HyperObject data model (HODM) overcomes this problem. An appropriate new query language, called HyperObject query language (HOQL) is also introduced to efficiently manipulate data stored using HODM. The paper also includes framework to implement the data model and its query language as a blade on top of the relational data model. This makes it easy to implement and solves the problem of handling heterogeneous semi-structured data at a higher level of abstraction on top of an underlying relational database.

Keywords Semi-structured · HyperFile · HyperObject · Middleware · Relational database

---

D. Pandit · S. Chattopadhyay
Information Technology, Jadavpur University, Kolkata, India
e-mail: diptangshu.max@gmail.com

S. Chattopadhyay
e-mail: samirancju@gmail.com

N. Chaki (✉)
Computer Science and Engineering, Calcutta University, Kolkata, India
e-mail: nchaki@gmail.com

S. Sengupta et al. (eds.), *Emerging Trends in Computing and Communication*, Lecture Notes in Electrical Engineering 298, DOI: 10.1007/978-81-322-1817-3_32,

32.1 Introduction

The semi-structured data is the information that does not fit well into relational databases or tables. The management of semi structured data is recognized as one of the major problems in the information technology industry as the variety and quantity of semi structured data found on World Wide Web has been increased dramatically in last few decades [1]. The conventional database query languages like SQL and OQL are inappropriate for querying semi-structured data as they are too constraining and rigid. Previously a few similar types of research work were done on modeling semi structured data and developing query languages. For example many query languages intending to manipulate semi-structured data are developed like GLASS [2], UMQL [3], WebSQL [4], WebOQL [5], WebLog [6], UnQL [7], W3QS [8], TQL [9], Lorel [10], Florid [11], etc. All those languages provide different ways of manipulating sub set of semi-structured data. Furthermore, some distributed approaches uses XML query languages and transformation sheet primarily for maintaining data warehouses [12, 13]. In the context of multiple databases and relational complex object some research is done with partially inconsistent information [14–16]. HyperFile data model [17] was introduced by Christopher Wade Clifton in context of manipulating partially structured data. We intend to solve the problem of storing partially structured data in relational database with the help of flexibility provided by HyperFile data model. Table 32.1 shows a brief summary of different existing data models [18] and our proposed HyperObject data model in terms of abstraction level, applicability and limitations. The reason why we choose to work on HyperFile data model is its simplicity and flexibility in storing diverse information for multiple object instances of the same entity type.

The reason why we choose to work on HyperFile data model is its very simple compared to the other works described above and yet the model is very flexible.

However, there exist severe limitations of HyperFile data model as we discovered while implementing it by developing a middleware layer on top of relational database. In this paper, we have proposed a simple data model, called Hyper Object data model (HODM) and implemented the same. HODM is implemented over existing relational database rather than creating a new file management system. This model also supports most of the basic operation of HyperFile and solves the difficulties of data representation and implementation faced while using original HyperFile data model.

Rest of the paper is organised as follows. Section 32.2 describes HyperFile data model and its problem that we found while implementing. In Sect. 32.3, the newly proposed HODM is introduced along with a suitable query language. Section 32.4 illustrates how it can be easily implemented on top of existing relational database and finally Sect. 32.5 concludes with problems still to be solved.

Table 32.1 Abstraction level, applicability and limitations of existing models

Data models	Abstraction level	Applicability and advantage	Limitations
Relational	Logical	Best suited for well-formed data	Unable to handle semi structured data
Object oriented	Physical/ logical	Flexible and handles nonconventional complex object interactions	Complex and time consuming search matching values
Graph	Logical/user	Well suited for complex connected data	Not useful for representing distinct sets of data
Network	Physical	Simple and easy to implement like hierarchical structures	Lacks structural dependence
HyperFile	Physical	Very simple and flexible	Lacks proper pointer representation and indexing mechanism
HODM (proposed)	Physical	Very simple and flexible	No indexing mechanism

32.2 The HyperFile Data Model and Its Limitations

The key idea in HyperFile data model is to see data as collection of objects which are linked in different ways. Data are browsed by traversing links. This key concept is same as Hypertext but the main difference is Hypertext is built on top of file system which usually do not provide data management facilities like indexing, concurrency control, recovery whereas HyperFile pledged to add those facilities either by hard coding on existing system or implementing HyperFile database on top of existing database management system that supports them.

The prime element in HyperFile data model is called a Triplet (/tuple/triple) which has a form <tuple_type, key, data>. Each tuple represents a property where tuple_type being type of property, data being the actual value of the property and key is the information that is used to isolate the property from same tuple type. Set of these tuples is called object/node. Objects can also be null set. The tuple schema will be stored in a table so that for each type of tuple the type of key and allowed data may differ. A brief summary of the HyperFile data model is:

```
Object <-- {Triple}
Triple <-- (Tuple_type, Key, Data)
Tuple_type <-- identifier
Key <-- Date | Numeric | String | Pointer
Data <-- Date | Numeric | String | Pointer| BLOB
```

The set above is an object (between two curly braces) which is composed of triplets (in between circular brackets). Triplets inside any object may link another object using pointer. Thus the database is formed like a structure of linked nodes/

objects which can be browsed traversing pointers. An example of such model is shown below

```
{
    (String , "Title", "MainProgramforSortRoutine" )
    (String , "Author", "3oeProgrammer" )
    (Text, "Description", <Arbitrarytextdescription>)
    (Text, "Code", <TextoftheProgram> )
    (Text, "ObjectCode", <Executableformodule> )
    (Pointer, "CalledRoutine", <Pointertoanotherobject>)
    (Pointer,                                  "Library",
"<Pointertoalibraryusedbythisroutine>)
}
```

While implementing HyperFile data model, the main problems are identified as:

Object Representation: Representing the object was the main problem in implementation. In HyperFile data model a tuple is represented by combining Tuple_type, Key and Data. This information does not give us any idea how the tuples are organized in an Object. That means if we consider a single tuple we will not find which Object it belongs to from the data that the tuple has. For example the triplet (Text, "Code", <TextoftheProgram>) does not specify the object it belongs to.

In the HyperFile data model, the triplets are grouped together in between two curly braces ({}). To implement the model in such a way we need to put all the triplets in order starting from start of the object to the end of object. The main problem of that idea is that if we need to change the object (i.e. insert some triplets or delete some triplet from the Object or just delete the Object) all the other triplets residing after the object that we are modifying must be shifted and thus the information regarding the Objects residing after the current object (which is being modified), must be changed. This creates a huge overhead in manipulation of triplets also insertion and deletion of triplets and destroying operation of objects become far more complex than it should be. In this representation if we do not shift the triplets after modification it will create fragmentation in the storage which is another major problem.

Object Linking: Without proper representation of Object structure it is impossible to use pointer. As described if we used a serial storage structure where the triplets are stored serially staring with a '{' and ending with a'}' the pointer must represent the location of the starting of the object that the pointer is pointing to. Now, if we update the object and suppose the object before it, was deleted we must shift all the objects after. If we do that the pointers will be invalid or we have to change each and every single pointer in the whole database which will also generate a lot of overhead. So, from here we can see the serial storage structure cannot be implemented efficiently. Linking can only be done when the object has proper representation and which is not dynamic in nature while the content of object may be dynamic.

The solution of the problems discussed above can be done adding some extra information to the triplets and changing the definition of tuples. The object information in each triplet will not only help the DBMS to organise objects but also will help to easily manipulate and link objects together. We introduce Hyper Object data model (HODM) which overcomes the difficulties that we have faced while implementing HyperFile data model as described above. The following section illustrates the proposed HODM along with its components and our proposed query language namely Hyper Object Query Language (HOQL) for efficient data manipulation best suited for HODM.

32.3 Hyper Object Data Model

32.3.1 The Model

HODM is built on the foundations of HyperFile data model. The extension over HyperFile model is aimed to make it simple and ready for implementation. The detail is explained below.

Quadruplet: The primary element in HyperFile data model is called a 'Quadruplet' is defined as the set {<Object_id>, <Data_type>, <Key>, <Data>}.

Where Object_id is id of object which the Quadruplet belongs to Data type is type of data to be stored. Key is some information about the data which will be used for searching purpose. Data is the actual data to be stored.

Object: Collection of Quadruplets forms an 'Object'. An object can have 0 or any number of Quadruplets. Here each Quintuplet stores some information about an object and as long as the type of data to be stored is supported by database. There is no restriction in adding large number of Quintuplet for a given object. One object can point to another object by using pointer.

By adding the extra information to each tuple it becomes easy now to represent object. Now as each of the tuple (quadruplet) has Object_id it describes which object it belongs to. So now it is not necessary to group the tuples or shift when any modification is made.

Pointing to another object becomes very easy as we can just create a tuple inside an object with pointer Data_type and put the Object_id of the object it is pointing to as Data of the tuple.

32.3.2 HyperObject Query Language

As we modified the HyperFile database we need to create a new query language which can efficiently manipulate data in HyperObject database.

HOQL: HyperObject Query Language (HOQL) is special type of query language developed for HyperObject database for ease of access of data stored in the database which has is using HyperObject data model. There are two types of query available on HyperObject database.

Create Command

The first type of query is to create objects or destroy objects. These are for defining the structure of the database or changing the same. The syntax of this type of query is given below

Create: +<object_id>;

The above syntax is used for creating an object. If the object already exists, it will notify with error message. Object id of a particular object is unique. If same object id already exists it will give an error.

Destroy Command

Destroy: −<object_id>;

The above syntax is used for removing an object. If the object does not exist, it will notify with error message.

Add Command

Another type of query is to add remove information stored in an object. It can also be used to change and search information inside an object or objects. The syntax of this type of queries is given below

Add: +<object_id> (<data_type>, <key>, <data>) [, (<data_type>, <key>, <data>)];*

This syntax can be used to add data to a specific object. The difference is here if the object does not exist or the same record exists in that object it will notify with error message.

Update Command

Update: ++<object_id> (<data_type>, <key>, <s_data>, <data>) [, (<data_type>, <key>, <s_data>, <data>)];*

This syntax can be used to add data to a specific object and will return the number of entries. If the object or the data to be updated do not exist it will notify with error message.

Delete Command

Delete: −<object_id> (<data_type>, <s_key>, <s_data>) [, (<data_type>, <s_key>, <s_data>)];*

This will delete the corresponding record and return no of triplets affected. If the object doesn't exist, it will notify with error message.

Union Command

Union: <object_id> + <object_id>;

This syntax will find union of the two objects. The output will be set of quadruplets. If the objects do not exist, it will notify with error message.

Intersection Command

*Intersection: <object_id> * <object_id>*;

This syntax will find intersection of the two objects. The output will be set of quadruplets. If the objects do not exist, it will notify with error message.

Difference Command

Difference: <object_id> – <object_id>;

This syntax will find difference of the two objects.The output will be set of quadruplets. If the objects do not exist, it will notify with error message.

Browse Command

Browse object: <object_id> [<level>];

This syntax will show all containing quadruplets of the object having the object_id and also add all containing quadruplets of linked objects up to <level>th level.

Filter Command

Filter: <object_id> (<s_data_type>, <s_key>, <s_data>);

The result from the filter operation will be an object which can be used as object in any other operation.

Key: It is a string which will be used to search the related data and also for indexing purpose.

s_key: Type of data to be stored with Special_operators.

data: Depends on data_type. According to data_type data is defined.

s_data: data with corresponding Special_operators.victim pass through the attacker's interface before getting forwarded to the victim. This allows the man-in-the-middle to change the traffic using filters.

32.4 Implementing HODM

We have implemented HyperObject database using a common relational database namely MySQL database.

32.4.1 Project Hierarchy

The hierarchy of project components is shown in Fig. 32.1. This has three main parts.

Front End: This fires HOQL queries. Also this will provide a graphical user interface to execute queries and display the result graphically.

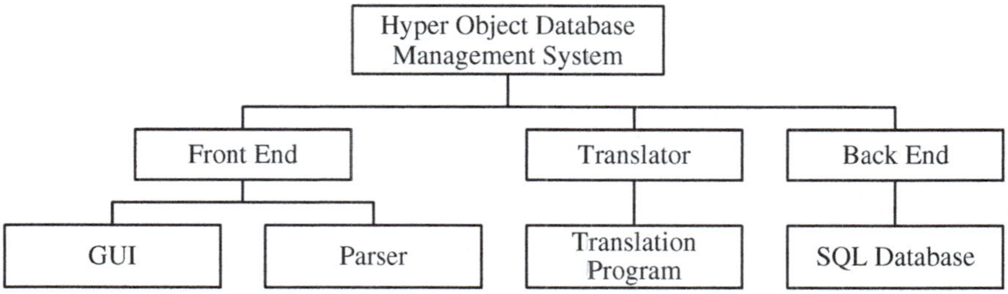

Fig. 32.1 Hierarchy of project components

Translator: Translator has a Translation program which translates input HOQL into SQL to execute on existing database and send the result from SQL in HOQL output format to rendering engine to display result.

Back End: Back End is the SQL engine (MySQL) to store and manipulate Hyper File data in traditional database format.

The Front End and Translator needs to reside in same node. But database can be in different node. Front End and Translator needs to be in application server in case of Web Applications.

32.4.2 Use Case Diagram

The use case diagram in Fig. 32.2 describes the interactions between the user and the system. User can either directly execute HOQL query through front end or can operate an Application which in turn executes HOQL queries through front end application. Front end after receiving HOQL parses it and calls according functions and pass the corresponding arguments to translator. Translator translates HOQL queries to SQL and execute on backend which is MySQL database.

Also the translator changes the result to according form and sends the result to the front end. The front end sends the result to either user or the application which is using HOQL.

32.4.3 Illustrative Example

The major motivation towards proposing HODM is to improve on the limitations of HyperFile data model. In this section, an example has been considered to verify the effectiveness of the proposed HODM over HyperFile model. Let's consider the following simple example of HyperFile data model entry

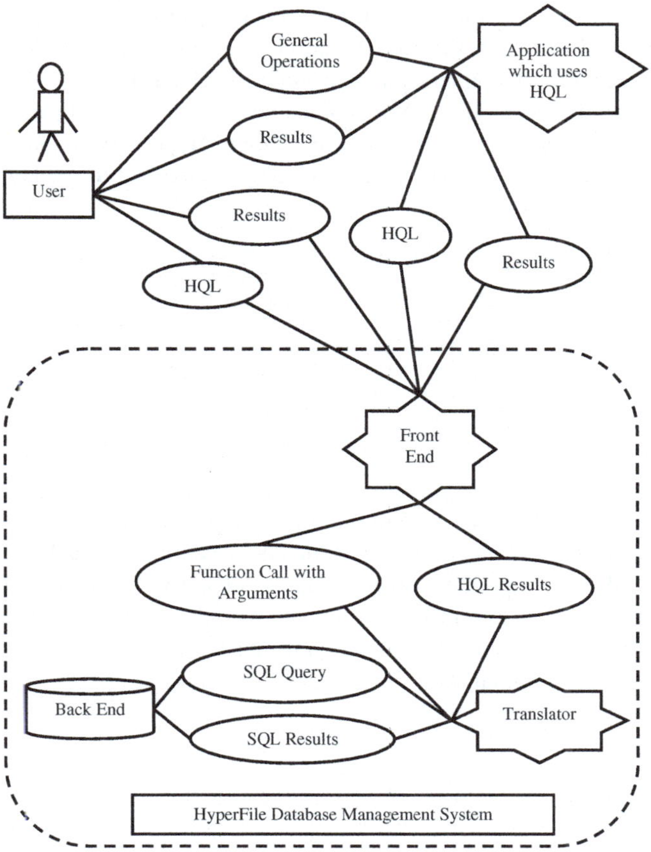

Fig. 32.2 Interaction between user and system

```
{
(String , "Title", "MainProgramforSortRoutine")
(String , "Author", "3oeProgrammer")
(Text, "Code", <TextoftheProgram>)
(Text, "ObjectCode", <Executableformodule>)
(Pointer,"CalledRoutine",<Pointertoanotherobject>)
}
{
(String, "Title", "AnotherSubroutine")
(String, "Author", "AnotherProgrammer")
(Text, "Code", <TextoftheProgram1>)
(Text, "ObjectCode", <Executableformodule>)
}
```

Looking at the above example, one can clearly see that two HyperFile objects are represented. The first object points to the second object (Pointer, "Called-Routine", <Pointertoanotherobject>). Implementation of the same is quite challenging. It is not possible to point to another object by pointing to its memory location. Specially, if the memory location is changed then it will be a huge problem to change all pointers that are affected. HODM solves this by adding unique object id for each object. Now pointers can point to any object even after its modified as object id is never changed according to HyperObject data model. Take a look at the following HODM example which is equivalent to the HyperFile data model entry stated above.

```
(obj01, String , "Title", "MainProgramforSortRoutine")
(obj01, String , "Author", "3oeProgrammer")
(obj01, Text, "Code", <TextoftheProgram>)
(obj01, Text, "ObjectCode", <Executableformodule>)
(obj01, Pointer, "CalledRoutine", obj02)
(obj02, String, "Title", "AnotherSubroutine")
(obj02, String, "Author", "AnotherProgrammer")
(obj02, Text, "Code", <TextoftheProgram1>)
(obj02, Text, "ObjectCode", <Executableformodule>)
```

It is easily understood from here that modifying location of an objection will not affect pointers. Another problem that might appear in HyperFile data model is if you delete the second object in HyperFile entry example. In such a scenario, the first object will have a pointer triplet but target object will not exist. This has not been mentioned or handled in HyperFile data model. HODM solves this problem by deleting pointer entries while destroying the target object so that no pointer object may appear in database.

32.5 Conclusion

The proposed HODM and its implementation on top of an underlying relational database that we have presented in this paper provide a simple solution to semi structured data manipulation using traditional relational databases. This work can be extended by developing proper indexing mechanism exploiting the power of relational database. Besides, query optimization is another issue that needs some research attention in future. The proposed model holds the potential for works like tweaking query structure or adding extended operations to get better result.

References

1. Chakraborty S, Chaki N (2011) A survey on the semi-structured data models. In: Springer proceedings of the 10th international conference on computer information systems and industrial management applications (CISIM-2011), Kolkata, 14–16 Dec 2011, pp 257–266
2. Ni W, Ling TW (2003) GLASS: a graphical query language for semi-structured data. In: Proceedings of international conference on database systems for advanced applications (DASFAA)

3. Cao Z, Wu Z, Wang Y (2007) UMQL: a unified multimedia query language. In: Proceedings of the 3rd international IEEE conference on signal-image technologies and internet-based system (SITIS 2007), Shanghai, pp 109–115
4. Mendelzon A, Mihaila G, Milo T (1996) Querying the world wide web. In: Proceedings of the 1st international conference on parallel and distributed information system, pp 80–91
5. Arocena G, Mendelzon A (1998) WebOQL: restructuring documents, databases and webs. In: Proceedings of the international conference on data engineering. IEEE Computer Society, pp 24–33
6. Lakshmanan LVS, Sadri F, Subramanian IN (1996) A declarative language for querying and restructuring the web. In: Proceedings of the 6th international workshop on research issues in data engineering
7. Buneman P, Davidson S, Hilebrand G, Suciu D (1996) A query language and optimization techniques for unstructured data. In: Proceedings of the ACM SIG-MOD international conference on management of data, pp 505–516
8. Shmueli O, Konopnicki D (1995) W3QS: a query system for the world-wide web. In: Proceedings of the international conference on very large data bases, Zurich, Switzerland. Morgan Kaufmann Publishers, Inc., pp 54–65
9. Cardelli L, Ghelli G (2004) TQL: a query language for semistructured data based on the ambient logic. Math Struct Comput Sci 14:285–327
10. Abiteboul S, Quass D, McHugh J, Widom J, Wiener JL (1997) The Lorel query language for semistructured data. Int J Digit Libr 1(1):68–88
11. Himmeroder R, Lausen G, Ludascher B, Schlepphorst C (1997) On a declarative semantics for web queries. In: Proceedings of the international conference on deductive and object-oriented databases, Switzerland. Springer LNCS, pp 386–398
12. Tseng F, Chen C (2005) Integrating heterogeneous data warehouses using XML technologies. J Inf Sci 31(3):209–229
13. Niemi T, Niinimäki M, Nummenmaa J, Thanisch P (2002) Constructing an OLAP cube from distributed XML data. In: Proceedings of 5th ACM international workshop data warehousing and OLAP (DOLAP 2002), pp 22–37
14. Motro A, Rakov I (1996) Estimating the quality of data in relational databases. In Proceedings of the 1996 conference on information quality, pp 94–106
15. Liu M, Ling TW (2000) A data model for semistructured data with partial and inconsistent information. In: Proceedings of the international conference on advances in database technology (EDBT 2000), Konstanz, Germany, 27–31 March 2000, pp 317–331
16. Bancilhon F, Khoshafian S (1989) A calculus for complex objects. J Comput Syst Sci 38(2):326–340
17. Clifton C, Garcia-Molina H, Bloom D (1995) HyperFile: a data and query model for documents. VLDB J 4(1):45–86
18. Angles R, Gutierrez C (2008) Survey of graph database models. ACM Comput Surv (CSUR) 40(1):1–39

Part V
Nanotechnology

Chapter 33
A Novel Carbon Nanotube Field Effect Transistor Based Analog Signal Processing Circuits for Low-power Communication Systems

P. A. Gowrisankar and K. Udhayakumar

Abstract There is a need to explore circuit application in new emerging technologies for their rapid commercialization as the CMOS technology is approaching its limits. Carbon Nanotube Field-Effect Transistor (CNFET) is a promising candidate for future electronic devices for low-power low-voltage digital or analog circuit application. In this paper, we presented a low-power, low-voltage CNFET operational amplifier (OPAMP) based signal processing circuits such as Half-wave rectifier, Full-wave rectifier, Clamper, Clipper, Comparator and Peak detector for a low-power communications system design application. The proposed signal processing circuits operation are studied by using HSPICE software for circuit simulation at 0.9 V input supply voltage. Simulation results show that the proposed signal processing circuits well suited for low-power low-voltage communication application for their lower power consumption, high speed operation and high frequency response.

Keywords Analog circuits · CNFET · OPAMP · Signal processing circuits

33.1 Introduction

CMOS technology is approaching its limits in the presence of challenges like extreme short-channel effects, lithographic limitations, process variations, leakage current and source-to-drain tunneling [1]. Many technological and device structure variations have been proposed in the literature like single electron transistor (SET) [2], FinFETs [3], and CNFETs [4, 5] etc. to provide improvements in electrostatics over CMOS. Among that, Carbon nanotubes (CNTs) field-effect transistors are one

P. A. Gowrisankar (✉) · K. Udhayakumar
Department of Electrical Engineering, College of Engineering Guindy, Anna University, Chennai 600025 Tamilnadu, India
e-mail: gowrisankar_eee@yahoo.co.in

S. Sengupta et al. (eds.), *Emerging Trends in Computing and Communication*, Lecture Notes in Electrical Engineering 298, DOI: 10.1007/978-81-322-1817-3_33,

of the new devices for designing low-power and high-performance circuits [5]. CNTs have special electronic and mechanical properties [6] that make them attractive for the future integrated circuit applications. Transistors with carbon nanotubes as their channel are called Carbon Nanotube Field Effect Transistor (CNFET). Carbon nanotube based transistor has significant potential to replace CMOS in the future due to its better electrostatics and higher mobility [5]. Utilizing the physics of nano-scale devices directly for circuit application is quite appealing. It not only enhances the functionality per device but also allows us to integrate a profusion of high-functionality devices in a small chip area.

Operational amplifiers (OPAMPs) are key elements of the analog and mixed signal circuit. Designing high-performance analog integrated circuits is becoming increasingly exigent with the relentless trend toward reduced supply voltages. The DC and AC performance of a CNFET OPAMP has already been analyzed and measured [7–9]. There has been a lot of work available in the literature on the digital circuit applications [10] of CNFET but its analog applications have not been explored. This paper present the first time CNFET operational amplifier (OPAMP) based analog signal processing circuits (such as Half-wave rectifier, Full-wave rectifier, Clamper, Clipper, Comparator and Peak detector) to propose their suitability in a wide range of future high performance, low-power analog system applications such as signal processing, remote sensing, portable bio-instrumentation. It begins with an overview of CNFET technology in Sect. 33.2. Section 33.3 covers the optimum design of CNFET operational amplifier. Section 33.4 presented the proposed analog signal processing circuits based on CNFET OPAMP. Section 33.5 presented the simulation results of CNFET OPAMP and the proposed analog signal processing circuits. In Sect. 33.6 conclude the paper.

33.2 CNFET Technology

Carbon Nanotube (CNT) is a sheet of graphite which is rolled up along a wrapping vector. The Single-walled carbon nanotube (SWCNT) could be metallic or semiconducting, depending on its chirality vector, which is determined by (n, m) indices and specify the arrangement angle of the carbon atoms along the nanotube. If $n - m = 3k$ ($k \varepsilon Z$), the SWCNT is conducting and otherwise it is semiconducting [5]. In Carbon Nanotube Field Effect Transistors (CNFETs) one or more semiconducting SWCNTs are used as the channel of the device as shown in Fig. 33.1.

To design a circuit with best performance based on an average power consumption and speed, it is very important to determine the threshold voltage because this affects the switching speed, the current and leakage power. Similar to a MOSFET device. A great advantage of CNFET is that its threshold voltage can be adjusted by changing the diameter of its CNTs. This practical characteristic makes CNFET more flexible than MOSFET [11] for designing digital circuits.

Fig. 33.1 Schematic diagram of a CNFET

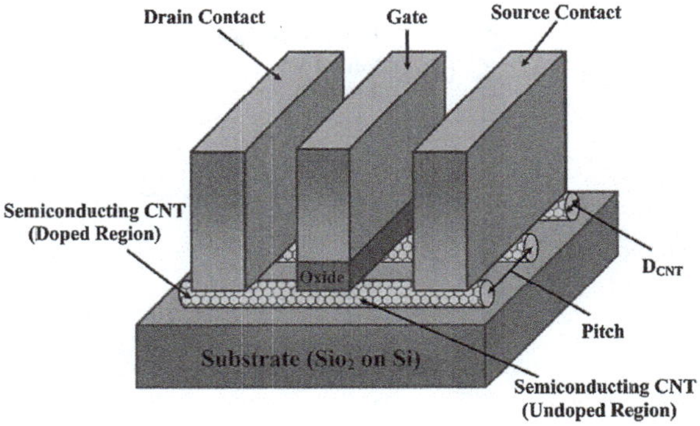

$$D_{CNT} = \frac{a * \sqrt{n_1^2 + n_2^2 + n_1 n_2}}{\pi} \qquad (33.1)$$

Where a = 2:49°A is the lattice constant. Since the bandgap of semiconducting CNTs is proportional to the diameter, then the threshold voltage of the intrinsic CNT channel can be approximated to the first order as the half bandgap (which is inversely proportional to the diameter). By adjusting the diameter, the threshold voltage can be controlled and is given by Eq. (33.2).

$$V_{th} = \frac{E_g}{2 * e} = \frac{\sqrt{3}}{3} \frac{a * V_\pi}{e * D_{CNT}} = \frac{0.43}{D_{CNT}(\text{nm})} \qquad (33.2)$$

D_{CNT} is the CNT diameter, e is the unit electron charge, $V\pi = 3.033$ eV is the carbon n–n bond energy in the tight bonding model, and a = 2.49 A is the carbon-to-carbon atom distance. For example, the threshold voltage of the CNFETs that use (19, 0) CNTs as channels is 0.289 V because the D_{CNT} of a (19, 0) CNT is 1.49 nm. The threshold voltage of the CNFET is inversely related to the CNT chirality vector. The current–voltage (I–V) characteristic of the CNFET are shows in the Fig. 33.2.

33.3 Design of Two Stage CNFET OPAMP Circuit

In this section, we implemented the operational amplifiers based on the CNFET technology at 32 nm node. CNFET based a two-stage Miller compensated operational amplifier is shown in Fig. 33.3. The proposed circuit is designed in terms of optimum structural device parameters namely Number of nanotubes (N), diameter of CNT, uniform inter-nanotube spacing pitch (S) and input supply voltage (V). The main design issue for the analog designer, when dealing with a CNFET technology, is the sizing of the transistors in order to achieve the desired electrical characteristics. The proposed operational amplifier optimum design

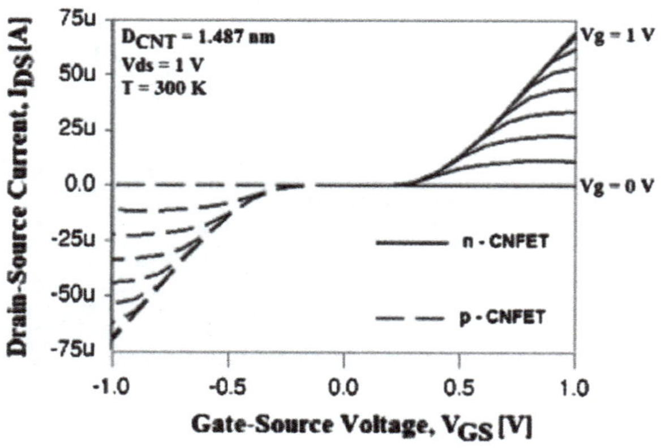

Fig. 33.2 I_{DS}–V_{GS} Characteristics of CNFET

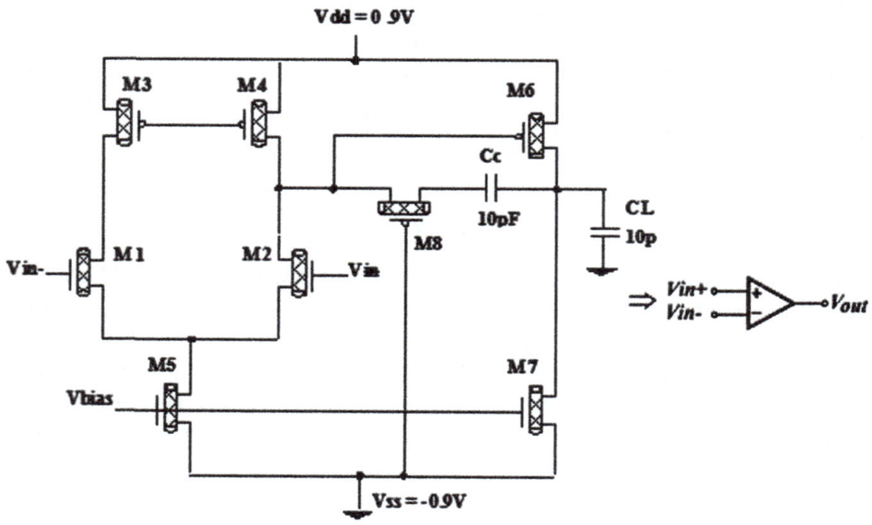

Fig. 33.3 Two stage CNFET OPAMP with miller compensation

parameter values are shown in Table 33.1. More detail about the CNFET OPAMP and its characteristics are discussed in the literature [7–9].

33.4 CNFET OPAMP Based Signal Processing Circuits

In this section, we designed the CNFET operational amplifier (OPAMP) based analog signal processing circuits such as half-wave rectifier, full-wave rectifier, clamper, clipper, comparator and peak detector. Most of the circuits presented in

Table 33.1 Optimized circuit parameters and its value of CNFET OPAMP [9]

Parameter	32 nm CNFET design	
	DCNT = 1.5 nm	
	N	S
M1 and M2	45	12
M3 and M4	22	12
M5	92	12
M6	257	8
M7	128	8
M8	3	12
IBias	1.81 μA	
VDD = [−VSS]	0.9 V	
Cc	3 pF	
C_L	10 pF	

this section serve to condition analog signals for subsequent input to another circuit. Many of them could be categorized as wave shaping or conditioning circuits. In this paper, we are examine circuits that can rectify low amplitude signals, limit the maximum excursion of signals, and change the DC level of waveforms. Many of the circuits are quite simple in terms of component count, but they play important roles in overall communication systems design. The designed analog signal processing circuits operate well at the input supply voltage V_{DD} to V_{SS} at room temperature T = 30 °C.

33.4.1 CNFET OPAMP Based Rectifier Circuit

33.4.1.1 Hall-Wave Rectifier

Figure 33.4a shows an ideal half-wave rectifier circuits. During the positive half cycle of the input waveform, the opamp goes positive and turns ON the diode (D1). The circuit then acts as a conventional non-inverting amplifier, and the positive half-cycle waveform appears across the output terminal (Vo). On the other hand, during the negative half cycle of the input waveform, the opamp output goes negative and turns OFF the diode (D1). Since the diode is open, no voltage appears across the output terminal (Vo). The overall result is perfect half-wave rectification, as represented by the transfer characteristic in Fig. 33.5a.

33.4.1.2 Full-Wave Rectifier

Figure 33.4b shows an ideal full-wave rectifier circuits. During the positive half cycle of the input waveform at A, the output of opamp A2 will go positive, mining diode D2 ON. A virtual short circuit will thus be established between the two input

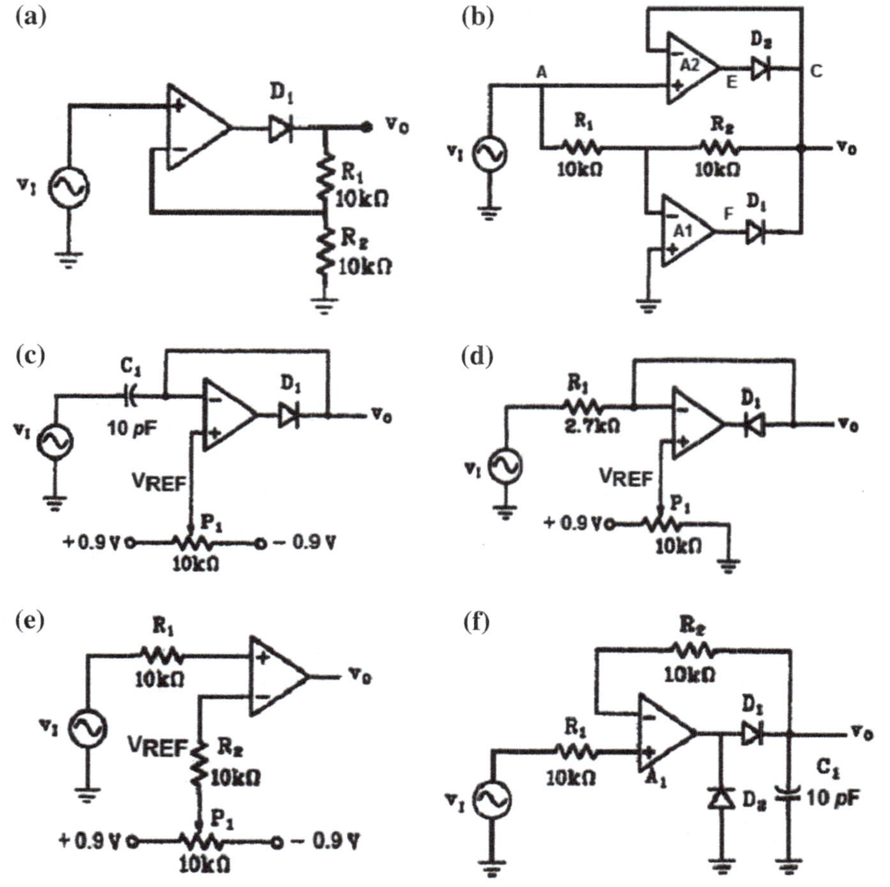

Fig. 33.4 CNFET OPAMP based analog signal processing circuits **a** Half-wave Rectifier, **b** Full-wave Rectifier, **c** Clamper, **d** Clipper, **e** Comparator, and **f** Peak Detector

terminals of opamp A2, and the voltage at the negative-input terminal, which is the output voltage of the circuit, will become equal to the input. Thus no current will flow through R1 and R2, and the voltage at the inverting input of opamp A1 will be equal to the input and hence positive. Therefore the output terminal (F) of opamp A1 will go negative until A1 saturates. This causes D1 to be turned OFF. Next consider, during the negative half cycle of the input waveform at A. The tendency for a negative voltage at the negative input of opamp A1, causes F to rise, making diode D1 conduct. Thus a virtual ground appears at the negative input of Al and the two equal resistances R1 and R2 force the voltage at C, which is the output voltage, to be equal to the negative of the input voltage at A and thus positive. The combination of positive voltage at C and negative voltage at A causes the output of A2 to saturate in the negative direction, thus keeping D2 OFF. The overall result is

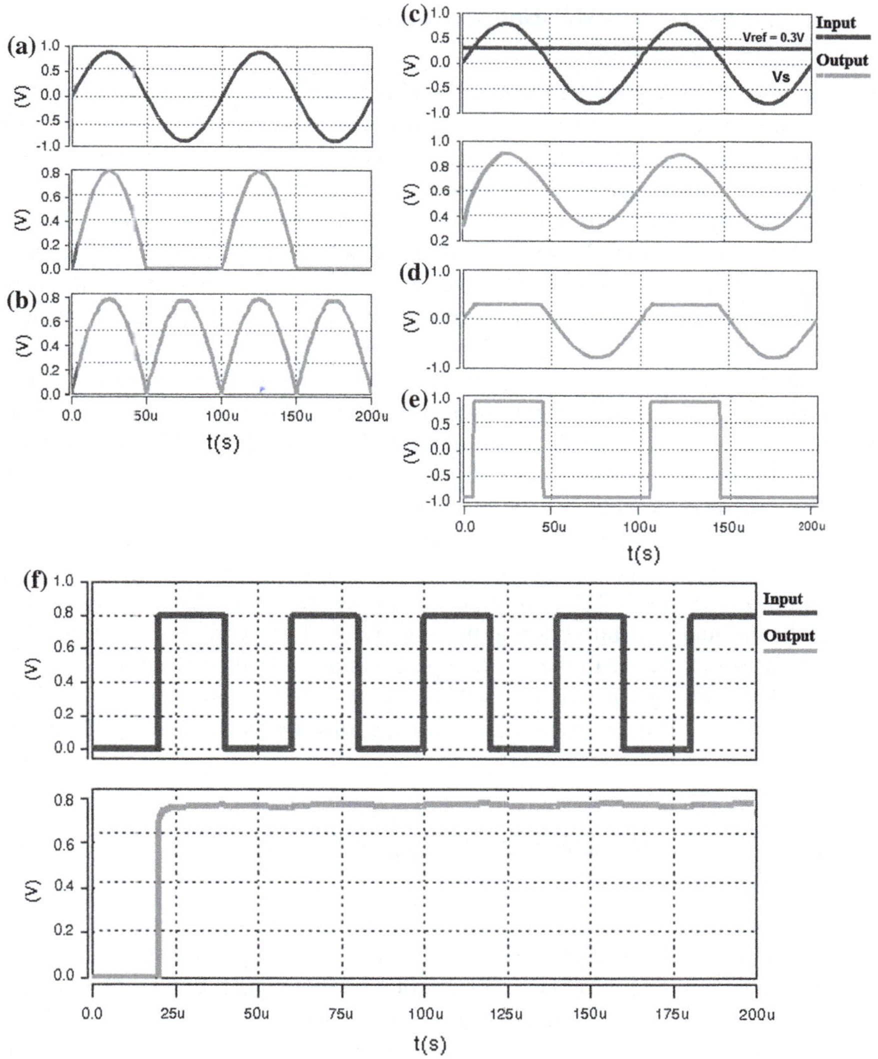

Fig. 33.5 Simulation output of the CNFET OPAMP based analog signal processing circuits **a** Half-wave Rectifier, **b** Full-wave Rectifier, **c** Clamper, **d** Clipper, **e** Comparator, and **f** Peak Detector

perfect full-wave rectification, as represented by the transfer characteristic in Fig. 33.5b. This precision is, of course, a result of placing the diodes in opamp feedback loops, thus masking their non-idealities. This circuit is one of the possible precision full-wave rectifier circuits.

33.4.2 CNFET OPAMP Based Clamper Circuit

In clamper circuits, a predetermined dc level is added to the input voltage. In other words, the output is clamped to a desired dc level. If the clamped dc level is positive, the clamper is called a positive clamper. On the other hand, if the clamped dc level is negative, it is called a negative clamper. The other equivalent terms for clamper are dc inserter or dc restorer. A clamper circuit with a variable dc level is shown in Fig. 33.4c. Here the input wave form is clamped at positive reference voltage (+Vref) and hence the circuit is called a positive clamper. The output voltage of the clamper is a net result of ac and dc input voltages applied to the inverting and non-inverting input terminals respectively. Therefore, to understand the circuit operation, each input must be considered separately. First, consider Vref at the non-inverting input. Since this voltage is positive, the output voltage (Vo) is positive, which forward biases diode D1. This closes the feedback loop and the op-amp operates as a voltage follower. This is possible because C1 is an open circuit for dc voltage. Therefore Vo = Vref. As for as voltage Vin at the inverting input is concerned during its negative half-cycle D1 conducts, charging C1 to the negative peak value of the voltage (VP). However, during the positive half-cycle of Vin diode D1 is reverse biased and hence the peak voltage (VP) across the capacitor acquired during the negative half-cycle is retained. Since this voltage VP is in series with the positive peak voltage VP, the output peak voltage VO = 2 VP. Thus the net output is Vo = Vref + VP, so the negative peak of 2 VPb is at Vref. The input and output wave forms are shown in Fig. 33.5c.

33.4.3 CNFET OPAMP Based Clipper Circuit

Clipper is a circuit that is used to clip off (remove) a certain portion of the input signal to obtain a desired output wave shape. In op-amp clipper circuits, a rectified diode may be used to clip off certain parts of the input signal. Figure 33.4d shows an active positive clipper, a circuit that removes positive parts of the input signal. The clipping level is determined by the reference voltage (Vref). When the Vref is ZERO and the non-inverting input is grounded. When Vin goes positive, the error voltage drives the op-amp output negative and turns ON the diode. This means the final output VO is ZERO (same as Vref) for any positive value of Vin. When Vin goes negative, the op-amp output is positive, which turns OFF the diode and opens the loop. When this happens, the final output Vo is free to follow the negative half cycle of the input voltage. This is why the negative half cycle appears at the output. To change the clipping level, all we do is adjust Vref as needed.

33.4.4 CNFET OPAMP Based Comparator Circuit

A comparator, as its name implies, compares a signal voltage on one input of op-amp with a known voltage called the reference voltage on the other input. Figure 33.4e shows an opamp comparator used as a voltage level detector. Consider a fixed reference voltage Vref of +0.3 V is applied to the inverting input and the other time varying signal voltage Vin is applied to the non-inverting input. Because of this arrangement, the circuit is of non-inverting type. When Vin is less than Vref, the output voltage Vo is −Vsat because the voltage at the inverting input is higher than Vin is greater than non-inverting input. On the other hand, when Vin is greater than Vref, Vo goes to +Vsat. This Vo changes from one level to another level whenever Vin = Vref as shown in Fig. 33.5e. At any given time, the circuit shows whether Vin is greater than or less than Vref. The circuit is hence called a Voltage level detector. The comparators are interface circuits between analog and digital domains, converting a continuous linear analog signal into a two-state digital signal. Comparators are used in circuits such as Digital interfacing, Schmitt triggers, Discriminators, Voltage level detectors, Oscillators, etc.

33.4.5 CNFET OPAMP Based Peak Detector Circuit

A Square, saw-tooth, pulse and triangular waves are typical examples of non-sinusoidal wave forms. A conventional AC voltmeter cannot be used to measure the RMS value of the pure sine wave. One possible solution for this problem is to measure the peak values of the non-sinusoidal wave forms. Figure 33.4f shows a peak detector that measures the positive peak values of the input. During the positive half-cycle of the input wave form Vin, the output of the opamp drives diode D1 ON charging capacitor C1 to the positive peak value VP of the input voltage Vin. Thus when diode D1 is forward biased, the opamp acts as a voltage follower. On the other hand, during the negative half-cycle of the input wave form Vin, the diode D1 is reverse biased, and the voltage across C1 is retained. The only discharge path for C1 is through output terminal Vo.

33.5 Simulation Results

In the first section simulation setup and its parameter values are discussed. In the second section the CNFET OPAMP characteristics are studied. In the third section the proposed CNFET OPAMP based analog signal processing circuits simulation output and their power consumption are presented.

Table 33.2 CNFET model parameters

Parameter	Description	Value
Lch	Physcial channel length	32 nm
Lgeff	The mean free path in the instrinsic CNT Channel	100 nm
Lss	The Length of doped CNT source-side extension region	32 nm
Ldd	The length of doped CNT drain-side extension region	32 nm
Kgate	The dielectric constant of high-k top gate dielectric material	16
Tox	The thickness of high-k top gate dielectric material	4 nm
Csub	The coupling capacitance between the channel region and the substrate	20 pF/m

33.5.1 Simulation Setup

In this section, the CNFET operational amplifier and analog signal processing circuits are comprehensively evaluated at operating input supply voltage. All the circuits are simulated using Synopsys HSPICE 2007 simulator tool with Compact SPICE model for CNFET [12] at 32 nm technology. The CNFET standard model has been designed for unipolar, MOSFET-like CNFET device [12, 13] in which each transistor may have one or more CNTs. This model also considers Schottky Barrier Effects, Parasitics, including CNT, Source/Drain, and Gate resistances and capacitances and CNT Charge Screening Effects. The parameters of the CNFET model and their values, with brief descriptions, are shown in Table 33.2.

33.5.2 Performance Characteristics of the CNFET OPAMP

The performances characteristics and robustness of the CNFET based operational amplifier are examined extensively in terms of the OPAMP circuit specifications, such as open loop gain (Av0), unity gain bandwidth, slew rate (SR), phase margin (PM), common-mode rejection ratio (CMRR), power supply rejection ratio (PSRR), slew rate (SR), output swing (OS) and power consumption as shown in the Table 33.3. More detail about the CNFET OPAMP and its characteristics are discussed in my previous paper [9].

33.5.3 Simulation Output of the Analog Signal Processing Circuits

The simulation result of the proposed CNFET OPAMP based analog signal processing circuits such as half-wave rectifier, full-wave rectifier, clamper, clipper, comparator and peak detector are presented in the Fig. 33.5. The performance of the proposed CNFET technology based signal processing circuits are compared

Table 33.3 CNFET OPAMP performance specifications [9]

Parameter	32 nm CNFET OPAMP
DC gain (dB)	49.12
−3 dB frequency (MHz)	198
Phase margin (degree)	48
CMRR (dB)	52.45
PSRR (dB)	54.35
Output swing (OS)	±0.9
Settling time (ns)	0.75
Slew rate (V/μs)	5841.2
Output resistance (kΩ)	67
Power dissipation (μW)	13

Table 33.4 Power consumption of the CNFET OPAMP based analog signal processing circuits

Analog signal processing circuits	0.18 μm CMOS technology Power consumption (μW)	32 nm CNFET technology Power consumption (μW)
Half-wave rectifier	1,256	42.44
Full-wave rectifier	1,421	78.09
Clamper	969	20.80
Clipper	1,011	50.82
Comparator	893	38.96
Peak detector	1,103	50.60

with 0.18 μm CMOS technology based analog signal processing circuits in term of power consumption. The proposed and conventional analog signal processing circuit's power consumption are shown in Table 33.4. Thus the proposed analog signal processing circuit characteristics are studied using state-of-the-art 32 nm CNFET technology. Simulation results show that the proposed signal processing circuits well suited for low-power low-voltage signal processing application for their lower power consumption. The conventional 0.18 μm CMOS technology based analog signal processing circuits and its parameters values are discussed already in [14].

33.6 Conclusion

In this paper, we presented a low-power, low-voltage CNFET operational amplifier (OPAMP) based analog signal processing circuit such as half-wave rectifier, full-wave rectifier, clamper, clipper, comparator and peak detector for low-power communication system design application. Proposed circuits are simple and much efficient than that their conventional CMOS based circuits especially for low-power analog circuit application. From the simulation result, an analog circuit application aspects of CNFET as a promising nano device. The CNFET motivates future works in the field of analog mixed signal integrated circuit application.

References

1. Thompson SE, Parthasarathy S (2006) Moor's law the future of Si microelectronics. Mater Today 9:20–25
2. Durrani ZAK (2009) Single-electron devices and circuits in silicon. World Scientific, Singapore, p 285
3. Kim JJ, Roy K (2004) Double gate-MOSFET subthreshold circuit for ultralow power applications. IEEE Trans on Electron Devices 51(9):1468–1473
4. McEuen PL, Fuhrer MS, Park H (2002) Single-walled carbon nanotube electronics. IEEE Trans Nanotechnol 1:78–85
5. Appenzeller J (2008) Carbon nanotubes for high-performance electronics—progress and prospect. Proc IEEE 96(2):201–211
6. Gowri Sankar PA, Udhayakumar K (2011) Mechanical and electrical properties of single walled carbon nanotubes: a computational study. Eur J Sci Res 60(3):342–358
7. Ali Usmani F, Hasan M (2010) Carbon nanotube field effect transistors for high performance analog applications: an optimum design approach. Microelectron J 41:395–402
8. Lewyn LL, Ytterdal T (2009) Analog circuit design in nanoscale CMOS technologies. Proc IEEE 97(10):1687–1714
9. Gowri Sankar, PA, Udhaya Kumar K (2013) Design and analysis of two stage operational amplifier based on emerging sub-32 nm technology. In: Advanced nanomaterials and emerging engineering technologies (ICANMEET 2013) international conference on, pp 664–668
10. Kim YB (2011) Integrated circuit design based on carbon nanotube field effect transistor. Trans Electr Electron Mater 12(5):175–188
11. Raychowdhury A, Roy K (2007) Carbon nanotube electronics: design of high-performance and low-power digital circuits. IEEE Trans Circ Syst 54:2391–2401
12. Deng J, Wong H-SP (2007) A compact SPICE model for carbon-nanotube field-effect transistors including nonidealities and its application–Part II: full device model and circuit performance benchmarking. IEEE Trans Electron Device 54(12):3195–3205
13. Stanford University CNFET model website (2008) Stanford University, Stanford. http://nano.stanford.edu/model_stan_cnt.htm
14. Terrell DL (2000) OPAMPS design, application, and troubleshooting, 2nd edn. Elsevier Science and Technology, USA

Chapter 34
Re-Programmable Logic Array for Logic Design and Its Reliability Analysis in QCA

Kunal Das, Debashis De, Sayantan Ghatak and Mallika De

Abstract Quantum dot cellular automaton is now considered as a strong alternative of Complementary Metal Oxide Semiconductor (CMOS) technology. In this paper, we demonstrate an empirical work for implementing Quantum dot Cellular Automata (QCA) based Re-Programmable two variables Re-programmable logic array. It is fully reprogrammable by exploiting the fact of bidirectional nature of QCA. AND or OR logic. In our proposal, we made a different aspect of designing PLA. We made a control word, which must be for both the plane i.e. AND plane and OR plane. The OR plane or AND plane is configured with Majority voter and orthogonal fully populated tile. The PLA cell designed for two variables PLA, Reprogrammable by means of altering control Inputs. In our proposal we can program AND plane as well as OR plane with the control word. The reliability of this Re-PLA is reported.

Keywords QCA tile · Orthogonal fully populated tile · MV · Re-PLA · Reliability

K. Das (✉)
Department of Information Technology, B.P. Poddar Institute of Management and Technology, 137, VIP Road, Kolkata 700052 West Bengal, India
e-mail: kunaldasqca@gmail.com

D. De · S. Ghatak
Department of Computer Science and Engineering, West Bengal University of Technology, BF-142, Sector-I, Salt Lake City, Kolkata 700064 West Bengal, India

D. De
School of Physics, University of Western Australia, M013, 35 Stirling Highway, Crawley, Perth, WA 6009, Australia

K. Das · M. De
Department of Engineering and Technological Studies, Kalyani University, Kalyani 741235 West Bengal, India

S. Sengupta et al. (eds.), *Emerging Trends in Computing and Communication*, Lecture Notes in Electrical Engineering 298, DOI: 10.1007/978-81-322-1817-3_34,

34.1 Introduction

The 'Complementary Metal Oxide Semiconductor' (CMOS) Technology in Very Large Scale Integration (VLSI) is reaching its limit due to high lithographic cost; leakage of current etc. Hence, we must ensure a strong alternative of CMOS technology for VLSI design. Carbon-nano tubes, Fin FET, Quantum computing etc. are becoming emerging technology. One of the emerging technologies in nano scale computing is Quantum dot Cellular Automata (QCA). Recently QCA has been in focus of research interest due to low power, high operating frequencies (T Hz) nano scale device architecture. Many researchers have focused on QCA during this nanotechnology era [1–5]. In early nineties Lent et al. [1] first introduced QCA as emerging technology for nano scale computing. The dynamic behavior of QCA cell is also being discussed [2]. It is becoming emerging Technology in nano scale computing very quickly as QCA cell was first fabricated in 1999 with GaAs. The four quantum dots are at the four corners of cell and a tunnel junction (made with metal) between two quantum dots to tunnel the electron from one dot to another dot [4]. Macucci et al. [5] demonstrated a theoretical and experimental approach to fabricate QCA cell in Silicon-On-Insulator (SOI) [4]. The QCA logic design has reported in [6–14]. The PLA design with QCA has also been reported in [15–17].

The fundamental theory of QCA is involved with coulomb interaction force. Due to this coulomb interaction force in square shaped four dot QCA cell, electrons (exact two consider for ideal fabrication) have only two choice to stay (see Fig. 34.1) in maximum distance of separation i.e. diagonal positions [1–4]. This is known as +1.00 (binary 1) and −1.00 (binary 0) polarization. One QCA cell is polarized with definite polarity and influences other cells to be polarized with same polarity. For example, three input Majority voter (MV) logic gate in QCA is considered here as shown in Fig. 34.1. The device cell is influenced by (three input cell) adjacent QCA cells and output polarity is hold as majority of influence. However, these polarities cannot hold long period. As a result QCA need four phases of adiabatic switching scheme [2, 3]. The four phases of clocking is required for QCA operation namely, Relax, Switch, Hold and Release. In the Relax phase the inter dot barrier hight is high and electron cannot able to tunnel the barrier (junction). In the switching phase the inter dot barrier becomes low and electrons are able to tunnel from one dot to other (with in cell only allowed due to its design of QCA) and hold a definite polarity. To hold the polarity, the clock is required is known as hold phase clocking. In the release phase, the polarity of QCA cell is released and inter do barrier becomes high [9].

Defect is uncertainties that are probable to occur during fabrication process [9–11, 18]. There are several possibilities in fabrication process (a) deposition phase defect (b) Synthesis phase defect. In the deposition phase defect, there are several type of defects (1) extra cell deposition (2) Missing cell deposition (3) Misplacement cell deposition. Extra cell deposition: the extra cells are deposited from exact design layout. Missing cell deposition: There is a possibility to miss the cell from the exact layout in QCA device fabrication. Misplacement cell deposition:

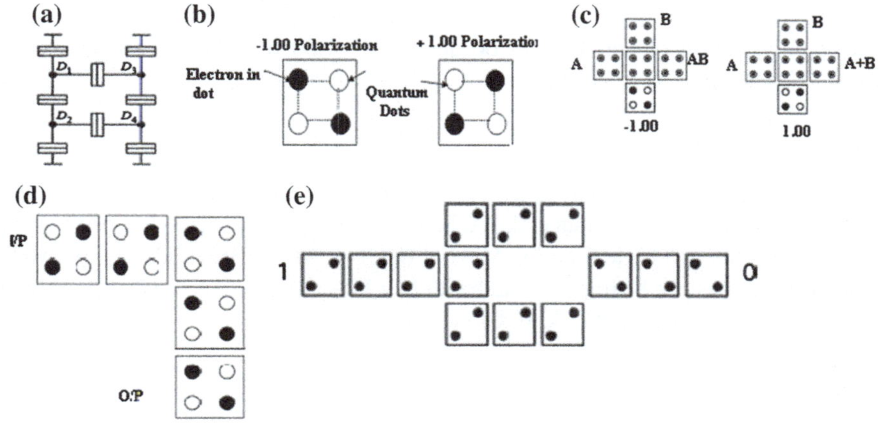

Fig. 34.1 **a** QCA cell design with tunnel junction, **b** QCA cell with two possible state polarizations, **c** three input majority voter, **d** QCA *L-Shape* wire, **e** inverter

the possibilities that the cell is misplaced from the exact design of QCA logic gate or circuit design. This probabilistic nature of QCA defect at fabrication phase is exploiting in this paper to defect analysis of QCA. In this paper, our aim is to implement Re-programmable Logic Array with QCA. There are few proposal have been reported on PLA [15–17], but here we concentrate to design a Re-programmable version of that PLA cell. In our proposal, we made a different aspect of designing PLA. In our work we made a control word, which must be for both the plane i.e. AND plane and OR plane. The OR plane or AND plane is configured with Majority voter and orthogonal fully populated tile [9]. The PLA cell designed for two variables PLA, Reprogrammable by means of altering control Inputs. There are two control inputs of each PLA cell. In our proposal we can program AND plane as well as OR plane with the control word. For two inputs PLA we required 8bits control word for AND PLA plane and 6-bits control word for OR-PLA plane. The probabilistic analysis of reliability test with information theory is also being reported for Re-PLA.

34.2 Basic QCA

Quantum dot cellular automata consist of four quantum dots positioned at four corners of cell and two extra electrons are confined within the cell [4, 5]. The electrons are placed at diagonals of square-cell by means of coulomb repulsive force. The two electrons used in QCA cell to store and transmit data by means of electron polarization. Electrons have the ability to tunnel from one quantum dot to the next; therefore the repelling force of electrons makes the charge to move to opposite corners of the quantum cells. In QCA, logic state is determined by the polarization of cell rather than voltage level as in CMOS technology. The two stable polarization of cell P = +1.00 and P = −1.00 of a QCA cell represents logic ‘1’ and logic ‘0’ respectively.

(a) QCA Majority Voter

A three input majority logic gate consists of an arrangement of five standard cells: a central device cell, three input cells (A, B, and C), and an output cell. Majority Voter (MV) is described as logic function MV (A, B, C) = AB + BC + CA. Logic AND and OR functions can be implemented with MV by setting one input permanently to “0” or “1”, respectively. M (A, B, 0) = AB, M (A, B, 1) = A + B

(b) QCA Wire

An array of QCA cells acts as a wire and is able to transmit information from one end to another, i.e., all the cells in the wire will switch their polarizations to follow that of the input or driver cell.

(c) QCA Inverter

The inverter or NOT gate is also simple to implement using QCA. It turns out that if two cells are placed at 45° with respect to each other they interact inversely.

34.3 Simulator Setup

The tile based Re-PLA simulation has been carried out with exhaustive simulation on simulation tool [19] QCAdesignerV2.0.3. The type of simulation engine is Coherence Vector. This type of simulation engine used due to its accuracy and detailed evaluation of QCA. The Coherence Vector engine is based on density matrix approach. We have used 20 nm × 20 nm cell size and 5 nm dot size. The radius of effect we consider is 41 nm. The number of sample we consider is 42,800. Convergence tolerance is 0.001000 and temperature we consider is 1 K. High temperature like 283 K can be performed with different setup. The relative permittivity 12.9 is considered and clock high is 9.8e-22, clock low is 3.8e-23.

34.4 Re-PLA

In this subsection, we demonstrated an empirical set up for implementing QCA based programmable two variables as well as three variables programmable logic array. It is fully reprogrammable by exploiting the fact of bidirectional nature of QCA. AND or OR logic can be implemented as mentioned in the previous section

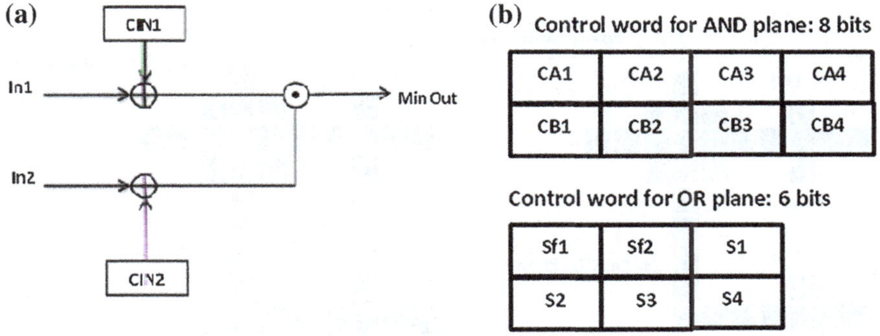

Fig. 34.2 **a** Symbolic logic diagram for AND plane/OR plane of Re-PLA, **b** control word for AND plane and OR plane where CA's, CB's are control input for AND plane and S's are control input for OR plane

by means of 1 × 1 or 3 × 3 tile. Now our big challenge is to design PLA cell. In early proposal, we found that there are two modes one is wire mode and another is logic mode. In our proposal, we made a different aspect of designing PLA. In our work, we made a control word, which must be equivalent to fuse like CMOS PLA. The control words are for both the plane i.e. AND plane and OR plane.

(a) Two variables Boolean Functions synthesis with PLA

The PLA cell is designed for two variables PLA. Reprogrammable by means of altering control Inputs. There are two control inputs of each PLA cell as shown in Fig. 34.2a, compared to early report we achieved four functionality in case of OR PLANE Re-PLA (a) OR operation with two input, while control input CIn1='1' and CIn2 = '1' (b) F = In1 acts like wire flow only In1 input while control input CIn1 = '1' and CIn2 = '1' (c) F = In2 acts like wire flow only In2 when control input CIn1 = '0' and CIn2 = '1' Finally, (d) F = '0' when both of the control inputs CIn1 = '0', CIn2 = '0', which is not in operation. The re-programmable PLA cell (shown in Fig. 34.3b) is known as Re-programmable–OR PLANE cell. The characterization of OR-PLANE Re-PLA is shown in Table 34.1. We also design another type of PLA cell known as Re-programmable –AND PLANE cell. This is shown in Fig. 34.3a, unlike the previous report [16, 17]. In case of AND PLANE Re-PLA (a) AND operation with two inputs, while control input CIn1 = '0' and Cin2 = '0' (b) F = In1 acts like wire flow only In1 input while control input CIn1 = '0' and CIn2 = '1' (c) F = In2 acts like wire flow only In2 when control input CIn1 = '1' and CIn2 = '0' Finally (d) F = '1' when both control input CIn1 = '1',CIn2 = '0', which is known as No operation. The characterization of AND- PLANE Re-PLA is shown in Table 34.2. In our proposal we can program AND plane as well as OR plane with the control word. For two inputs PLA we required 8 bits control word for AND-PLA plane and 6-bits control word for OR-PLA plane which is shown in Fig. 34.1b. The block diagram and

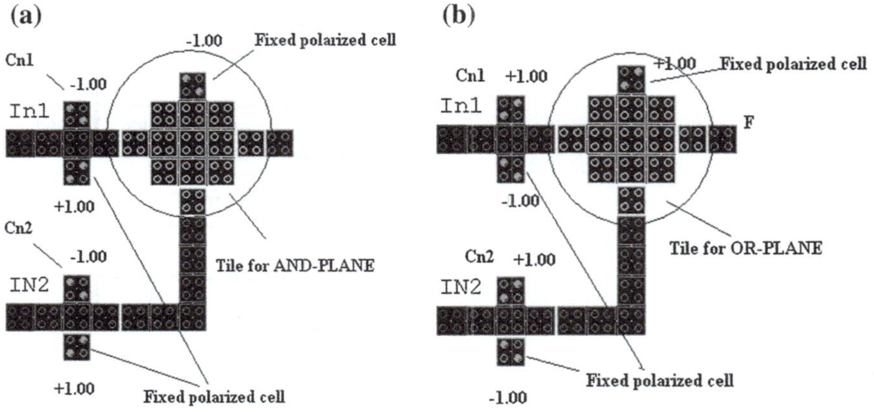

Fig. 34.3 **a** QCA layout representation of Re-PLA cell in AND-PLANE, **b** QCA Layout representation of Re-PLA cell in OR-PLANE, where Cn1, Cn2 are two control input, In1, In2 are two input

Table 34.1 Characterization table for OR-PLANE Re-PLA

Control input for OR-PLANE		Output functionality
Cn1	Cn2	F
−1.000	−1.000	No operation
−1.000	+1.000	In2
+1.000	−1.000	In1
+1.000	+1.000	In1 OR In2

Table 34.2 Characterization table for AND-PLANE Re-PLA

Control input for AND-PLA		Output functionality
Cn1	Cn2	F
−1.000	−1.000	In1 AND In2
−1.000	+1.000	In1
+1.000	−1.000	In2
+1.000	+1.000	No operation

symbols diagram and QCA Layout design by QCADesigner [19]. AND-PLANE Re-PLA and OR-PLANE Re-PLA cells are shown in Fig. 34.3a, b.

Let us consider the two variable Boolean function $F = AB' + A'B$, can be computed with Re-PLA for the three variable computer $\{x \leftarrow add(x, a, b)\}$. Now to implement this function we have to set up the control word. The control word of AND-PLA and OR PLA for above functions are {1,1,0,0,0,0,1,1} and

{1,1,0,1,1,0} respectively, where CIn3 = CIn4 = CIn5 = CIn6 = '0' and S1 = S3 = '0', rest remain '1' for selecting A′B and AB′ respectively.

(b) Re-PLA programming

The Programming technique of Re-programmable Logic array is explored in this subsection. For example, a fundamental logic function F = a′b + b′a is considered. The fundamental approach of assigning the control word is based on Directed Acyclic Graph (DAG) traversal [20]. DAG can represent the entire logic in RePLA. Each node of DAG shows either AND-plane or OR-plane cell. Figure 34.4d shows the DAG representation of entire logic circuit for Fig. 34.4a. Nodes can only pass up input or down input as switch or it can have computational part i.e. AND or OR operation. For AND-plane AND operation for 00 control input, up pass for 01 and down pass for 10 and similarly for OR-plane OR operation for 11 control input, up pass for 10 and down pass for 01 control inputs are shown in Fig. 34.4e, which is also shown in Tables 34.1, 34.2. Appropriate traversal of DAG will produce the AND-plane control word as well as OR-plane control word for programming the Re-PLA. Now for function F = a′b + b′a, we have the traversal node 2 AND operation, node 3 AND operation for AND-plane, node 5 down pass operation, node 6 up pass operation and node 7 OR operation for OR-plane.

34.5 Reliability of Re-PLA

QCA has found to suffer from high error rate during fabrication. It is reported in several report [10, 14, 18, 21,22, 23] the error may occur during fabrication of dot/cell and placement of cell, there may be extra cell, missing cell and misplaced cell displacement of cell in design phase. In this section, we describe the information-theoretic approach to investigate the reliability of Re-PLA cell. The fundamental theory behind this analysis is Shannon's joint source channel encoding theory [24–26]. The entropy H(X) can be defined as

$$\mathrm{H(X)} \equiv -\sum_{x\in X} p(x)\log(p(x)) \tag{34.1}$$

where H(X) is measured information content in the variable X, higher in entropy means larger information carried with the variable. If we consider that X suffers errors with the transformation function f: x → y, where f is erroneous mapping X∈ x, Y∈ y. Hence, the mutual information I(X; Y) can defined as

$$\begin{aligned} \mathrm{I(X;\,Y)} &= \mathrm{H(X) - H(X \mid Y)} \\ &= \mathrm{H(Y) - H(Y \mid X)} \end{aligned} \tag{34.2}$$

where H (Y|X) is conditional entropy of Y over X can be defined as

(a)

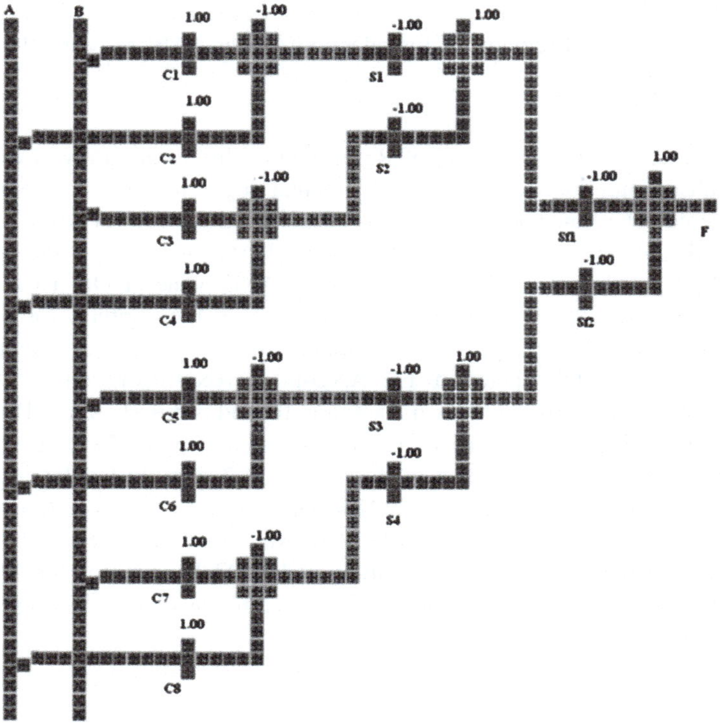

(b)

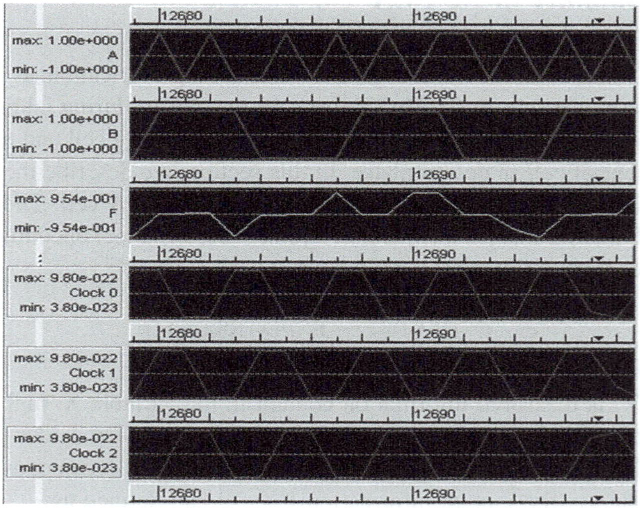

Fig. 34.4 **a** QCA layout of Re-PLA to functional output F = A′B + AB′, the control word for AND-PLANE Re-PLA is {1,1,0,0,0,0,1,1} and for OR-PLANE Re-PLA is {1,1,0,1,1,0}, **b** simulated result of Re-PLA for function F = A′B + AB′, **c** corresponding schematic diagram of Re-PLA, **d** equivalent DAG for Re-PLA, **e** each DAG node with functional representation

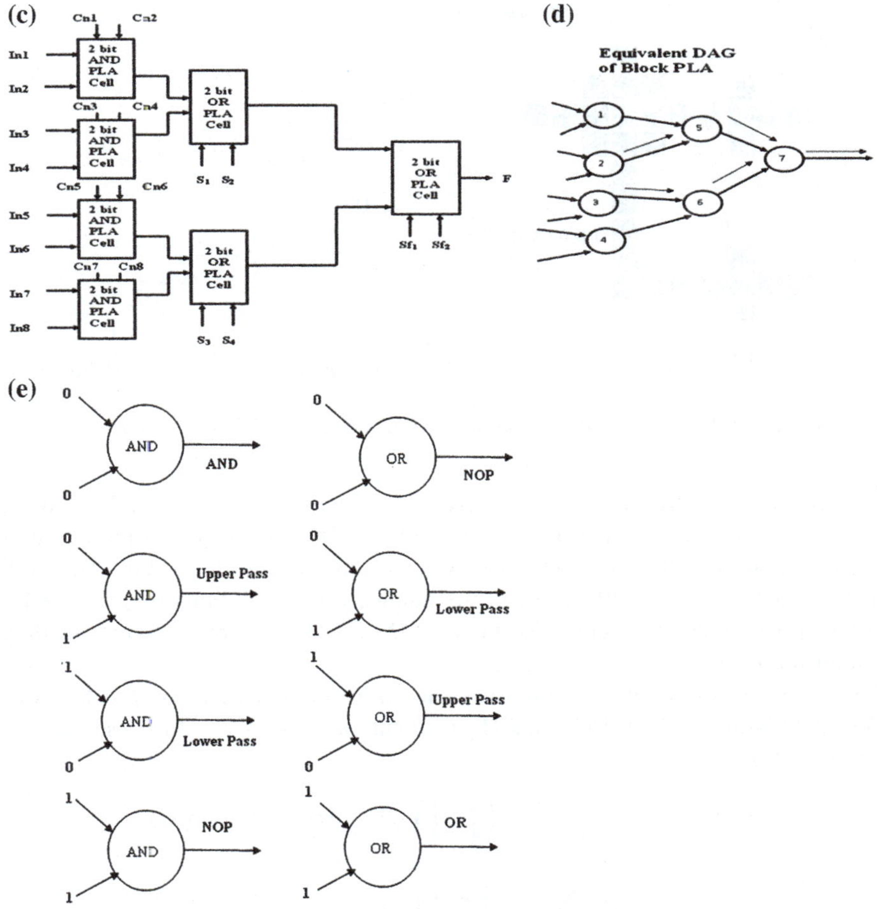

Fig. 34.4 (continued)

$$
\begin{aligned}
\mathrm{H}\ (\mathrm{Y} \mid \mathrm{X}) &= -\sum_{x \in X} \sum_{y \in Y} p(x, y) \log(p(y|x)) \\
&= -\sum_{x \in X} \sum_{y \in Y} p(x) p(y|x) \log(p(y|x))
\end{aligned}
\tag{34.3}
$$

where p(x, y) and p(y | x) are joint probability and conditional probability, respectively. The maximum information content with arbitrary error rate, gives the maximum capacity defined as

$$
C = \max_{\forall p(x)} I(X; Y) \tag{34.4}
$$

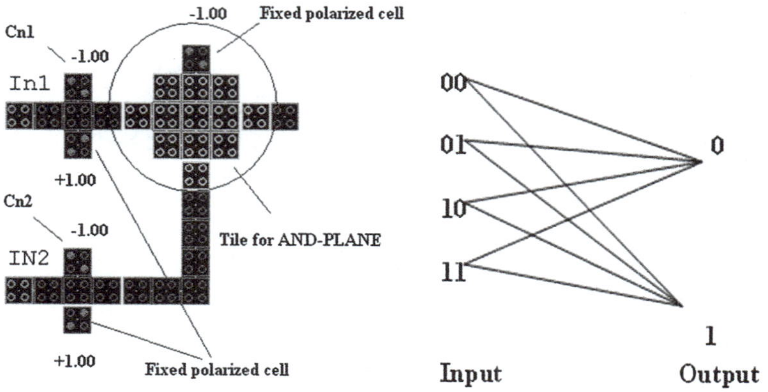

Fig. 34.5 Re-PLA AND PLANE cell and it's corresponding probabilities of output

Now in case of Re-PLA AND PLANE cell, There are possible outputs for input '00', '01', '10' and '11' are '0', '0', '0' and '1'. The corresponding probabilities are $\rho000$, $\rho010$, $\rho100$ and $\rho111$. The erroneous outputs for inputs '00', '01', '00' and '11' are '1', '1' '1', '0'. The corresponding error probabilities are $\rho001$, $\rho011$, $\rho101$, $\rho110$. Figure 34.5 shows Re-PLA AND PLANE cell and it's corresponding output probability.

The maximum Information transfer capacity of Re-PLA AND PLANE cell can be described with Eq. 34.4, H(Y) and H (Y|X) can be expressed with respect to ρi's are as follows

$$\mathrm{H(Y)} = -\sum_{\mathrm{Yi}\in\mathrm{y}}\sum_{\mathrm{Xj}\in\mathrm{x}} 0.5 \prod_{Ym=1Yi}\prod_{Yn=0}(1+\rho\mathrm{m,xi})(1-\rho\mathrm{n,xj}) \times \log\left(\sum_{Xj\in x} 0.5\prod_{Ym=1}\prod_{Yn=0}(1+\rho\mathrm{m,xi})(1-\rho\mathrm{n,xj})\right)$$

$$\mathrm{H(Y|X)} = -\sum_{\mathrm{Yi}\in\mathrm{y}}\sum_{\mathrm{Xj}\in\mathrm{x}} (\rho(X=Xj)) \times 0.5 \prod_{Ym=1Yi}\prod_{Yn=0}(1+\rho\mathrm{m,xi})(1-\rho\mathrm{n,xj}) \times \log\left(\sum_{Xj\in x} 0.5\prod_{Ym=1}\prod_{Yn=0}(1+\rho\mathrm{m,xi})(1-\rho\mathrm{n,xj})\right)$$

For $\rho000 = 0.99$, $\rho010 = 0.99$, $\rho100 = 0.99$ and $\rho111 = 0.99$, error probabilities are $\rho001 = 0.01$, $\rho011 = 0.01$, $\rho101 = 0.01$ and $\rho110 = 0.01$. We have the information capacity for Re-PLA AND PLANE cell is 0.99999 bits/use and Fig. 34.6 shows that information capacitates with average probabilities of erroneous output, which implies the design of this Re-PLA cell is reliable.

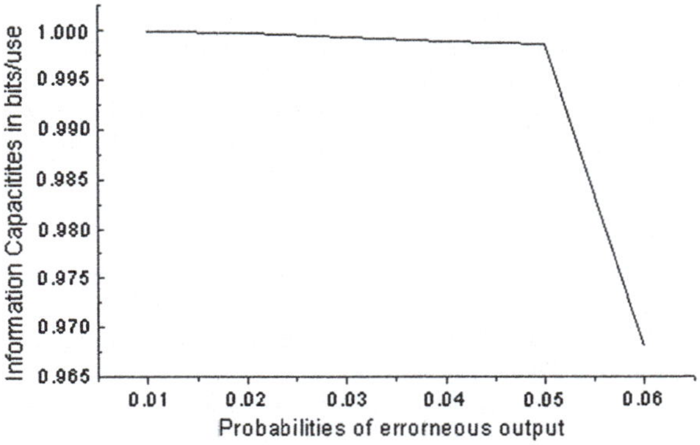

Fig. 34.6 Re-PLA AND PLANE information capacities corresponding to probabilities of output

34.6 Conclusion

In this paper, we have observed the Re-PLA design with Majority Voter and 3 $\times$ 3 orthogonal tile. The control words for AND-PLANE Re-PLA cell and OR-PLANE Re-PLA cell for two functional realizations have been reported. It can be concluded that the Re-programmability of Re-PLA is innovative with earlier report due to its control words. Programmers can program the Re-PLA repeatedly by altering the control words. The control words for different functionality are predefined to the Re-PLA programmer. The reliability of Re-PLA has also been discussed. The information theoretic approach has been adopted to show the functional reliability which concludes here if probability of error for different erroneous output is between 5 % the circuit design remains reliable.

Acknowledgments The authors are grateful to the University Grants Commission (UGC), India File No.: 41-631/2012(SR), under which this paper has been completed.

References

1. Lent CS, Taugaw PD, Porod W, Bernstein GH (1993) Quantum dot cellular automata. Nanotechnology 4:49–57
2. Lent CS, Tougaw PD, Porod W (1993) Bistable saturation in coupled quantum dots for quantum cellular automata. Appl Phys Lett 62:7–14
3. Amlani I, Orlov A, Snider G, Lent C, Porod W, Bernstein G (1999) Experimental demonstration of electron switching in a quantum-dot cellular automaton (QCA) cell. Superlattices Microstruct 25(1–2):273–278
4. Lent CS, Taugaw PD (1996) Dynamic behavior of quantum cellular automata. J Appl Phys 80(8):4722–4736
5. Macucci M, Gattobigio M, Bonci L, Iannaccone G, Prins FE, Single C, Wetekam G, Kern DP (2003) A QCA cell in silicon on insulator technology: theory and experiment. Superlattices Microstruct 34:205–211

6. Momenzadeh M, Huang J, Tahoori MB, Lombardi F (2005) Characterization, test, and logic synthesis of AND–OR-inverter (AOI) gate design for QCA implementation, IEEE Trans. Comput Aided Des Integr Circuits Syst 24:1881–1893
7. Das K, De D (2009) A study on diverse nanostructure for implementing logic gate design for QCA. In: Proceedings of the international conference ICANN-2009, IIT Guwahati, Guwahati, Assam
8. Das K, De D (2009) A novel approach of AND–OR-inverter (AOI) gate design for QCA. In: Proceedings of IEEE conference CODEC-09, Kolkata
9. Das K, De D (2011) Characterisation, applicability and defect analysis for tiles nanostructure of quantum dot cellular automata. Mol Simul 37(3):210–225
10. Das K, De D (2010) QCA defect and fault analysis of diverse nanostructure for implementing logic gate. Int J Recent Trends Eng Finl 3(1):1–5
11. Momenzadeh M, Huang J, Lombardi F (2005) Defect and fault tolerance in VLSI systems DFT 2005. In: 20th IEEE international symposium, Washington
12. Tougaw PD, Lent CS (1994) Logical devices implemented using quantum cellular automata. J Appl Phys 75(3):1818–1825
13. Wang W, Walus K, Jullien GA (2003) Quantum-dot cellular automata adders. In: IEEE Nano2003 conference, San Francisco
14. Jha N, Gupta S (2003) Testing of digital system. Cambridge University Press, Cambridge
15. Crocker M, Hu XS, Niemier M, Yan M, Bernstein G (2008) PLAs in quantum-dot cellular automata. IEEE Trans Nanotechnol 7(3):376–386
16. Dysart TJ, Kogge PM (2008) Comparing the reliability of PLA and custom logic implementations of a QCA adder In: IEEE international workshop on design and test of nano devices, circuits and systems, pp 53–56
17. Crocker M, Hu XS, Niemier M (2007) Fault models and yield analysis for QCA-based PLAs. In: International conference on field programmable logic and applications, pp 435–440
18. Thoori M, Huang J, Momenzadeh M, Lombardi F (2004) Testing of quantum Cell automata. IEEE Trans Nanotechnol 3(4):432–442
19. Walus K, Dysart TJ, Jullien GA, Budiman RA (2002) ATIPS laboratory QCA designer. ATIPS laboratory, University of Calgary, Canada. http://www.atips.ca/projects/qcadesigner
20. Tarjan RE (1972) Depth-first search and linear graph algorithms. SIAM J Comput 1(2):146–160
21. Fijany A, Toomarian BN (2001) New design for quantum dot cellular automata to obtain fault tolerant logic gates. J Nanopart Res 3:27–37
22. Huang J, Momenzadeh M, Lombardi F (2007) On the tolerance to manufacturing defects in molecular QCA tiles for processing-by wire. J Electron Test Theory Appl 23(2):163–174
23. Momenzadeh M, Ottavi M, Lombardi F (2005) Modeling QCA defects at molecular-level in combinational circuits. In: Proceedings of 20th IEEE international symposium on DFT, pp 208–216
24. Krishnaswamy S, Viamontes GF, Markov IL, Hayes JP (2008) Probabilistic transfer matrices in symbolic reliability analysis of logic circuits. ACM Trans Des Autom Electron Syst 13(1):8–35
25. Han J, Taylor E, Gao J, Fortes J (2005) Reliability modeling of nanoeltronic circuits. In: Proceedings of 5th IEEE conference on nanotechnology, Nagoya, July 2005
26. Wang L, Jain F, Lombardi F (2011) Information-theoretic modeling and analysis of stochastic behaviors in quantum-dot cellular automata. Intech Open, Croatia, pp 1–22

Chapter 35
Realization of Bi-Quinary Coded Decimal Adder in Quantum Dot Cellular Automata

Dipannita Podder, Kunal Das, Debashis De and Mallika De

Abstract Bi-quinary coded parallel adder design with Quantum Cellular Automata is presented in this brief contribution. The nano-electronic computer architecture using QCA technology is in infancy stage. It requires more advancement with new approaches. In this paper the design of parallel decimal adder is proposed using bi-quinary encoding techniques with algorithm. The circuits are implemented using QCA designer tool and analyzed using simulation result. The signal propagation delay, complexity, required area, hardware cost are calculated and compare with previously proposed decimal QCA adders.

Keywords QCA basic · 3 × 3 tile · Decimal digit encoding · Bi-quinary code · Parallel decimal adder

D. Podder (✉) · K. Das (✉)
B. P. Poddar Institute of Management and Technology, 137, VIP Road, Kolkata 700052, India
e-mail: dip.237@gmail.com

K. Das
e-mail: kunaldasqca@gmail.com

D. De
Department of Computer Science and Engineering, West Bengal University of Technology, BF-142, Sector-I, Salt Lake City, Kolkata 700064, India
e-mail: dr.debashis.de@gmail.com

K. Das · M. De
Department of Engineering and Technological Studies, Kalyani University, Kalyani 741235 West Bengal, India
e-mail: demallika@yahoo.com

S. Sengupta et al. (eds.), *Emerging Trends in Computing and Communication*, Lecture Notes in Electrical Engineering 298, DOI: 10.1007/978-81-322-1817-3_35,

35.1 Introduction

Quantum dot Cellular Automata (QCA) is an emerging technology in nanotechnology era that has been recognized as one of future alternative of Complementary Metal Oxide Semiconductor (CMOS) technology in Very Large Scale Integration (VLSI) design by providing high density, high switching speed, and ultralow power dissipation and reducing the drawbacks of CMOS technology like very high leakage of current and high lithographic cost in fabrication [1–3]. In modern era computers process large volumes of decimal information in financial, commercial, Internet-based, and automatic control applications, this cannot tolerate errors from converting between decimal and binary formats reported in [4]. This problem can be resolved using decimal coded arithmetic units. Several QCA-based decimal encoded adder, the main component of arithmetic unit designs, have been proposed, i.e. BCD adder [5], JMC adder [4, 6, 7] etc. In this paper we have proposed and designed parallel decimal adder using the bi-quinary encoding technique. In this design we have used modified XOR gate, proposed in earlier work [8].

QCA cell has four quantum dots positioned at the corners of the square shaped cell and two free electrons. Each dot can be occupied by one of the two hopping electrons. The mutual behavior of the electrons is based on the columbic interaction; they arrange themselves diagonally in order to reach to the maximum distance. In QCA logic states are defined in terms of polarization of electrons rather than the voltage level. When electrons are polarized $P = +1.00$ then it represents binary '1' equivalent and $P = -1.00$ then it represents binary '0', shown in Fig. 35.1a. QCA requires four-phased clocking signal. The four phases are relaxed, switch, hold, and release. In the relax phase there is no inter dot barrier. In the switch phase, barrier slowly becomes high and cell attends definite polarity depending on the input. The polarity of electron is retained in the hold phase. The barrier slowly gets lowered and the cell releases polarity in the release phase. Majority Voter is the basic building block of QCA gates [9, 10], shown in Fig. 35.1b. The logic function of MV is explained by Boolean function MV (A, B, C) = AB + BC + CA. When C = 0, MV acts as AND gate, and for C = 1 MV act as OR gate. Inverter is designed in a special manner, the 45° of interaction between two QCA cell are involved in inverter design. Figure 35.1c shows the structure of five input MV using cascading 3×3 tile [8]. The logic function of five input majority voter is MV (A, B, C, D, E) = ABC + ABD + ABE + ACD + ACE + ADE + BCD + BCE + BDE + CDE; for the values C = 0, D = 0 it acts as AND gate and for C = 1, D = 1 it acts as OR gate.

35.2 Proposed Decimal Adder

Previously Binary Coded Decimal adder [5], Jonson Mobius coded serial and parallel adders are designed in QCA technology [4, 6, 7]. The BCD adder followed the fundamental design principal of CMOS technology. In recent years, operations

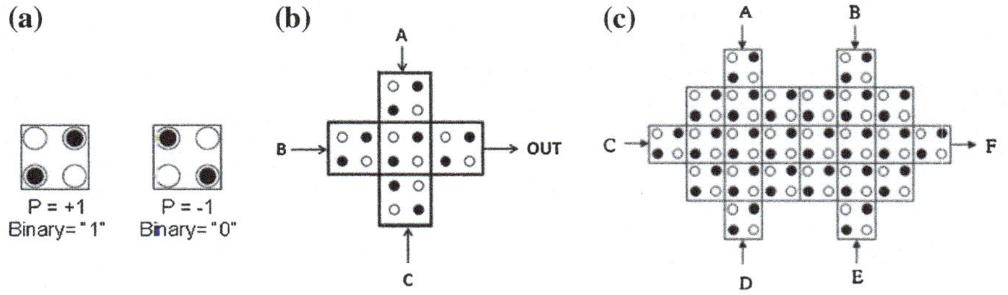

Fig. 35.1 **a** QCA cell with polarization, **b** QCA layout of three input MV, **c** QCA layout of five input MV after cascading two 3 × 3 tile

Table 35.1 6 bit bi-quinary code

Decimal digit	5	4	3	2	1	0
0	0	0	0	0	0	1
1	0	0	0	0	1	0
2	0	0	0	1	0	0
3	0	0	1	0	0	0
4	0	1	0	0	0	0
5	1	0	0	0	0	1
6	1	0	0	0	1	0
7	1	0	0	1	0	0
8	1	0	1	0	0	0
9	1	1	0	0	0	0

are performed upon large amount of data, so parallel computing approach executes the job faster with less complexity than serial computing. So, Parallel decimal adder using 6 bit bi-quinary encoding techniques [4] is proposed in this paper. The bi-quinary representations of decimal numbers are shown in Table 35.1. It is observed that the decimal equivalent quinary output is simplified like the output of a decoder and the binary component act as a select line. The quinary components repeat themselves when the binary component inverts.

The algorithm of the suggested adder is as follows.

```
1) CALCULATE D(A4,A3,A2,A1,A0,B4,B3,B2,B1,B0);
2) D5c = CORRECT D5(A5, B5, D5);
3) CALCULATE Co(A5, B5, and D5c);
4) SHIFT();
END,
PROCEDURE SHIFT ()
1) IF (Cin = 1), THEN
LEFT ROTATE D4, D3, D2, D1, AND D0;
2) IF (D4 = 1 AND Cin = 1), THEN
INVERT D5c;
3) IF (D4 = 1 AND D5c = 1), THEN
INVERT Co;
END,
```

Table 35.2 Truth table of sum generator

A4	A3	A2	A1	A0	D5	D4	D3	D2	D1	D0
0	0	0	0	1	0	B4	B3	B2	B1	B0
0	0	0	1	0	B4	B3	B2	B1	B0	B4
0	0	1	0	0	B4 + B3	B2	B1	B0	B4	B3
0	1	0	0	0	B4 + B3 + B2	B1	B0	B4	B3	B2
1	0	0	0	0	B4 + B3 + B2 + B1	B0	B4	B3	B2	B1

where A5 A4 A3 A2 A1 A0 is the first operand, B5 B4 B3 B2 B1 B0 is the second operand, D = {D5, D4, D3, D2, D1, D0} is the result of sum, produced by the quinary components of operands. CI is the carry-in bit. D5c is the binary component of sum after correction.

According to step 1 of the algorithm the sum, D5, D4, D3, D2, D1, D0 are calculated by processing the quinary components of the second operand (addend), B4 B3 B2 B1 B0, under the control of the first operand (augends) bits, A4 A3 A2 A1 A0. The MSB of D will be corrected according to step 2, by processing the binary component of the addend B4 and binary component of the augends A5 and D5. Then according to step 3 the carry out will be calculated and finally the bits can be shifted ones under the mentioned conditions.

For example: A = 7, B = 7, Cin = 1. Bi-quinary representation of A, B are 100100, 100100. In step 1, D5, D4, D3, D2, D1, D0 are calculated, using the functions established from Table 35.2, i.e. 010000. After step 2 and 3, D5c and Co become 0 and 1, respectively. In step 4 the shift operation is done.

Cin = 1; so, D4, D3, D2, D1, D0 rotate towards left and produce the output 00001. Again according to step 2 of procedure shift( ) D5c is inverted. So, the final output becomes 1100001.

35.2.1 Block Diagram

The adder consists of two main blocks, i.e., Operational Unit and Bit Shifter, shown in Fig. 35.2.

Operational Block:

Sum Generator.
The Sum generator block is expressed with the help of concept multiplexer; where A will act as select input and output of five multiplexer produce D. The truth table of Sum generator is shown in Table 35.2. It can be noted that the quinary components of the second operand (addend) B4 B3 B2 B1 B0 act as Inputs which will be selected by the quinary components of the first operand (augends) bits, A4 A3 A2 A1 A0. Each output of D = {D5, D4, D3, D2, D1, D0} can be expressed in Boolean expressions as shown in Table 35.2.

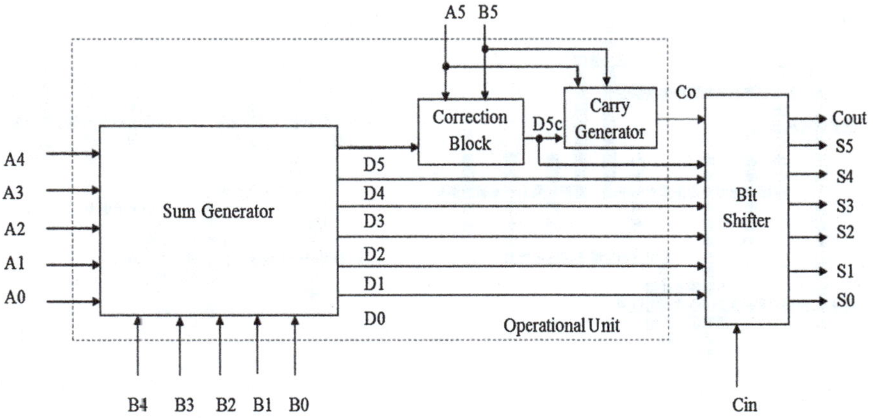

Fig. 35.2 Block diagram of proposed bi-quinary parallel decimal adder

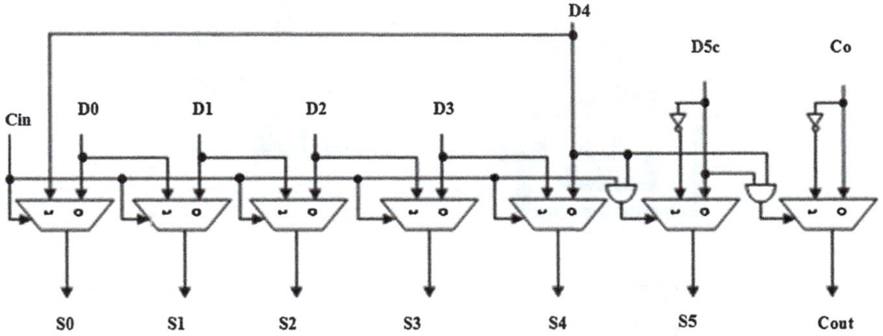

Fig. 35.3 Block diagram of bit shifter

Correction Block.
The binary components of the operands play a role in the formation of output. These bits correct and produce the binary component of the sum D5c. The function is $D5c = A5 \oplus B5 \oplus D5$. Here we have used a modified three input XOR gate in QCA [8] to implement this correction Block.

Carry generator.
The carry output can be calculated using the binary components operand and sum. The Boolean function is $Co = \left((A5 + B5)\overline{D5c}\right) + A5B5)$.

Bit Shifter:

This block consists of five multiplexer and two AND gates and two inverters. If Cin = 1, the previously calculated sum and carryout need to shift in a special manner to produce the proper output. The shifting operations operate in a different

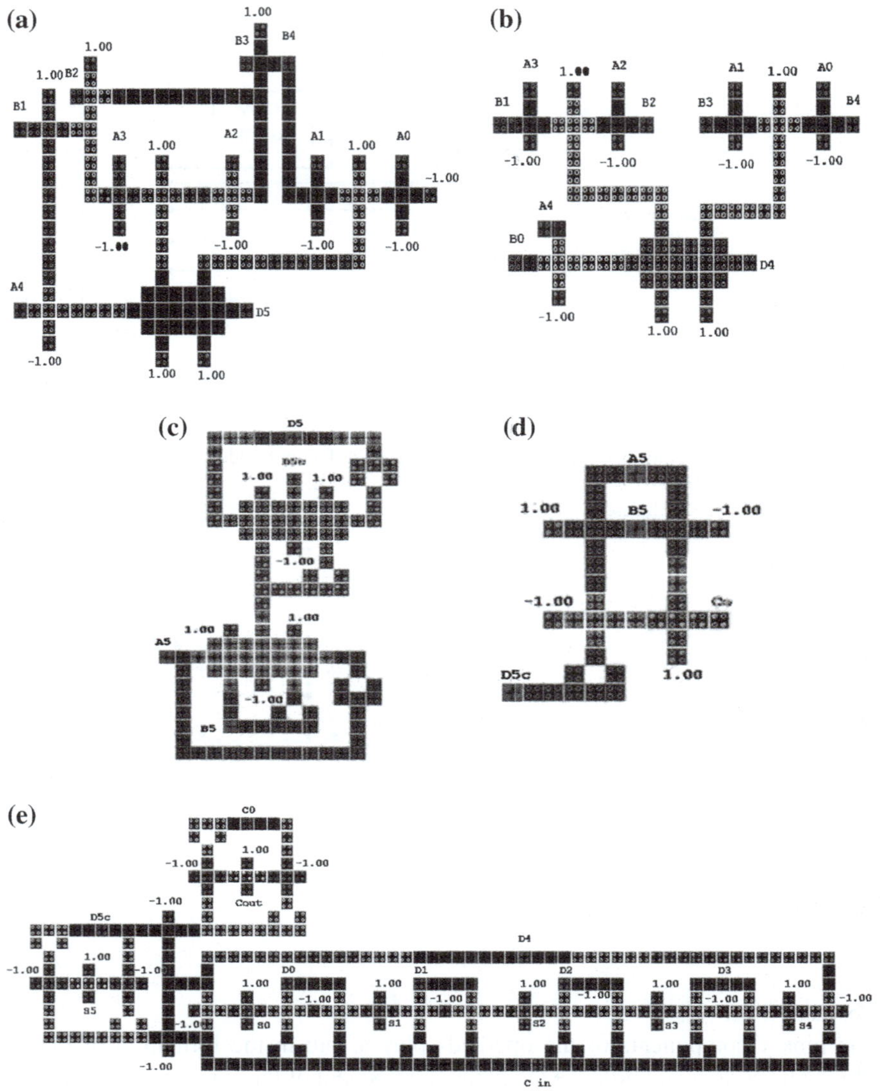

Fig. 35.4 QCA layout: **a** and **b** are output of sum generator, D5 and D4 respectively, **c** correction block, **d** carry generator, **e** bit shifter

way for the binary component of Sum, quinary component of sum and carry out. The block diagram is shown in Fig. 35.3. If Cin = 1, then the quinary components of the sum shift towards left. Again D5c will be inverted at the condition Cin = 1 and D4 = 1 and Co will be inverted at the condition D5c = 1 and D4 = 1.

35.2.2 QCA Layout

The QCA layouts of the function D5 and D4 are shown in Fig. 35.4a and b. Similarly D3, D2, D1, D0 are designed. The correction block, carry generator and bit shifter are implemented, and the QCA layouts are shown in Fig. 35.4c, d and e respectively.

35.3 Simulation and Result

The various input combinations of A,B, and Cin are simulated with QCA Designer [11, 12], among them one combination is explained, i.e. A = 8, B = 4, Cin = 1. Bi-quinary representation of A, B are 101000, 010000. D5, D4, D3, D2, D1, D0 are calculated by processing the quinary components of operands A and B. The simulation result of Sum generator is shown in Fig. 35.5a. The simulation result of correction block and carry generator are shown in Fig. 35.5b and c. Finally the shifting block executes, and provides the output 1001000 is shown in Fig. 35.5d. All simulation results show that the proposed adder in QCA performed faster addition.

35.3.1 Comparisons and Discussions

All blocks consist of 1,266 cells with required area 2.14 μm^2. The total propagation delay of all blocks of the proposed adder is six clock cycles. We made a comparison with decimal adder designed in QCA although we also compared with Brent-Kung adder [13] which is binary adder while bi-quinary adder is decimal adder. The comparison with previously proposed adders is mentioned in Table 35.3. A complexity (number of cells), required area (size of QCA cell = 18 × 18 nm, center-to-center distance = 20 nm), propagation delay (latency), and cost function are given for each variant [4]. In [14] reports that for QCA circuit *Power* ≡ *Complexity*. The cost function of a QCA circuit can be calculated with the product of an area, delay and power, as well as the same function of a microelectronic circuit Cost = Area × Delay × Power [15]. It is seen from Table 35.3 that all blocks of proposed adder require lowest complexity and minimal area among all adders designed in QCA. It can reduce the delay also. The bi-quinary technique requires less number of bits than the conventional decimal system to represent the decimal digit as well as the self-complementing property of the bi-quinary code is very useful in arithmetic unit design [16]. The problem of binary to BCD conversion is not required in this adder design; a decoder can easily convert the binary/decimal to bi-quinary code.

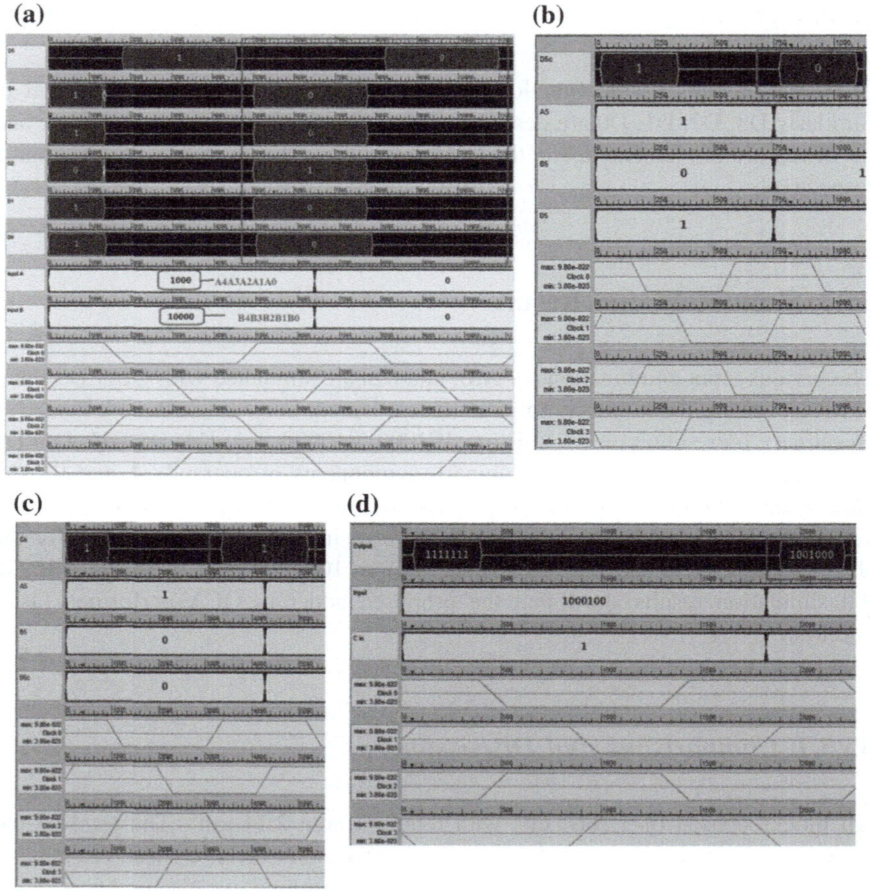

Fig. 35.5 Simulation result for QCA layout designed with QCA Designer **a** sum generator, **b** correction block, **c** carry generator, **d** bit shifter

Table 35.3 QCA decimal adder comparison

Variant	Different adder	Complexity	Area (μm^2)	Delay	Cost
1	Parallel BCD adder [5]	1,348 cells	2.28	8 CLK	24,588
2	Parallel JMC adder [7]	3,017 cells	4.00	12 CLK	144,816
3	Parallel JMC adder [6]	3,560 cells	4.00	7 CLK	99,680
4	Serial JMC adder [4]	1,130 cells	1.77	10 CLK	20,001
5	[a]Brent–Kung adder [13]	1,782 cells	1.498	2.5CLK	6,673.6
6	This work	1,266 cells	2.14	6 CLK	16,255

[a] Binary adder, while other adders are decimal adder

35.4 Conclusion

The adder is an extremely important part of arithmetic logic unit, as well as in nano-computing system. This paper presents the bi-quinary coded parallel decimal adder. The new algorithm design and implementation is presented in QCA platform. The proposed adder executes with a lesser amount of complexity and delay than the other QCA decimal adders. Each block of this adder is verified using QCA designer tool. In future decimal adder-subtractor and error checking strategy can be implemented with the help of this logic.

Acknowledgments The authors are grateful to the University Grants Commission (UGC), India File No.: 41-631/2012(SR), under which this paper has been completed.

References

1. Lent CS, Taugaw PD, Porod W, Bernstein GH (1993) Quantum dot cellular automata. Nanotechnology 4:49–57
2. Tougaw PD, Lent CS (1996) Dynamic behavior of quantum cellular automata. J Appl Phys 80(8):4722–4736
3. Lent CS, Tougaw PD, Porod W (1993) Bistable saturation in coupled quantum dots for quantum cellular automata. Appl Phys Lett 62:7–14
4. Gladshtein M (2011) Quantum-dot cellular automata serial decimal adder. IEEE Trans Nanotechnol 10(6): 1377–1382
5. Kharbash F, Chaudhry GM (2008) The design of quantum-dot cellular automata decimal adder. In: IEEE international multi-topic conference, INMIC 2008. IEEE (bcd adder)
6. Gladshtein MA (2010) The signal propagation delay reduction of the combinational adder of decimal digits encoded by the Johnson-Mobius code. Autom Control Comput Sci 44(2):103–109
7. Gladshtein MA (2009) Algorithmic synthesis of a combinational adder of decimal digits encoded by the Johnson-Mobius code. Autom Control Comput Sci 43(5):233–240
8. Das K, De D (2011) Characterisation, applicability and defect analysis for tiles nanostructure of quantum dot cellular automata. Mol Simul 37(03):210–225
9. Das K, De D (2009) A Novel approach of And-Or-Inverter (AOI) gate design for QCA. In: Proceedings of IEEE conference CODEC-09 (14–16 Dec 2009), pp 1–4
10. Das K, De D (2011) A study on diverse nanostructure for implementing logic gate design for QCA. Int J Nanosci 10(1–2) (Feb and April 2011):263–269 (World Scientific)
11. Walus K et al (2004) QCA designer: a rapid design and simulation tool for quantum-dot cellular automata. IEEE Trans Nanotechnol 3(1):26–31
12. Walus K et al (2002) ATIPS laboratory QCADesigner homepage. ATIPS laboratory, University of Calgary, Canada. http://www.atips.ca/projects/qcadesigner
13. Pudi V, Sridharan K (2012) Low complexity design of ripple carry and Brent–Kung adders in QCA. IEEE Trans Nanotechnol 11(1):105–119
14. Srivastava S, Sarkar S, Bhanja S (2009) Estimation of upper bound of power dissipation in QCA circuits. IEEE Trans Nanotechnol 8(1):116–127
15. Oklobdzija V (ed) (2002) The computer engineering handbook. CRC press, Florida
16. Kashio T (1962) TOSHIO KASHIO. U.S. Patent No. 3,015,445. 2 Jan 1962

Chapter 36
Synthesis of ESOP-Based Reversible Logic Using Positive Polarity Reed-Muller Form

Chandan Bandyopadhyay and Hafizur Rahaman

Abstract The development of efficient techniques for reversible quantum circuit synthesis has received significant attention now-a-days due to recent emphasis on low power circuit design. This work presents two new deterministic methods which evaluate the PPRM structure of logic functions. After extracting the structure, the synthesis of ESOP based reversible logic is performed using PPRM form. The first approach is based on transformation technique, whereas the second method is based on iterative reduction procedure. In both the approaches, we have derived the PPRM expression from an input truth table. Based on this expression, the design of ESOP-based reversible circuit is achieved.

Keywords ESOP · Quantum cost · Gate count · Reversible circuit · Cube list

36.1 Introduction

Landauer's [1, 2] stated that the minimum energy of amount *KT log* 2 Joules/bit (where K is Boltzman constant, T is absolute temperature) is dissipated in traditional logic computation technique, due to loss of every bit of information. If VLSI technology follows Moore's law [3], the energy loss in conventional circuit design is likely to become more dominant. Advances in VLSI technology and the use of new fabrication processes over the last few decades have rendered the heat loss

C. Bandyopadhyay (✉) · H. Rahaman (✉)
Bengal Engineering and Science University, Shibpur, Howrah 711103 West Bengal, India
e-mail: chandanb.iiest@gmail.com

H. Rahaman
e-mail: rahaman.h@yahoo.co.in; rahaman.h@gmail.com

S. Sengupta et al. (eds.), *Emerging Trends in Computing and Communication*, Lecture Notes in Electrical Engineering 298, DOI: 10.1007/978-81-322-1817-3_36,

and dissipation problem more complex in deep-submicron integrated circuits (IC). With the exponential growth of packing density, the traditional technologies like CMOS are reaching to a limit. So, some alternative technology is required to overcome from this stagnancy.

Reversible logic may provide a potential solution of such problems. Bennet [4] proposed that zero energy dissipation is possible if the circuit is reversible in nature and the inherent energy loss resulting from the irreversibility of information processing may be mitigated by implementing reversibility (in ideal condition), which is information lossless.

In present days, EXOR-based logic synthesis has been proved to be highly effective in the field of reversible computing [5]. It also been noticed that EXOR-based realizations are very efficient for arithmetic functions, error detecting and correcting functions, and symmetric functions. Testing of multi-level tree of XOR gates is easy as single fault detection [6] does not depend on input test vector. There exist several canonical families of AND/XOR form. Shannon, Positive Davio and Negative Davio representations [7] are the examples of canonical binary forms (AND/XOR forms). Canonical trees formed by the combination of above mentioned expansion technique are Kronecker, Pseudo Kronecker and Reed-Muller trees. Among these forms, Reed-Muller form has received extensive attention from researchers as logic implemented based on Reed-Muller form allows to generate a number possible representations of Boolean functions. Another aspect of Reed-Muller form is its easily testable characteristics. An efficient technique for conversion of minterms to positive polarity Reed-Muller coefficients and vice versa is introduced by Khan and Alam [8].

Apart from this, representing the PPRM expression by means Binary Decision Diagram (BDDs) have reported in [9]. The algorithm presented in [10–14] calculates PPRM coefficients by applying fixed polarity Reed-Muller coefficients generation algorithm. But the main drawback of algorithm presented in [10–14] is the use of a matrix data structure of size ($2^n \times 2^n$) to calculate PPRM coefficients, involves tedious computation. An idea to formulate optimal fixed polarity Reed-Muller expansions of Boolean functions is introduced in [15, 16]. But the use of their approach to obtain optimal FPRM form became very difficult when a Boolean function contains a higher number of variables.

Perkowski [6] has reported the ease of circuit testability based on Reed-Muller representation. He also has shown that Circuit constructed from Reed-Muller representation requires minimum number of tests for fault detection.

In this work, we have proposed two deterministic algorithms that produce PPRM cover corresponding to a Boolean logic function. The first approach is based on transformation technique, which produces a canonical positive control cube list. Eventually, this cube list is equivalent to the PPRM form of input function. The second approach first designs an initial circuit having 2^n-1 (n-bit function) Toffoli gates and then uses unwanted gate removal strategy that ensures the design of canonical ESOP-based circuit which is the circuit realization of PPRM expression in reversible domain.

The rest of the paper is organized as follows: Sect. 36.2 describes reversible preliminaries. The Proposed technique has been explained in Sect. 36.3. Examples on proposed technique have been presented in Sect. 36.4. The discussion is concluded in Sect. 36.5.

36.2 Background

36.2.1 Reversibility

Definition 1 *A fan-out free circuit (C_{nf}) with circuit depth (d) over the set of input lines $X = \{x_1, x_2, \ldots, x_n\}$ is said to be reversible (R_c) if the circuit is consist of reversible gates (g_i) only and the number of inputs (m) is equal with number of outputs (n) i.e. $C_{nf} = g_0 \cdot g_1 \cdot g_2 \cdot \ldots \cdot g_{(d-1)}$ where g_i represent ith reversible gate of the circuit.*

36.2.2 Reversible Gates

Definition 2 *For the domain variables $\{x_1, x_2, \ldots, x_n\}$, the multiple control Toffoli gate has the form $TOF(C; t)$, where $C = \{x_{i_1}, x_{i_2}, \ldots, x_{i_k}\}$, $t = \{x_j\}$ and assures $C \cap t = \phi$. It maps the input vector $(x_1^0, x_2^0, \ldots, x_n^0)$ to Boolean pattern $\left\{x_1^0, x_2^0, \ldots, x_{j-1}^0, x_j^0 \oplus x_{i_1}^0 x_{i_2}^0 \ldots x_{i_k}^0, x_{j+1}^0, \ldots, x_n^0\right\}$. The set C that controls the change of jth bit is the set of controls and t is called the target.*

Some of the basic reversible gates are as follows:

1. 1-input/1-output NOT $(x_1 \rightarrow \overline{x_1})$
2. 2-input/2-output controlled NOT (CNOT) gate: $(x_1, x_2) \rightarrow (x_1, x_1 \oplus x_2)$; and
3. 3-input/3-output Toffoli [5] gate $(x_1, x_2, x_3) \rightarrow (x_1, x_2, x_1x_2 \oplus x_3)$;

The basic reversible gates are shown in Fig. 36.1 where NOT, CNOT and Toffoli gate corresponds to Fig. 36.1a–c respectively.

36.2.3 Reed-Muller Form

Boolean algebra operators play a crucial role in logic synthesis and testing of digital circuits. The simplest case of the Galois field algebra is modulo-2 algebra and the representation of any Boolean function is possible by modulo-2 algebra. In practical field of application Reed-Muller expansion is nothing but modulo-2 sum-of products expression. Construction of reversible logic circuit by modulo-2

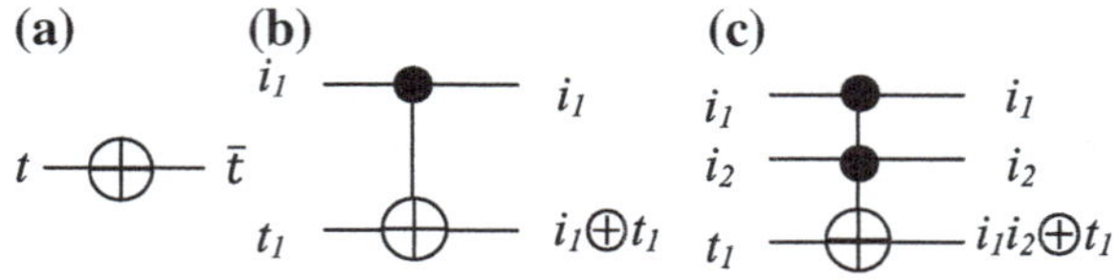

Fig. 36.1 Basic reversible gates. **a** NOT. **b** CNOT. **c** Toffoli

operations can be realized by means of exclusive OR (EXOR) gates. Any Boolean function $f(x_1x_2...x_n)$ of n variables can be represented by following forms of the Reed-Muller expansion [10, 11].

PPRM: The positive polarity Reed-Muller form is an EXOR sum of products, where each variable has positive polarity (not complemented form). PPRM is a canonical expression, so further minimization is not possible. An n variable PPRM expression can be represented as

$$f(x_1x_2...x_n) = a_0 \oplus a_1x_1 \oplus a_2x_2 \oplus a_3x_1x_2 \oplus \ldots \oplus a_{2^n-1}x_1x_2...x_n$$

where $a_i \in \{0, 1\}$. The variable a_i represent coefficient vector, where x_i are input variable terms. If the coefficient is 0, then the associated product term does not appear in the PPRM expression. If the coefficient is 1, then the associated product term appears in the PPRM expression.

The pD(positive Davio) expansion on all variable of an n-variable Boolean function $f(x_1x_2...x_n) = f_0 \oplus x_if_2$, where $f_0 = f(x_1, ..., x_{i-1}, 0, x_{i+1}, ...x_n)$ and $f_2 = f_0 \oplus f_1$ results PPRM expression.

FPRM: FPRM is similar to PPRM form, where each literal will appear as complemented or un-complemented form throughout the expression i.e. polarity of each variable is fixed. For an n variable Boolean function there exists 2^n number of different FPRM expression. An n variable FPRM expression can be represented as

$$f(x_1x_2...x_n) = a_0 \oplus a_1\dot{x}_1 \oplus a_2\dot{x}_2 \oplus a_3\dot{x}_1\dot{x}_2 \oplus \ldots \oplus a_{2^n-1}\dot{x}_1\dot{x}_2...\dot{x}_n$$

where $a_i \in \{0, 1\}$ and $\dot{x} \in \{x, \bar{x}\}$.

GRM/MPRM: The mixed polarity Reed-Muller expression is the generalized form of FPRM expression where there is no restriction in polarity of each variable. For an n variable Boolean function there exists $2^{n2^{n-1}}$ number of GRM expression. An n variable MPRM expression can be represented as

$$f(x_1x_2...x_n) = a_0 \oplus a_1\dot{x}_1 \oplus a_2\dot{x}_2 \oplus a_3\dot{x}_1\dot{x}_2 \oplus \ldots \oplus a_{2^n-1}\dot{x}_1\dot{x}_2...\dot{x}_n$$

where $a_i \in \{0, 1\}$ and $\dot{x} \in \{x, \bar{x}\}$.

NPRM: The negative polarity Reed-Muller form is an EXOR sum of products, where each variable has negative polarity (complemented form). NPRM is also a canonical expression, so further minimization is not possible. An n variable NPRM expression can be represented as

$$f(x_1x_2\ldots x_n) = a_0 \oplus a_1\bar{x}_1 \oplus a_2\bar{x}_2 \oplus a_3\bar{x}_1\bar{x}_2 \oplus \ldots \oplus a_{2^n-1}\bar{x}_1\bar{x}_2\ldots\bar{x}_n$$

where $a_i \in \{0, 1\}$.

The relations between the Reed-Muller forms are as follows:

$$\text{PPRM / NPRM} \subset \text{FPRM} \subset \text{GRM / MPRM}$$

36.2.4 ESOP

An exclusive-or-sum-of-product (ESOP) [17, 18] form is a variation of sum of products (SOP) form in boolean function representation. In ESOP, OR (+) and exclusive-or ($\oplus$) operator are associative in nature. Any boolean function can be converted to the equivalent ESOP form. For example, $f = xy + yz$ is in SOP form. Now, if we convert it to ESOP form, then the equivalent function becomes $f = xy \oplus yz \oplus xyz$. The ESOP-based circuit representation of an n bit single output function requires total $(n + 1)$ circuit lines. First n lines are termed as control lines and the last $(n + 1)$th line is the target line, which provides the functional output.

36.2.5 Cube List

A cube list is represented by a matrix (M) of size $(2^n \times n)$, where each row represents a cube (C_i). The cube list generates the function $f(x_1, x_2,\ldots, x_n)$, which is a collection of product term separated by exclusive-or ($\oplus$) operator. Each product term is represented by a cube (C_i), which is a vector in a vector space of dimension 2^n. An ESOP cube C_i is represented as $C_i = \langle a_1\, a_2\, a_3 \ldots a_n \rangle$, where $a_i \in \{0, 1, -\}$ for function $f(x_1, x_2,\ldots, x_n) = C_1 \oplus C_2 \oplus \ldots \oplus C_k$. Each cube C_i always mapped to a Toffoli gate T_i, where a defined bit value (0 or 1) of respective bit position in cube C_i corresponds to a control node in Toffoli gate T_i. If the bit value at kth position in cube C_i is "zero" then the control value of kth position in Toffoli gate T_i is a negative control. An example of cube list and its equivalent Toffoli representation corresponding to the MPRM (Mixed Polarity Reed-Muller) logical expression $f(x_1, x_2, x_3, x_4) = x_2x_3 \oplus x_4 \oplus x_1 \oplus \bar{x}_1x_2x_4 \oplus \bar{x}_1x_2x_3\bar{x}_4$, is shown in Fig. 36.2a, b.

(a)

x_1	x_2	x_3	x_4
–	1	1	–
–	–	–	1
1	–	–	–
0	1	–	1
0	1	1	0

(b)

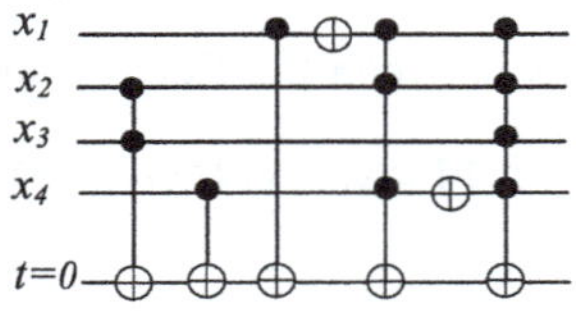

Fig. 36.2 **a** Cube list for $f(x_1, x_2, x_3, x_4) = x_2x_3 \oplus x_4 \oplus x_1 \oplus \bar{x}_1x_2x_4 \oplus \bar{x}_1x_2x_3\bar{x}_4$. **b** Equivalent Toffoli network

36.3 Proposed Technique

The algorithm for calculating PPRM covers using transformation technique is stated in algorithm1. Input specification for PPRM cover extraction algorithm is a *truth table*. A cube list which is equivalent to the PPRM form of the input specification is generated from output section of this algorithm. Steps related to the construction of PPRM cube list is stated here.

36.3.1 PPRM Cover Extraction Using Transformation Technique

- ***Step 1***: Initially the *PPRMCubelist* is empty. Our aim is to make all output value of the corresponding input function by all 0, using top-down approach. The function *Read_Spec_File()* reads the input (*truth table*).
- ***Step 2***: We set the constant line coefficient (a_0) for PPRM expression by $a_0 = 0$ or $a_0 = 1$ according to the function output value *Get_Output()* is 0 or 1. This function checks the output value of the input bit string [x_1, x_2,..., x_n] is either 0 or 1, where $x_i = 0$ and $i \in \{1, 2,..., \text{n}\}$ in *TTList*.
- ***Step 3***: If the constant line coefficient a_0 is set to 1 then toggle all the output column value of *TTList* using function *Flip Output Element();*
- ***Step 4***: Function *Get_Output()* retrieve the output bit from TTList and locate the first bit string whose output value is 1 using *Get_Bit_String()* function.
- ***Step 5***: We add the recently located bit string into the cube list by replacing all 0s by "–" (don't care).
- ***Step 6***: We form a *k*-CNOT gate where *k* value represents total number of 1 and each control node of *k*-CNOT gate corresponds to the position of 1 value present in that bit string.
- ***Step 7***: Now apply all the bit strings starting from the encountered bit position to 2^n-th position through positive-control *k*-CNOT gate.
- ***Step 8***: The TTList remains updated using function *Replace()*.

Steps 1–7 are repeated until all the output bit value in *TTList* become 0. Finally we obtain the equivalent PPRM cube list corresponding to the input specification (*truth table*).

Algorithm1: Algorithm for generation of equivalent PPRM cover cube list

PPRM_Cover_Generation( )
Input : Truth table ("*example.spec*")
Output: Equivalent PPRM Cubelist(*PPRMCubelist*)

```
begin
    PPRMCubelist = NULL;
    TTList = Read_Spec_File("example.spec");
    if Get_Output(TTList,0) = = 1 then
       Flip_Output_Element(TTList);
    end
    index = 1;
    while index ≠ Size(TTList) do
          output_element = Get_Output(TTList, index);
          if output_element = = 1 then
            string_element = Get_Bit_String(TTList,index);
            Add(string element, PPRMCubelist);
             one_index _list = Get_One_Index_ Psition(string_element);
            k = Size(one_index_list);
            k_CNOT_Gate = Create_k_CNOT_Gate(one_index_list);
            for i=index to Size(TTList) do
                element = Get_Bit_String(TTList);
                output = Apply_CNOT(k_CNOT Gate, element);
                Replace(TTList, i, element, output);
            end
          end
    end
    Replace_Zero_Dash(PPRMList);
end
```

36.3.2 Construction of ESOP Based Circuit for PPRM Expression Using Iterative Reduction Process

Another deterministic way of deriving PPRM form and designing equivalent ESOP-based canonical Toffoli network is presented here. This very approach first design a dynamic circuit having totaled $(n + 1)$ circuits line, where first n lines corresponds to n variables of an input function i.e. control lines and the $(n + 1)$th line corresponds to function output line. Next, all possible 2^n-1 input bit strings are applied individually over the dynamic circuit. While processing the bit strings, the dynamic circuit intelligently removes unwanted gate from the circuit and make the circuit updated. After processing all the input bit patterns over the circuit, a canonical Toffoli network with positive control Toffoli gates is obtained.

The algorithm for designing ESOP based circuit for PPRM expression using Iterative Reduction Process is stated in algorithm2.

- ***Step 1***: First read the input table to determine the number of input variable (let n) using function *Read_Truth_Table()*.

- ***Step 2***: Next, draw a network consisting of n control line and one target line i.e. having total $(n + 1)$ circuit line.
- ***Step 3***: Set the target line value by 0 or 1 according to the function output value in the truth table for the input bit string $[x_1x_2...x_n]$ is 0 or 1, where $x_i = 0, \forall i = 1, 2, \ldots, n$
- ***Step 4***: *Retrive_Bit_String(),* retrieves a string from the *TTList* in top-down manner and records the position of 1's in the input bit string and number of 1s (say k) present in the ith bit string in *string_element.*
- ***Step 5***: Use function *Design_Toffoli_Gate()* and draw a k-CNOT gate from this information on $(n + 1)$th line in such a way that the position of positive control values (1 values) in the ith bit string is same as that of controls of this k-CNOT gate.
- ***Step 6***: Apply all the bit string starting from position $i = 1$ to 2^n while scanning the truth table in top-down fashion to this positive-control k-CNOT gate individually and collect the output value from $(n + 1)$th line using function
  Retrive_Output().

Algorithm2:

PPRM_Circuit_Generation( )
Input : Truth table
Output: ESOP-based canonical Toffoli networks

```
begin
   Boolean decession;
   TTList = Read_Truth_Table(truth table);
   if Get_Output(TTList,0) = = 1 then
      set_target_line=1;
   end
   else
      set_target_line=0;
   index = 1;
   while index ≠ Size(TTList) do
        string_element = Retrive_Bit_String(TTList, index);
        toffoli_network = Design_Toffoli_Gate(string_element);
   end
   for i=1: Size(TTList) do
        string_element = Retrive_Bit_String(TTList, i);
        output_value = Retrive_Output(toffoli_network, string_element);
        decession = Compare_Output
                      (output_value, string_element,TTList);
        if decession = = False then
        gate_position = Retrive_Position(output_value,
                         toffoli_network, string_element);
        toffoli_network = Delete_Gate(toffoli_network. gate_position)
        end
        else
        No Operation:
        end
   end
end
```

- ***Step 7***: Now, compare the truth table output value with recently collected output value for each bit pattern. Function *Compare_Output()* performs this operation and return true value if both the outputs are same.
- ***Step 8***: If it returns false value then find out the unwanted gate where the target line bit value toggles first while scanning the Toffoli network from right to left.
- ***Step 9***: Delete the unwanted gate using function *Delete_Gate()* and update the circuit. This process continues until all the input bit patterns are passed by the network.

Obtain the ESOP-based Toffoli network which corresponds to the PPRM form of input specification. The details have been illustrated in algorithm2.

36.4 Experimental Results

36.4.1 Framework

The proposed algorithms have been implemented using JAVA SWING on an Intel(R) Core Duo CPU T6500, 2.10 GHz computer with 3 GB memory.

36.4.2 Illustration with Examples

In this section we have discussed the proposed deterministic models for PPRM cover generation. Each of the models has been illustrated here with examples.

Illustration of algorithm1

Example 1: Input specification to algorithm1 is a *truth table* which is shown in Fig. 36.3a, b.

- Start the process by reading the input *table* in top-down manner and find out the first bit string for which the function output is 1. Our goal is to make this initial table's output column all 0.
- It is found that for bit string [001], the function output (f_{out}) value is 1. Now, calculate the *k* value which represents number of 1's present in bit string [001] that is 1 and also remember the position of 1's.
- We obtain the cube [- - 1] from [001] bit string by replacing 0 by "–" and we store it in *PPRM cube list*.
- Consequently, a *k*-CNOT gate is defined by *Tof(b,c; t)*, where a control node appears in *b* and *c* variable line and target line (function output line) is denoted by variable *t*. The designed positive-control *k*-CNOT gate is then applied to all the bit string of the *TTList* starting from 2nd to 8th position.

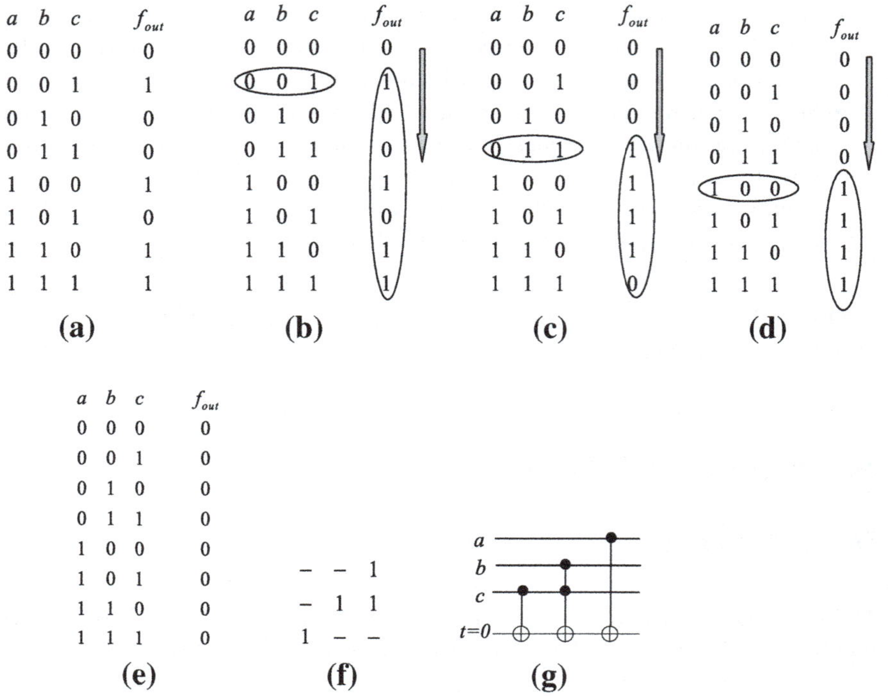

Fig. 36.3 **a** Input truth table T_1. **b** Creation of cube [- - 1] from table T_1. **c** Creation of cube [- 1 1] from table T_2. **d** Creation of cube [1- -] from table T_3. **e** Destination table T_4. **f** Equivalent PPRM cube list. **g** ESOP-based representation of PPRM cube list

- Now the *TTList* is modified. The modified *TTList* (T_2) is shown in Fig. 36.3c, d, e.
- The same procedure is repeated by reading the recently modified *TTList* in top-down manner and finding out the first bit string which has output value one, until all the output values of *TTList* are made 0.
- In this way all the cubes namely, [- - 1], [-11], and [11-] are generated. Finally, the PPRM cover corresponding to the input specification is extracted and has been depicted in Fig. 36.3f.

The PPRM expression corresponding to the input specification is $f(a,b,c) = c \oplus bc \oplus a$. Resulting ESOP-based Toffoli network corresponding to PPRM expression is designed and has been shown in Fig. 36.3g.

Illustration of algorithm2

The operating principle of the proposed technique which is stated in algorithm2 has been illustrated here with an example.

Example 2: A truth table, which is tabulated in Table 36.1, is the input specification here.

Table 36.1 Input truth table

a	b	c	f_{out}
0	0	0	0
0	0	1	1
0	1	0	1
0	1	1	0
1	0	0	1
1	0	1	0
1	1	0	0
1	1	1	1

- As the input truth table has 3 inputs and 1 output, initial circuit contains three control lines namely *a, b,* and *c* and one target line (*t*).
- Set the target line value (*t*) to 0 as in the given truth table, the output value for bit string 000 is 0.
- Next, generate all possible $2^n - 1$ Toffoli gate connection on target line *t* in such a way that bit value "1" indicate a positive control node and number of "1" (say *k*)in a input bit string build *k*-CNOT gate on target line *t*. Initial circuit is shown in Fig. 36.4a.
- Apply the input pattern [001] on the initial circuit at the input level and observe the output value of the target line at each level.

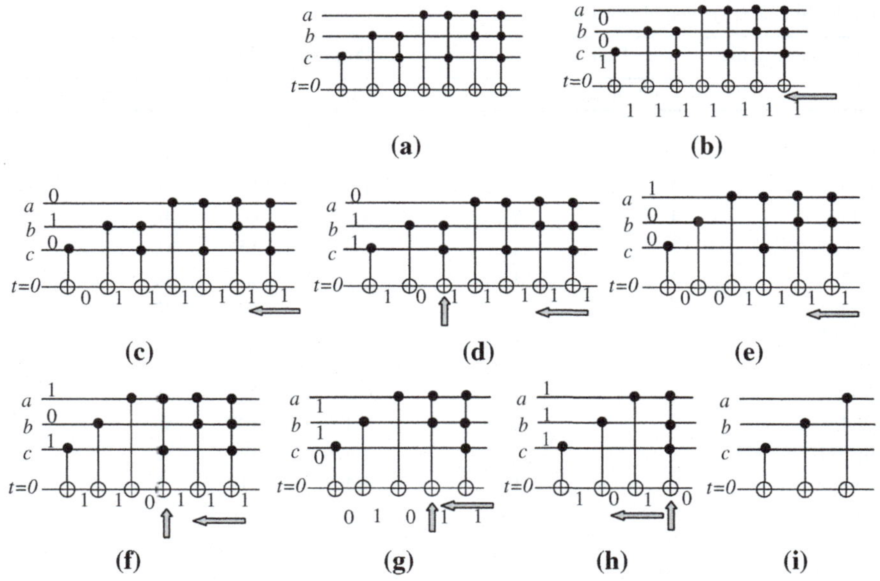

Fig. 36.4a **a** Initial circuit. **b** Checking with bit pattern [001]. **c** Checking with bit pattern [010]. **d** Checking with bit pattern [011]. **e** Checking with bit pattern [100]. **f** Checking with bit pattern [101]. **g** Checking with bit pattern [110]. **h** Checking with bit pattern [111]. **i** Final circuit

- As the output of the truth table for bit string [001] and the final output of the target line of the circuit is same i.e. is 1, so the initial circuit remain unaltered. Checking with bit string [001] has been shown in Fig. 36.4b.
- The same fact also happens for bit string [010] and circuit remain unaltered. Checking with bit string [010] has been shown in Fig. 36.4c.
- Now apply bit pattern [011]. It is found that the output of the truth table for bit string [011] and the final output of the target line of the circuit are different. So, scan the circuit from right to left and identify the first unwanted Toffoli gate where the bit value toggles first.
- Checking with bit string [011] is presented in Fig. 36.4d, where the unwanted Toffoli gate has been marked by an upward arrow. Hence, remove that Toffoli gate and update the circuit as shown in Fig. 36.4e.
- Similarly, bit pattern {100,101, 110, 111} one by one are passed through the circuit and update the circuit in each pass.
- Finally, we obtain the ESOP-based reversible circuit which realizes the PPRM expression i.e. $f(a,b,c) = c \oplus b \oplus a$, of input logic function as shown in Fig. 36.4e, f, g, h, i.

Table 36.2 Experimental results

Benchmark name	No. of circuit lines		Proposed technique		
	Control lines	Target lines	QC	GC	Execution time (s)
3_17_6	3	3	27	11	0
4gt12_24	4	1	55	3	0
4mod7	4	3	129	13	0
wim	4	7	330	46	0
Ex2	5	1	219	15	0
xor5	5	1	5	5	0
squar5	5	8	490	30	0
C7552_119	5	16	2,257	81	0
con1	7	2	239	18	0
Z4ml	7	4	448	32	0
rd84	8	4	2,477	107	0
misex1	8	7	2,576	164	0
9symml	9	1	4,368	210	0
apex4	9	19	166,275	3,472	1
x2	10	7	1,978	86	0
ex1010	10	10	237,841	4,961	2
add6	12	7	5,084	132	1
alu1	12	8	277	37	1
misex3c	14	14	466,912	6,302	29
pcler8	16	5	776	32	20

36.4.3 Implementation Details

Experimental results are presented in Table 36.2. We have used the benchmark circuits from [19] to test our algorithm. The first two column of the table show the name of the different benchmarks, number of primary inputs respectively. The third, fourth and fifth column denote the gate count (denoted by GC), quantum costs [19] (denoted by QC) and the execution time respectively.

36.5 Conclusions

In this work, we have presented two deterministic algorithms, which initially evaluate PPRM form of input specification and after that design ESOP-based circuit corresponding to the evaluated form and perform synthesis of reversible logics.

References

1. Landauer R (1961) Irreversibility and heat generation in the computing process. IBM Res Dev 5:183–191
2. Keyes RW, Landauer R (1970) Minimal energy dissipation in logic. IBM J Res Dev 152–157
3. Moore GE (1965) Cramming more components onto integrated circuits. Electronics 38(8)
4. Bennett CH (1973) Logical reversibility of computation. IBM J Res Dev 17:525–532
5. Toffoli T (1980) Reversible computing. In: Technical Memo-MIT/LCS/TM-151, MIT Lab for Computer Science
6. Perkowski M (1996) A unified approach to EXOR-based representation of Boolean functions. In: Proceedings of the XIX national conference circuit theory and electronics circuits, vol 1, Krynica, Poland, pp 27–41
7. Sasao T (1995) Representation of logic functions using EXOR operators. In: Proceedings of workshop applications of the Read-Muller expansion in circuit design, Makuhari, Japan, pp 308–313
8. Khan MMHA, Alam MS (1997) Algorithms for conversion of minterms to positive polarity Reed-Muller coefficients and vice versa. Inf Process Lett 62(5):223–230
9. Purwar S (1991) An efficient method of computing generalized Reed-Muller expansions from binary decision diagram. IEEE Trans Comput 40(11):1298–1301
10. Zhang YZ, Rayner PJW (1984) Minimization of Reed-Muller polynomials with fixed polarity. IEE Proc Comput Digit Tech 131(5):177–186
11. Harking B (1990) Efficient algorithm for canonical Reed-Muller expansion of Boolean function. IEE Proc Comput Digit Tech 137(5):366–377
12. Saluja KK, Ong EH (1979) Minimization of Reed-Muller canonic expansion. IEEE Trans Comput 28:535–537
13. Besslich PhW (1983) Efficient computer method for EXOR logic design. IEE Proc Comput Digit Tech 130(6):203–206
14. Habib MK (1992) Efficient algorithm for Reed-Muller expansions of completely and incompletely specified functions. In: Proceedings of the international symposium on logic synthesis and microprocessor architecture

15. Falkowski BJ, Chang CH (1995) An exact minimizer of fixed polarity Reed-Muller expansion. Int J Electron 79(3):389–409
16. Falkowski BJ, Chang CH (2000) Minimisation of k variable mixed-polarity Reed-Muller expansions. VLSI Des 11(4):311–320
17. Bandyopadhyay C, Roy D, Kole DK, Dutta K, Rahaman H (2013) ESOP-based synthesis of reversible circuit using improved cube list. In: International symposium on electronic system design
18. Fazel K, Thornton M, Rice JE (2007) ESOP-based Toffoli gate cascade generation. In: PACRIM, Victoria, BC, Canada, 22–24 Aug 2007. IEEE Press, pp 206–209
19. Wille R, Grosse D, Teuver L, Dueck GW, Drechsler R (2008) Revlib: an online resources for reversible functions and reversible circuits. In: 38th international symposium on multiple-valued logic, vol 24, May 2008, pp 220–225

Chapter 37
The Structural Studies of Luminescent Vapour Phase Etched Porous Silicon

Madhumita Das Sarkar, Debashis Jana and Kuntal Ghosh

Abstract Porous Silicon (PS) layers have been fabricated on p-type crystalline silicon (c-Si) using Reaction Induced Vapour Phase Stain Etching (RIVPSE) for different growth condition. The morphological properties of the porous Silicon samples have been investigated by using Scanning Electron Microscope (SEM). The Scanning electron micrographs indicate that these samples have structures of predominantly small size clusters having dual nature of pores instead of the postulated columns. Bonding structures for the samples have been investigated by using Fourier Transform Infrared Spectroscopy (FTIR) and compared for different metal induced (Zn, Al + Si, Si) vapour phase stain etching. The study reveals that Zn induced vapour phase stain etched porous silicon is most reactive surface and may be of greater life.

Keywords Porous silicon · Vapour phase stain etching · Fourier transform infrared spectroscopy · Scanning electron microscope

37.1 Introduction

Porous Silicon with various structure and morphology is widely used for solar cell and other optoelectronics device applications. Stain etching is being used recently in contrast to conventional anodization method because of its less setup costs and

M. Das Sarkar (✉) · D. Jana (✉)
Department of Computer Science Engineering, West Bengal University of Technology, Salt Lake, Kolkata, India
e-mail: dassarkar.madhumita@gmail.com

D. Jana
e-mail: debashis006jana@gmail.com

K. Ghosh
Indian Institute of Technology, Bombay, India
e-mail: kuntalgh24@gmail.com

S. Sengupta et al. (eds.), *Emerging Trends in Computing and Communication*, Lecture Notes in Electrical Engineering 298, DOI: 10.1007/978-81-322-1817-3_37,

especially its ability of batch processing. High efficiency LED, good antireflection property, better sensitivity in sensor have already been reported by using either pure stain etching or by using modified stain etched method [1].

In vapour phase etching porous silicon has been fabricated by exposing Si substrate to the vapour of HF and HNO_3 mixture. However this process takes several hours to etch the substrate. To reduce the incubation period some metal is added in the mixture to enhance the reaction rate. Aluminium, Silicon are reported to use as an induced material and is called Reaction induced Vapour Phase Stain Etching. In this work the Porous Silicon has been fabricated by using a metal Zn in mixture of acid solution. SEM image is done to visualize the porous structure. The bonding structures have been investigated using FTIR. A comparative analysis has been done for different material induced (Zn, Al + Si, Si) vapour phase stain etching to predict the stability of the surface layer according to the chemical bonding. The formation mechanism and even some of the simplest properties of the porous silicon material are still the matter of investigation [2–11].

37.2 Experimental

37.2.1 RIVPSE Etching

A Porous Silicon layer was fabricated by Reaction Induced Vapour Phase Stain Etching (RIVPSE) process of a P-type c-Si (orientation: 100, boron doped, double side polished surface) wafer with a resistivity of 3–10 Ohm-cm and a thickness of 300 μm. Initially, the c-Si wafer was cleaned by solvent method using and heavy organic contaminants were removed using Standard-clean method (RCA). The c-Si samples were then dipped in $HF:H_2O$ (1:10) solution at normal temperature to remove native oxide as well as to maintain the homogeneity of surface. The cleaned sample mounted on top of the beaker is exposed to HF: HNO_3 vapours. Here the white vapour arising from the dissolution of sacrificial Zn pieces transformed into dense reddish brown vapours which reacted with the silicon substrate supported on the top of the beaker.

Porous silicon samples have been prepared for different oxidant ratio, etching time and height of sample surface from solution. Optimization of growth for fabrication of porous silicon is done by taking three parameters in RIVPSE, such as oxidant ratio, etching time, height of sample from liquid level.

Best bright luminescence come under the UV light from PS substrate which was exposed 4 min to the acidic vapor with oxidant ratio ($HF:HNO_3$) 6:1 and 8.4 cm distance between silicon surface and electrolyte solution. Different materials (Al + Si, Si) have been induced in acid mixture of HF and HNO_3 (6:1) keeping other parameters same. Here Sample N1, N2, N3 have been prepared by RIVPSE adding Zn continuously during etching. Continuous addition of Zn offers less control over the fabrication, so the next sample N4 has been fabricated by adding fixed amount of Zn for only one time.

37.3 Result and Discussion

37.3.1 Scanning Electron Microscope Picture

The surface of Zn induced RIVPSE based PS film have been investigated by scanning electron microscope (VEGA II TESCAN). Before loading the samples into SEM chamber, the samples were coated with a thin layer of gold film to avoid charging effect. Figures 37.1, 37.2 and 37.3 shows the scanning microscope image of a Porous silicon surface with oxidant ratio 4:1, 6:1 and 8:1 etching time 4 min and 8.4 cm height. Here Zn as a metal to induce the reaction is added (1.68 g) for one time only during the RIVPSE. There is a high possibility to form a single etched PS layer by adding Zn for only one time. There are some crystallite structures in some places throughout the surface (Fig. 37.4).

37.3.2 FTIR Study

The bonding structure in porous Si layers was studied using Bruker 3,000 Hyperion Microscope with vertex 80 FTIR system equipped with 15x, 20x transmission/reflection objectives, automated x–y scanning stage for spectroscopic mapping; single element MCT-A detector. The readings were taken for sample N-1, N-2 and N-3 in single point mode.

The FTIR absorption peaks and the corresponding bonds were shown in the Table 37.1.

With increasing HF concentration from sample N1 to N3 ($HF:HNO_3$::4:1, 6:1 and 8:1) the possibility of having (Si–Hx) bonds increases in N3 ($HF:HNO_3 = 8:1$) sample. The presence of these hydride bonds is more prone to aging and therefore may be of more instable in nature. The degree of oxidation is more in N1 sample in comparison N2 and N3 with already confirmed by the presence of Si–OH bond with sharp peak found in the N1 sample.

The relative intensity of Si–OH to Si–Hx peaks are increasing with increasing oxidant ratio. The presence of intense Si–OH peak may be treated as the best chemically active surface platform for any kind of protein attachment. The shift of the peak with respect to inverse wavelength may be correlated with pore diameter to the depth of the pore (Figs. 37.5, 37.6, and 37.7).

The FTIR absorption peaks and the corresponding bonds were shown in the Table 37.2

Available oxidizing agent ($NO/(NO)_2$, NO_2) in wavenumber 1,625.63, 16,660.06 cm^{-1} is much more in Zn induced vapour phase etched PS in comparison to Al + Si and Si induced vapour phase etched PS. For this gaseous

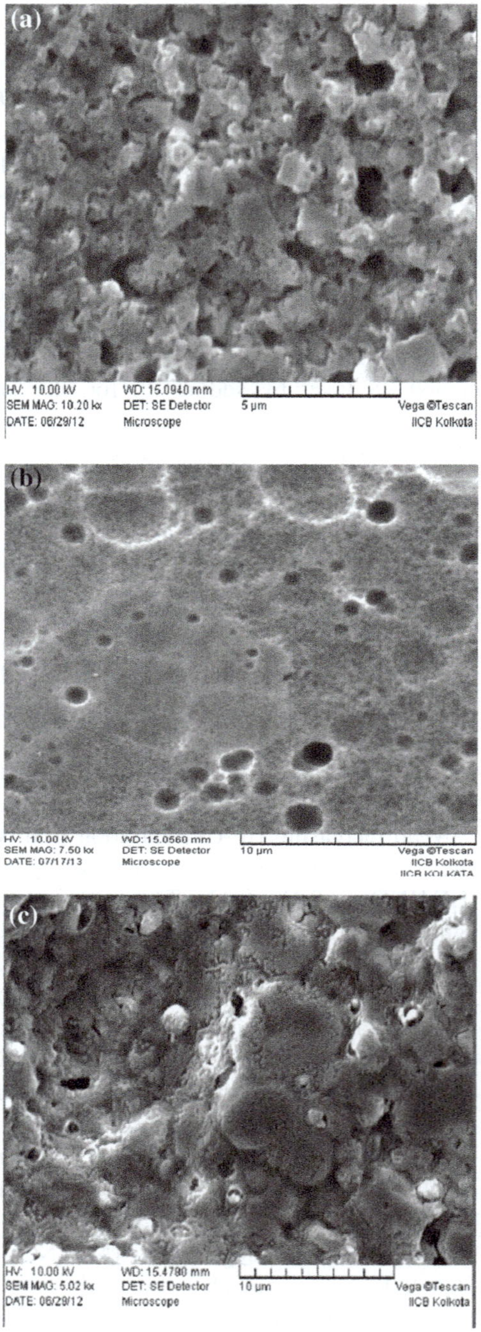

Fig. 37.1 SEM image of RIVPSE PS of **a** HF:HNO_3 (4:1) **b** HF:HNO_3 (6:1) **c** HF:HNO_3 (8:1) (*top view*)

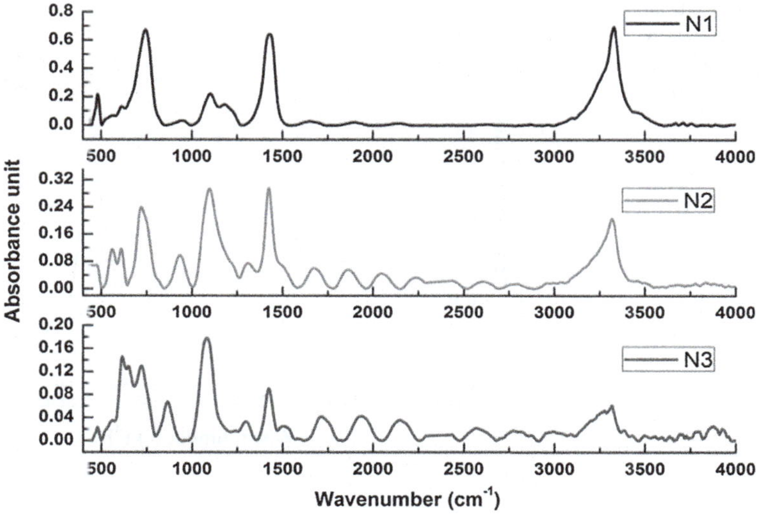

Fig. 37.2 FTIR single point image of sample N-1 (HF/HNO_3 4:1), N-2 (HF/HNO_3 6:1) and N-3 (HF/HNO_3 8:1)

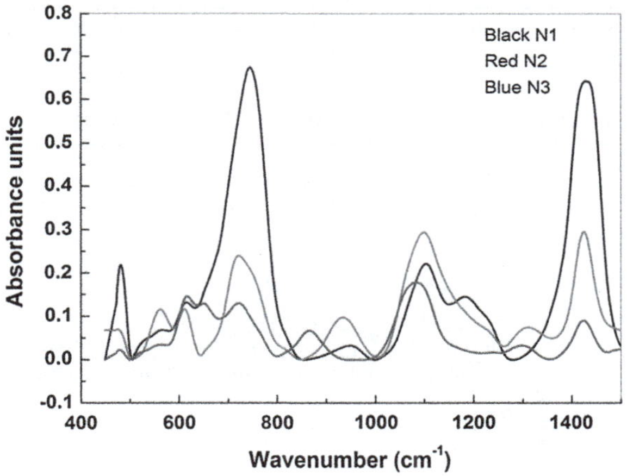

Fig. 37.3 FTIR single point image of sample N-1, N-2 and N-3 (wave number: 400–1,400 cm^{-1})

etching SiF_4 is the main gaseous product keeping SiF_6 as a complex formed on the surface. The competitive nature of Si–OH to Si–H_x is also maximum in case of Zn induced vapour phase etched PS and gradually decreases to Al + Si induced vapour phase etched PS and least in Si induced vapour phase etched PS. The self passivating nature of Zn may lead to the less incubation time to the etching and therefore Si–H_x bonds are less in case of Zn induced vapour phase etched PS. Here partial oxidation is more in Si induced vapour phase etched PS.

Fig. 37.4 FTIR single point image of sample N-1, N-2 and N-3 (wave number: 1,400–4,000 cm^{-1})

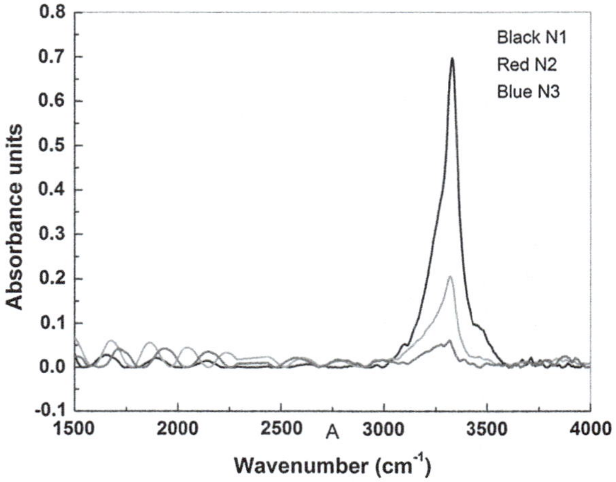

Table 37.1 The FTIR absorption peaks and the corresponding bonds

Sample	Si–H bonding (cm^{-1})	Si–OH bonding (cm^{-1})	Si–O–Si asymmetric stretching (cm^{-1})	Si=O (cm^{-1})	Si–H stretching (cm^{-1})	Si–OH stretching (cm^{-1})
N-1 (4:1, 4 min)	614.56	744.59	1,103.86	1,428.58	2,137.90	3,328.12
N-2 (6:1, 4 min)	611.56	724.84	1,100.15	1,426.59	2,021.57, 2,196.81	3,314.05
N-3 (8:1, 4 min)	616.08	721.80	1,084.15	1,424.78	2,147.83	3,316.41

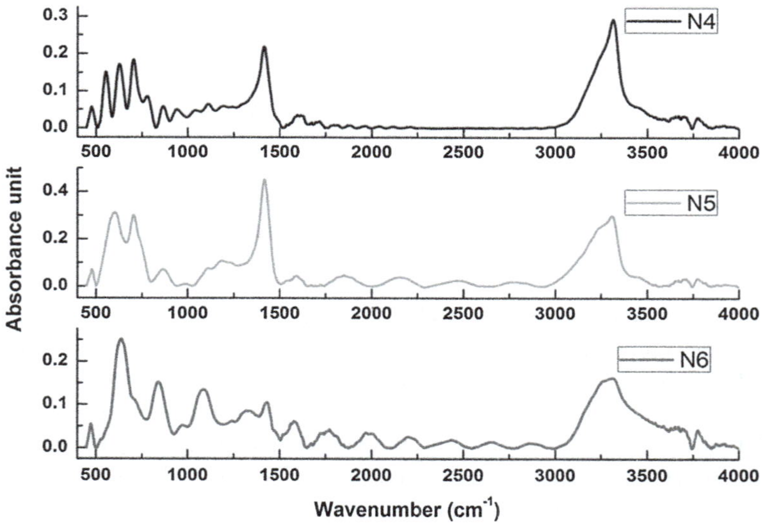

Fig. 37.5 FTIR multi point image of sample N-4, N-5 and N-6 (HF/HNO_3 6:1)

Fig. 37.6 FTIR multi point image of N-4, N-5 and N-6 (wave number: 400–1,400 cm^{-1})

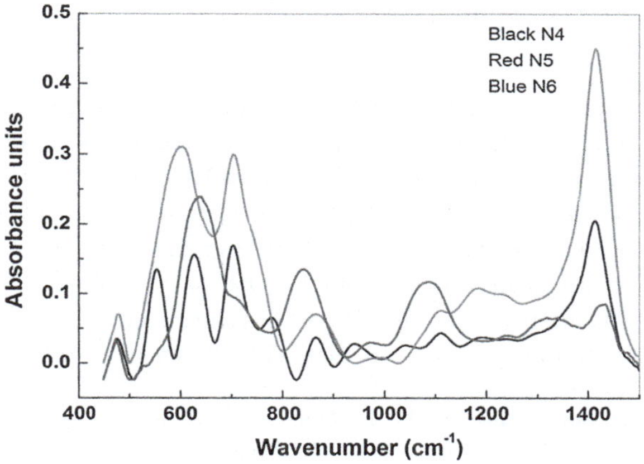

Fig. 37.7 FTIR multi point image of N-4, N-5 and N-6 (wave number: 1,500–4,000 cm^{-1})

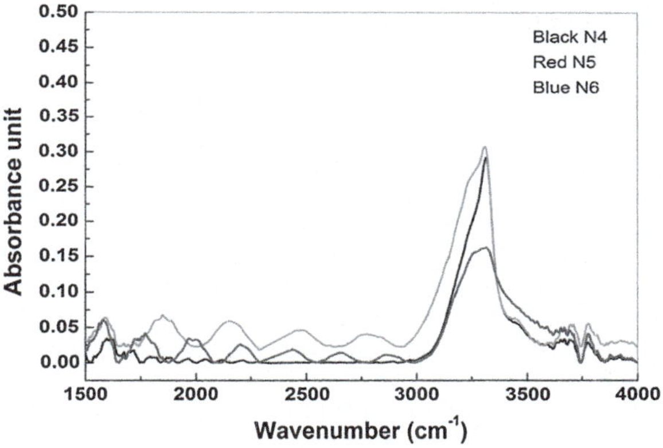

Table 37.2 The FTIR (multi point) absorption peaks and the corresponding bonds

Sample	Si–H bonding (cm^{-1})	Si–OH bonding (cm^{-1})	Si–O–Si asymmetric stretching (cm^{-1})	Si=O (cm^{-1})	Si–H stretching (cm^{-1})	Si–OH stretching (cm^{-1})
N-4 (6:1, 4 min Zn induced)	627.43	703.33	1,111.26	1,413.75	Not found	3,315.45
N-5 (6:1, 4 min, Si + Al induced)	603.51	704.01	1,110.65, 1,183.13	1,415.35	2,021.57, 2,196.81	3,309.61
N-6 (6:1, 4 min, Si induced)	635.94	721.80	1,088.22	1,433.46	2,147.83	3,314.52

37.4 Conclusion

The morphology of porous silicon layers produced using vapour phase of an HF:HNO_3 mixture depends on the technology parameters. The intensity of peaks responsible for Si–O–Si and Si=O bonding are lower and more sharp for Zn induced vapour etched porous silicon surfaces than Al or Al + Si induced etched surface. In the Zn induced etching the reaction is more selective to the target surface. Peak intensity for Si–OH was also found good. It might be removed by annealing the porous silicon in future study. Further studies are required to correlate the bonding structure with EDXS from SEM.

Acknowledgments We thankfully acknowledge Dr. Rupak Bhadra, IICB Kolkata for Scanning Electron microscope images. We also heartily acknowledge Mr. Kuntal Ghosh, IIT Bombay for FTIR data.

References

1. Xu J, Steckl AJ (1995) Stain etched porous silicon visible light emitting diodes. J Vac Sci Technol B 13(3):1221–1224
2. Palsule LC, Liu S, Gangopadhyay S, Holtz M, Lamp D, Kristiansen M (1997) Electrical and optical characterisation or crystalline silicon/porous silicon heterojunctions. Solar Energ Mater. Solar Cells 46:261–269
3. Pickering C, Beale MJJ, Robbins DJ, Pearson PJ, Greef R (1984) Optical studies of the structure of porous silicon films formed in p-type degenerate and non-degenerate silicon. J Phys CI7(35):6535–6552
4. Deal BE, Grove AS (1965) General relationship for the thermal oxidation of silicon. J Appl Phy 36(12):3770–3778
5. Akiyama T, Kageshima H (2003) Microscopic mechanism of interfacial reaction during Si oxidation. App Surf Sci 216:270
6. Koynov S, Brandit MS, Stutzmann M (2006) Black nonreflecting silicon surfaces for solar cells. Appl Phys Lett 88:203107
7. Pickering C, Beale MJJ, Robbins DJ, Pearson PJ, Gree R (1984) Optical studies of the structure of porous silicon films formed in p-type degenerate and non-degenerate silicon. J Phys CI7:6535–6552
8. Duttagupta SP et al (1998) Photovoltaic device applications of porous microcrystalline silicon. Solar Energ Mater. Selar Cells 52:271–283
9. Schirone L, Sotgiu G, Parisini A, Montecchi M (1997) Optical and morphological properties of stain-etched porous silicon films for anti-reflection coatings of photovoltaic devices. Solid State Phenom 54:59–64
10. Ben Ranha M, Bessais B (2010) Enhancement of photovoltaic properties of multicrystalline silicon solar cells by combination of buried metallic contacts and thin porous silicon. Sol Energ 84:486–491
11. Salman KA, Omar K, Hassan Z (2012) Effective conversion efficiency enhancement of solar cell using ZnO/PS antireflection coating layers. Sol Energ 86:541–547

Chapter 38
Online Testable Conservative Adder Design in Quantum Dot Cellular Automata

Arijit Dey, Kunal Das, Debashis De and Mallika De

Abstract Garbage count minimization and low power, lossless conservative full adder design and its online testing in Quantum dot Cellular Automata is prime research interest of this work. Parity preserving reversible logic design as well as conservative logic design is a lossless paradigm in Nanotechnology. Errors can be detected by means of parity in conservative logic design. We introduce a conservative logic gate to design full adder with zero garbage count. The proposed two conservative logic gate (PCLG) is universal in nature. A tester reversible logic gate (TRLG) is designed to perform online test of proposed conservative logic gate (PCLG). We demonstrate the most promising two PCLG and a TRLG to design full adder and to online test of PCLG respectively. We compared our PCLG with well-known Fredkin gate in terms of implementation of thirteen standard functions.

Keywords Parity preserving · Conservative logic gate · Reversible logic gate · 5—input majority voter · Online testing

A. Dey (✉) · K. Das (✉)
B. P. Poddar Institute of Management and Technology, 137, VIP Road, Kol 700052, India
e-mail: ad.computerapplication@gmail.com

K. Das
e-mail: kunaldasqca@gmail.com

D. De
Department of Computer Science and Engineering, West Bengal University of Technology, BF-142, Sector-I, Salt Lake City, Kolkata 700064, India
e-mail: dr.debasis.de@gmail.com

D. De
School of Physics, University of Western Australia, M013, 35 Stirling Highway, Crawley, Perth WA 6009, Australia

K. Das · M. De
Department of Engineering and Technological Studies, Kalyani University, Kalyani 741235 West Bengal, India
e-mail: de.mallika@yahoo.com

S. Sengupta et al. (eds.), *Emerging Trends in Computing and Communication*, Lecture Notes in Electrical Engineering 298, DOI: 10.1007/978-81-322-1817-3_38,

38.1 Introduction

In conventional computers, the computation is irreversible. Irreversible means once the output is generated from the logic block the input bits are lost, so that the power is retain in the system. Reversible computing is a possible solution to compute with almost zero power dissipation. Launder [1] has proved that each bit information lost produce K_BT ln 2 joules of heat energy for irreversible logic computation, where K_B is Boltzmann's constant and T is the absolute temperature at which computation is performed. Bennett [2] has proved zero power dissipation in case of reversible logic computation. The key feature of a reversible logic is to recover bit loss. But reversibility feature does not able to identify bit error in the circuit. The fault tolerant circuit can be obtained by means of parity. In earlier several proposal have been introduced to design fault tolerant circuit by parity preserving reversible gate or conservative logic gate [3, 4].

In this paper, we are presenting a proposed two conservative logic gate namely PCLG which shows parity preserving approach to design a full adder circuit. We also proposed a reversible logic gate namely TRLG for computing the input–output parity that will test the PCLG. Here we focus to introduce a full adder circuit design using PCLG with zero power dissipation and minimum garbage count. Here we introduced a reversible logic gate as a tester block to test the proposed conservative logic gate.

38.2 Preliminaries

38.2.1 Reversible Logic Gate

The mapping of input vector I_V and output vector O_V is bijective, i.e., each input yields a distinct output (one- to -one mapping) and number of inputs in I_V and outputs in O_V are same. This type of gate is known as Reversible Logic Gate [5, 6].

Feynman Gate, Toffoli Gate, Fredkin Gate are commonly performed as reversible logic gate Fig. 38.1 shows the truth table of Feynman gate, Toffoli gate, and Fredkin gate. Reversible logic gate must have the following four key features [7].

1. Minimum number of garbage output;
2. Minimum input constants;
3. Minimum circuit level;
4. Minimum number of gates;

Fig. 38.1 Truth table of **a** feynman gate, **b** toffoli gate, **c** fredkin gate

(a)

INPUT	A	0	0	1	1
	B	0	1	0	1
OUTPUTS	P	0	0	1	1
	Q	0	1	1	0

(b)

INPUT	A	0	0	0	0	1	1	1	1
	B	0	0	1	1	0	0	1	1
	C	0	1	0	1	0	1	0	1
OUTPUTS	P	0	0	0	0	1	1	1	1
	Q	0	0	1	1	0	0	1	1
	R	0	1	0	1	0	1	1	0

(c)

INPUT	A	0	0	0	0	1	1	1	1
	B	0	0	1	1	0	0	1	1
	C	0	1	0	1	0	1	0	1
OUTPUTS	P	0	0	0	0	1	1	1	1
	Q	0	0	1	1	0	1	0	1
	R	0	1	0	1	0	0	1	1

38.2.2 Conservative Logic Gate

Conservative logic gate is input vector I_V and output vector O_V are mapped in a way where parity of inputs in I_V and outputs in O_V are preserved i.e., number of 1's present in each input and number of 1's present in output must be same. In earlier an algorithm for K × K conservative logic has been introduced [3], where the inputs are defined in (K + 1) number of sets in terms of parity.

38.2.3 QCA Basics

QCA is becoming an emerging technology in VLSI design. It was first introduced by C.S.Lent in the year 1993 [8, 9]. Each QCA cell consists of 4 dots in a square and two extra electrons are confined within the cell. The maximum distance of electron gives the polarization. The electron gives two states of polarization: P = +1.00 gives logic 1 and P = −1.00 gives logic 0, shown in Fig. 38.2a. Information is flowing by means of polarization from one QCA cell to next QCA cell in influence of knik energy. The knik energy is inversely proportional to the distance between two charges qi qj defined as

$$E_{i,j}^{kink} = \frac{1}{4\pi\varepsilon_0\varepsilon_r|r_i - r_j|} q_i q_j \tag{38.1}$$

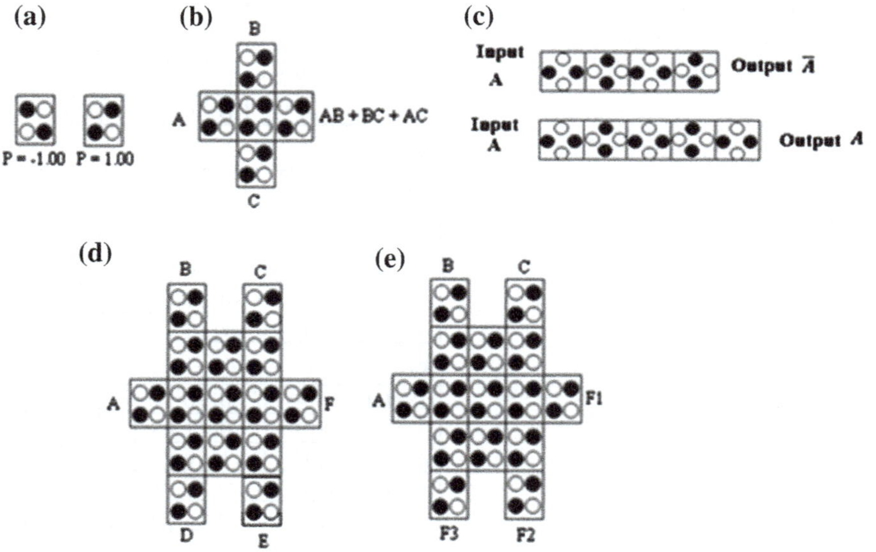

Fig. 38.2 **a** QCA cell polarization. **b** Majority voter. **c** Inverter chain. **d** Five input majority voter. **e** Tripple fan—out butterfly tile

where ε_0 the permittivity of is free space and ε_r is the relative permittivity. In earlier, several proposals have been reported regarding gates in quantum dot cellular automata [10–14]. 3 input Majority Voter (MV) is described as MV (A, B, C) = AB + BC + AC shown in Fig. 38.2b. We also rotate each quantum cell by 45° and make a binary wire, which creates an Inverter chain shown in Fig. 38.2c [15]. In previous, the cascading effect has been described to design more complex gate in quantum dot cellular automata [16]. The five input majority voter is shown in Fig. 38.2d, the two inputs are −1.00 polarized to produce the 3 input AND functionality. Butterfly tile is utilized to produce multiple fan-outs, shown in Fig. 38.2e.

38.3 Proposed Conservative Full Adder

The testable reversible logic gate to design reversible full adder is reported in [10, 17]. The garbage count optimization or zero garbage, lossless in terms of bits circuit design becomes prime research problem. In this paper, our attempt is to introduce a conservative logic gate to design a fault tolerant [18, 19], lossless zero garbage full adder. We have designed the architecture of full adder, which requires one PCLG block shown in Fig. 38.3a, which is more effective than other previously reported full adder in terms of size of the architecture [20]. Here we use a 4 × 4 mapping technique to design the effective architecture of the full adder.

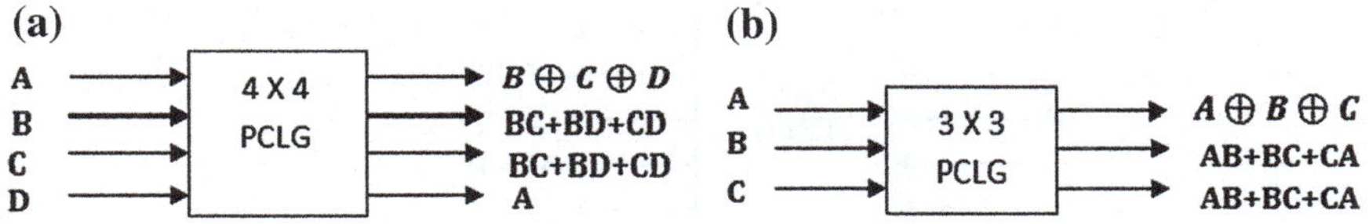

Fig. 38.3 Block diagram of proposed **a** 4 × 4 conservative logic gate (*PCLG*), **b** 3 × 3 conservative logic gate (*PCLG*)

Table 38.1 Input vector showing set for 4 × 4 PCLG

I_V	Inputs	I_V set	Outputs	O_V	O_V set
	ABCD		PQRS		
X0	0000	S0	0000	X0	S0
X1	0001	S1	1000	X8	S1
X2	0010	S1	1000	X8	S1
X3	0011	S2	0110	X6	S2
X4	0100	S1	1000	X8	S1
X5	0101	S2	0110	X6	S2
X6	0110	S2	0110	X6	S2
X7	0111	S3	1110	X14	S3
X8	1000	S1	0001	X1	S1
X9	1001	S2	1001	X9	S2
X10	1010	S2	1001	X9	S2
X11	1011	S3	0111	X7	S3
X12	1100	S2	1001	X9	S2
X13	1101	S3	0111	X7	S3
X14	1110	S3	0111	X7	S3
X15	1111	S4	1111	X15	S4

The input vector I_V (A, B, C, D) are mapped to the outputs in output vector O_V ($P = B \oplus C \oplus D$, $Q = BC + BD + CD$, $R = BC + BD + CD$, $S = A$) shown in Fig. 38.3a. We classify the inputs in I_V in the concept of set [3]. We define the sets in terms of presence of 1's in I_V. We use 4 × 4 mapping technique to design the proposed conservative logic (PCLG), so that the number of sets are five (S0, S1, S2, S3, S4) [3]. The following Table 38.1 shows the classification of set. I_V represents numbers of input vectors of 4 × 4 PCLG, which is decimal equivalent of input 'ABCD'. I_V Set represents 1's count of input 'ABCD'. Another 3 × 3 PCLG is designed to implement zero garbage full adder. In the Fig. 38.3b, the block diagram of 3 × 3 PCLG is shown. The input vector I_V (A, B, C) are mapped to the outputs in output vector O_V ($P = A \oplus B \oplus C$, $Q = AB + BC + CA$, $R = AB + BC + CA$). The proposed PCLG is universal i.e., we can implement the three basic gates from our PCLG block. A single PCLG block is needed to implement conservative full adder that is generate zero garbage count in production, shown in Fig. 38.3b. We also explored a tester—reversible logic gate (TRLG) as a tester block to test our proposed conservative logic gate shown in Fig. 38.4a.

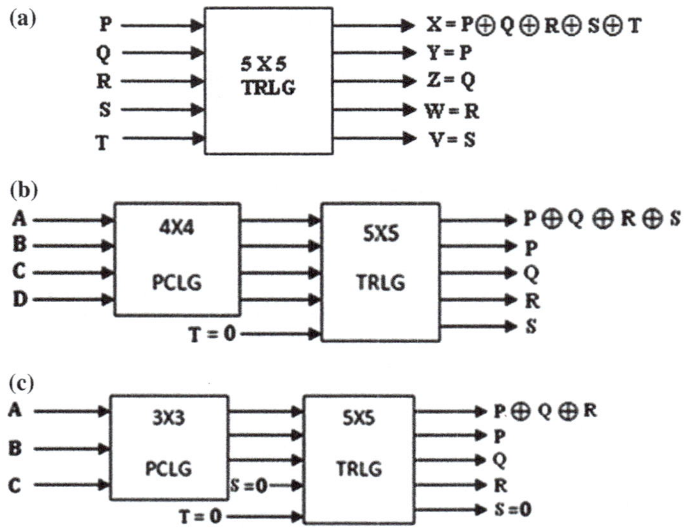

Fig. 38.4 **a** Block diagram of TRLG. **b** Block diagram online testing of 4 × 4 PCLG using 5 × 5 proposed TRLG. **c** Block diagram online testing of 3 × 3 PCLG using 5 × 5 proposed TRLG

38.4 Online Testing

In this section, we made an attempt to explore the online testing strategy. The online testing strategy is proposed for reversible logic gate test in Quantum computing [10, 17, 21]. In this regards, we proposed a 5 × 5 reversible logic gate known as TRLG, the tester gate which will test PCLG without any other computation i.e. online test will performed. The TRLG is also designed with QCADesigner [22] and simulated. The input vector I_V (P, Q, R, S, T) of the reversible logic is mapped to the output vector O_V ($X = P \oplus Q \oplus R \oplus S \oplus T$, $Y = P$, $Z = Q$, $W = R$, $V = S$) shown in Fig. 38.4a. The two pair rail of PCLG and TRLG is shown in Fig. 38.4b and c to test PCLG online. When TRLG input $T = 0$, the tester will perform the online testing if we railed with PCLG as shown in Fig. 38.4b. The tester output will produce same parity output as input parity i.e. if input I_V is even parity or odd parity, the output 'X' of TRLG will produce same parity and corresponding input vector I_V (A, B, C, D) of 4 × 4 PCLG will reflect with other output (Y, Z, W, V). Similarly, we can test any 4 × 4 Conservative logic gate (CLG) as well as 3 × 3 CLG. In Fig. 38.4c shows the 3 × 3 PCLG testability with two pair railed with 5 × 5 TRLG, In this test we have to put input S as well as T input to zero i.e. for $S = T = 0$., the tester will perform the online testing.

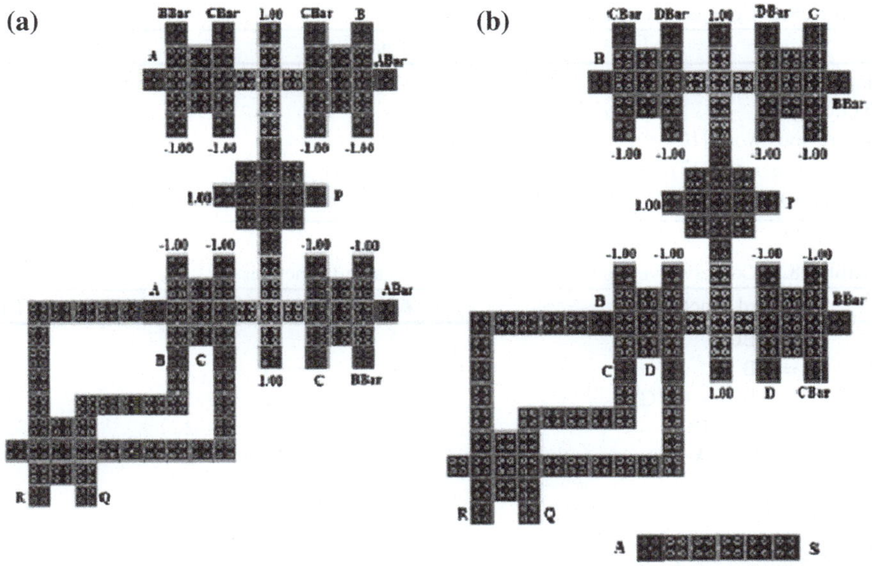

Fig. 38.5 **a** 3 × 3 PCLG QCA layout designed with QCADesigner. **b** 4 × 4 PCLG QCA layout designed with QCADesigner

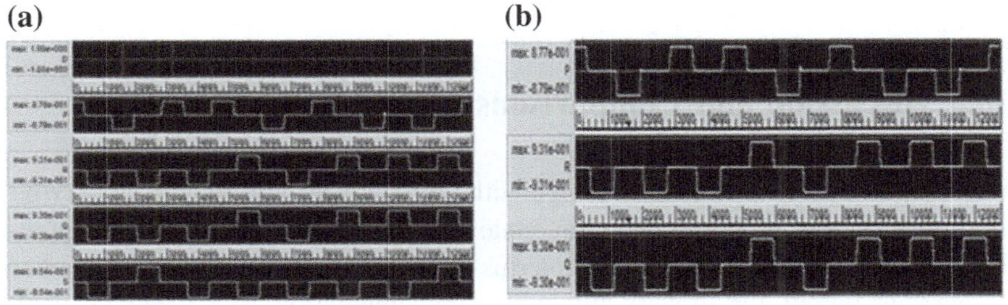

Fig. 38.6 **a** simulation result of 4 × 4 PCLG. **b** Simulation result of 3 × 3 PCLG

38.5 Simulation

We have designed and simulate our proposed PCLG and TRLG by QCADesigner [22]. The Signal Distribution Network is used to generate ABAR, BBAR and CBAR from input A, B and C [23], for simplicity, we have not shown the Signal Distribution Network in this design, Vector table in QCADesigner tool is utilized for Signal Distribution Network. In Fig. 38.5a and b, the designing architecture of our 3 × 3 PCLG and 4 × 4 PCLG with QCADesigner tool are shown respectively. The simulated result of 4 × 4 PCLG and 3 × 3 PCLG are shown in Fig. 38.6a and b respectively.

Table 38.2 Comparison of lossless full adder design

	No. of gates	Garbage output
Fredkin gate	5	4
Toffoli gate	4	2
In previous work [10]	2	1
In previous work [17]	1	2
In this work	1	0

Table 38.3 Conservative logic design of thirteen standard functions

	Standard functions	#Fredkin	#CLK	#4 × 4PCLG	#CLK
1	F = ABC	2	8	2	2
2	F = AB	1	4	1	1
3	F = ABC + AB'C'	3	12	1	3
4	F = AB + BC	2	8	2	3
5	F = AB + A'B'C	5	16	2	2
6	F = AB + A'B'C'	4	12	2	3
7	F = ABC + A'BC' + AB'C'	6	16	1	3
8	F = A	1	4	1	1
9	F = AB + AC + BC	5	16	1	1
10	F = AB + AB'	1	4	2	2
11	F = A'B + BC + A'B'C'	6	16	4	3
12	F = AB + A'B'	2	8	1	1
13	F = ABC + A'B'C + AB'C' + A'BC'	3	12	1	1
	Total	41	136	21	26

38.6 Discussion and Comparisons

We compared lossless full adder design with Fredkin gate, Toffoli gate, in previous work [10] and our proposed PCLG in context of number of gates, garbage outputs. Table 38.2 shows the comparisons in terms of numbers of CLG/RLG and numbers of garbage output for lossless full adder implementation. We generate all 13 standard functions with our proposed PCLG and we compare it with the Fredkin gate shown in Table 38.3. Table 38.3 shows 13 standard functions implement [3]. Fredkin uses total 41 gates with 136 clocks where as our work present the functions with only 21 gates which produce 50 % reduction in terms of number of gates and with 26 clocks which produce 80 % reduction in clocking zones.

38.7 Conclusion

This work presents a new lossless and effective architecture of full adder using conservative logic design. A new reversible logic TRLG is demonstrated for online testing of the proposed advanced PCLG. In this work, we made an attempt to optimize the garbage count of the conservative full adder design. The online testing strategy is also described with a tester TRLG. This tester can also be applicable to test online any 2×2, 3×3, 4×4 conservative logic gate. Finally, we can conclude that zero garbage count, testable lossless conservative full adder design can be applicable to implement advanced cryptographic architecture.

References

1. Landauer R (1961) Irreversibility and heat generation in the computing process. IBM J Res Dev 5(3):183–191
2. Bennett CH (1973) Logical reversibility of computation. IBM J Res Dev 17(6):525–532
3. Das K, De D (2010) Characterization, test and logic synthesis of novel conservative and reversible logic gates for Qca. Int J Nanosci 9(03):201–214
4. Das K, De D (2010) Novel approach to design a testable conservative logic gate for QCA implementation. In: 2010 IEEE 2nd international advance computing conference (IACC). IEEE
5. Feynman R (1985) Quantum mechanical computers. Opt News 11:11
6. Toffoli T (1980) Reversible computing. Tech Memo MIT/LCS/TM-151, MIT Lab for Computer Science
7. Fredkin E, Toffoli T (2002) Conservative logic. Collision-based computing, vol. 1 Springer, London pp 47–81
8. Lent CS et al (1993) Quantum cellular automata. Nanotechnology 4(1):49
9. Lent CS, Tougaw PD, Porod W (1993) Bistable saturation in coupled quantum dots for quantum cellular automata. Appl Phys Lett 62:7–14
10. Vasudevan A, Dilip P et al (2006) Reversible-logic design with online testability. In: IEEE transactions on instrumentation and measurement 55.2
11. Sen B, Ganeriwal S, Sikdar BK (2013) Reversible logic-based fault-tolerant nanocircuits in QCA. ISRN Electronics 2013
12. Das K, De D (2011) A study on diverse nanostructure for implementing logic gate design for QCA. Int J Nanosci 10(01n02):263–269
13. Das K, De D (2009) A novel approach of and-or-inverter (AOI) gate design for QCA. In: 2009 4th international conference on computers and devices for communication (CODEC 2009). IEEE
14. Lent CS, Taugaw PD, Porod W, Bernstein GH (1993) Quantum dot cellular automata. Nanotechnology 4:49–57
15. Tougaw PD, Lent CS (1996) Dynamic behavior of quantum cellular automata. J Appl Phys 80(8):4722–4736
16. Das K, De D (2011) Characterisation, applicability and defect analysis for tiles nanostructure of quantum dot cellular automata. Mol Simul 37(03):210–225
17. Thapliyal H, Ranganathan N (2010) Reversible logic based concurrent error detection methodology for emerging nanocircuits. In: 2010 10th IEEE conference on nanotechnology (IEEE-NANO). IEEE
18. Parhami B (2006) Fault-tolerant reversible circuits. In: 2006 Fortieth Asilomar conference on signals, systems and computers (ACSSC'06). IEEE
19. Haghparast M, Navi K (2008) A novel fault tolerant reversible gate for nanotechnology based systems. Am J Appl Sci 5(5):519
20. Fredkin E, Toffoli T (1982) Conservative logic. Int J Theor Phys 21:219–253
21. Ma X et al (2006) Testing reversible 1D arrays for molecular QCA. In: 2006 21st IEEE international symposium on defect and fault tolerance in VLSI systems (DFT'06). IEEE
22. Walus K (2002) ATIPS laboratory QCADesigner homepage. ATIPS Laboratory, University of Calgary, Canada
23. Tougaw D, Khatun M (2013) A scalable signal distribution network for quantum-dot cellular automata, pp 1–1

Chapter 39
Calculation of Bridge Function and Thermodynamic Properties of Lennard-Jones Fluid Using Integral Equation Theory

Rupa Pal

Abstract The integral equation theory is nowadays one of the most widely used approaches for prediction of thermodynamic behaviour of homogeneous liquid system based on Ornstein–Zernike equation together with Closure relation. For improvement of correlation functions these closure properties are replaced by bridge function expansion. In this paper, the bridge function is first calculated from soft sphere mean spherical model approximation (SMSA) theory. Then, for systematic study of phase behavior of L-J fluid, the thermodynamic properties of interest like isothermal compressibility and chemical potential are derived from state of equations based on integral equation theory.

Keywords Bridge function · Lennard-Jones · Potential · Isothermal compressibility · Chemical potential

39.1 Introduction

The current approach of phase behaviour of simple liquid systems along with their critical parameters is based on integral equation theory as it is able to describe the structural and thermodynamic properties of both liquids and vapour phases. A large amount of work has been devoted on integral equation theory because of two reasons:

(1) To obtain analytically representable results about the structure and thermodynamic properties of liquids and (2) To solve the inverse problem and reconstruct the form of intermolecular potential $u(r)$; if(r) is known.

R. Pal (✉)
Department of Engineering Chemistry, B. P. Poddar Institute of Management and Technology, Kolkata, India
e-mail: rupray@gmail.com

S. Sengupta et al. (eds.), *Emerging Trends in Computing and Communication*, Lecture Notes in Electrical Engineering 298, DOI: 10.1007/978-81-322-1817-3_39,

For a homogeneous isotropic fluid of number density ρ, the basic integral equation is the well-known Ornstein–Zernike (OZ) equation [1] that establish a relation between the direct correlation function $c(r)$ and the total correlation function $h(r)$ between two atoms separated by a distance r; these are connected by a suitable closure relation

$$h(r) = c(r) + \rho \int c(|r - r^1|)h(r^1)dr^1 = c(r) + \gamma(r) \tag{39.1}$$

Here $\gamma(r)$ is the indirect correlation function via all possible chains of atoms correlated directly. The closure relation depends on the so-called bridge function $B(r) = \sum_{n=0}^{1} \{\varepsilon_n(r)\}$ that represents an infinite sum of n points elementary diagrams [2] and thus is practically incomputable.

To build tractable expressions for B(r), different routes of empirical closure have been explored. Some of them are:

The Percus–Yevick (PY) closure [3]:

$$B(r) = \gamma(r) + \ln\{1 + \gamma(r)\} \tag{39.2A}$$

The Martynov–Sariskov (MS) closure [4]:

$$B(r) = \{1 + 2\gamma(r)\}^{1/2} - 1 - \gamma(r) \tag{39.2B}$$

The extension of above MS equation given by Vompe and Martynov (VM) [5]:

$$B(r) = \frac{1}{2A}\left[(1 + 4A\{\gamma(r) - \beta u_2(r)\})^{1/2} - 1 - 2A\{\gamma(r) - \beta u_2(r)\}\right] \tag{39.2C}$$

Here A is the fitting parameter and β is inverse temperature [$\beta = 1/k_B T$, here k_B is Boltzmann's constant]. The potential energy function is considered to be composed of a short ranged repulsive part $u_1(r)$ and a weak long-ranged attractive part $u_2(r)$.

$$u(r) = u_1(r) + u_2(r) \tag{39.3}$$

According to Lee [6] $B(r)$ is a function of $\gamma(r)$ and despite its empirical nature, very accurate results are obtained in various situations compared to simulation data. Thus, the VM equation can be simplified as,

$$B(r) = \frac{-\gamma(r)^2}{2[1 + \alpha\gamma(r)]} \tag{39.4}$$

In original VM bridge function α has the value 0.8 [5]. This VM bridge function, although originally developed and applied for hard sphere systems, can be extended and modified for Lennard-Jones liquid systems [7]. Thus in L-J liquid the above bridge function is reformed as soft sphere Mean-Spherical model approximation (SMSA).

In the study of phase diagram the thermodynamic properties such as pressure (P), isothermal compressibility (β_{Tcm}) and excess chemical potential (μ^{ex}) may be calculated for each state point using the pair distribution functions $g(r)$ where

$$g(r) = h(r) + 1 \tag{39.5}$$

These are given in literatures in the following forms [8]:

- The pressure obtained from the virial equation of states:

$$P = \rho k_B T - \int_0^{\alpha} r \frac{du(r)}{dr} g(r) r^2 dr \tag{39.6}$$

- The isothermal compressibility (β_{Tcm}) obtained by deriving P with respect to ρ and takes the form:

$$\frac{1}{\beta_{Tcm}} = \rho k_B T - \frac{4\pi\rho^2}{3} \int_0^{\alpha} r \frac{du(r)}{dr} \left\{ g(r) + \rho/_2 \frac{\partial g(r)}{\partial \rho} \right\} r^2 dr \tag{39.7}$$

- The excess chemical potential calculated from direct correlation function:

$$\beta\mu^{ex} = 4\pi\rho \int \left[\gamma(r) + B(r) - h(r) + \frac{1}{2} h(r) \left\{ \gamma(r) + \frac{4}{3} B(r) \right\} \right] r^2 dr \tag{39.8}$$

However the purpose of the present work is to calculate and apply the bride function in L-J fluid by SMSA; also to investigate alternative method for the above thermodynamic properties and compare the report found with those obtained from the simulation of above equations derived.

39.2 Calculation and Observation

39.2.1 Calculation of Bridge Function

To solve the closure equation, considering a solute of hard sphere diameter (σ), density (ρ) and packing fraction (η)[$\eta = (\pi/6)\rho\sigma^3$], the simplified form of VM approximation as per Eq. (39.4) be: $B(r) = \frac{-\gamma^2(r)}{2[1+\alpha\gamma(r)]}$

Here;

$$\alpha = \frac{17}{120\,\eta} + 0.5150 - 0.2210\,\eta \tag{39.9}$$

By solving OZ equation for limit condition $h(r) = -1$ [1] the indirect correlation function $\gamma(r)$ can be obtained from the direct correlation function $c(r)$:

$$\gamma(r) = -1 - c(r) \tag{39.10}$$

Now, the calculation of direct correlation function consists of two parts.

(a) Inside the hard core: where $r < \sigma$, the exact analytical solution of $c(r)$ is;

$$-c(r) = \alpha + \beta(r/\sigma) + \gamma(r/\sigma)^3 \tag{39.11}$$

Here,

$$\alpha = \frac{(1+2\eta)^2}{(1-\eta)^4} \tag{39.12}$$

$$\beta = -6\eta\frac{(1+0.5\eta)^2}{(1-\eta)^4} \tag{39.13}$$

$$\gamma = 0.5\eta\alpha \tag{39.14}$$

(b) Outside the hard core:

In case where $r \geq \sigma$, the mean spherical model approximation theory has been applied. As the L-J potential contains two components: attractive and repulsive; as per Eq. (39.3) the direct correlation function can be written as:

$$c(r) = \beta u_1(r) + \beta u_2(r) \tag{39.15}$$

where [9],

$$\beta u_1(r) = \frac{4\left[(\sigma/r)^{12} - (\sigma/r)^6\right]}{T^*} \tag{39.16}$$

and [10],

$$\beta u_2(r) = \frac{4(\sigma/r)^6 \exp\left[-{}^1\!/_\rho(\sigma/r)^6\right]}{T^*} \tag{39.17}$$

Now, the well-known L-J (12-6) potential has the form

$$U_{Lj} = 4\varepsilon\left[\left(\frac{\sigma}{r}\right)^{12} - \left(\frac{\sigma}{r}\right)^6\right] \tag{39.18}$$

ε and σ are characteristic energy and hard sphere diameter respectively.

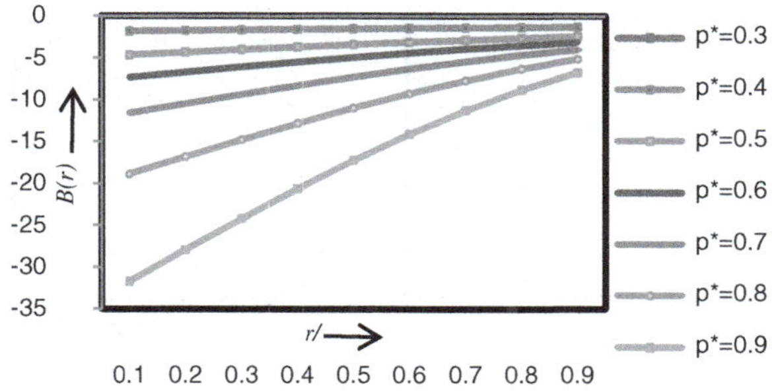

Fig. 39.1 Change in bridge function values for L-J fluid inside the core region ($r < 0.9\sigma$) at different reduced density ranges from $\rho* = 0.1$ to $\rho* = 0.9$. This plot is temperature independent

39.2.1.1 Observation

The objective here is to calculate the bridge function values of L-J fluid over wide range of densities $\rho* = 0.1$–0.9 for three different temperatures $T* = 1.35$, 2.74 and 5.00 ($T^* = k_B T/\varepsilon$ and $\rho^* = \rho\sigma^3$). At first the direct correlation functions have been calculated differently for inside and outside the hard-core diameters as described above. It has been found that inside the hard core, the direct correlation functions are temperature independent. Thus the bridge function values have been calculated separately for inside the hard core and plotted in Fig. 39.1. It has been found at low density up to $\rho* = 0.3$, $B(r)$ remains almost constant. At medium or higher density $B(r)$ values deviate a lot and slowly converge at $r \rightarrow \sigma$. However outside the hard core both c(r) and B(r) functions are temperature dependent. Thus these have been calculated at three temperature zone $T* = 1.35$, 2.74 and 5.00 and reported in Fig. 39.2. This figure shows that all values are convergent. This observation is quite agreeable with the work done by Charpentier and Jakse [8].

39.2.2 Calculation of Thermodynamic Parameters

Once the values of bridge functions are found, it can be possible to propose a simple method to calculate excess chemical potential, isothermal compressibility and entropy of the system in a thermodynamic consistent manner.

Now, for soft sphere interaction through a pair potential $u(r)$, the essential thermodynamic properties as excess chemical potential (μ_{ex}), the isothermal compressibility (β_T) etc. can be determined using Eqs. (39.7, 39.8) by solving the corresponding equations. However, it has already been reported in Fig. 39.2 that $B(r)$ values do not vary so much at outside the hard core. So to avoid unwieldy thermodynamic integration which requires either typical program or software and increases the complicacy and difficulty for coding, it has been considered that the excess chemical potential depends on $B(r)$ inside the hard core and this dependency can be neglected at outside the hardcore (98 % in the case of hard sphere

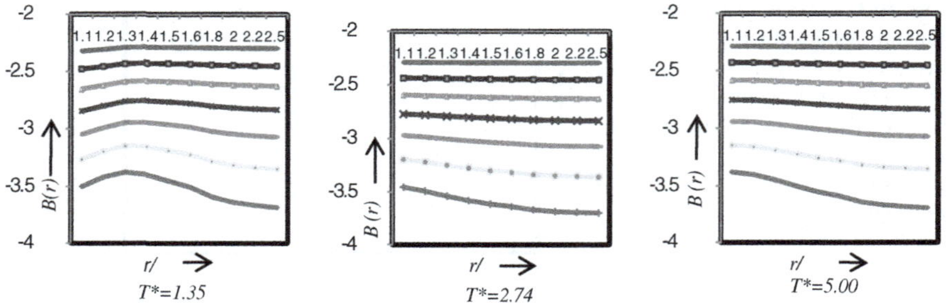

Fig. 39.2 Plot of bridge function versus diameter ratio for same L-J fluid outside the core region ($r > 1.1\sigma$) over wide density ranges from $\rho^* = 0.1$ to $\rho^* = 0.9$ as obtained at temperature 1.35, 2.74, 5.00 respectively

fluid as of Ref. [11]). So to determine the chemical potential for L-J fluid the subsequent method has been applied using series of equations as:

Lomba et al. [12] gave an equation for the entropy S of the system for L-J pair potential $u(r)$,

$$S = u(r) + \beta_T - 1 - \beta\mu_{ex} \tag{39.19}$$

If a system of non-interacting particles of the excluded volume V_0 be considered then g(r) equals to zero and unity inside and outside the excluded volume respectively. Therefore, its entropy S can be written as S_0,

$$S_0 = -2\pi\rho/3 \tag{39.20}$$

Goin et al. [13] derived an analytical expression for compressibility factor for L-J fluid,

$$\beta_T = \beta_{T\,HS} - \frac{1}{T^* d_B^3}\left(14.85\eta - 26.31\eta^2 + 154.5\eta^3 - 182.4\eta^4\right) \tag{39.21}$$

Verlet–Weis [14] formula can be used to calculate d_B, density factor, as:

$$d_B = \frac{1.0683 + 0.3813T^*}{1 + 0.4293T^*} \tag{39.22}$$

The term $\beta_{T\,HS}$ is the compressibility factor of hard sphere system and that can be calculated using the Carnahan–Starling [15] equation,

$$\beta_{T\,HS} = \frac{1 + \eta + \eta^2 - \eta^3}{(1 - \eta)^3} \tag{39.23}$$

Table 39.1 Isothermal compressibility and chemical potential values of L-J fluid in terms of β_T and $\beta\mu_{ex}$ respectively over wide density ranges at temperatures $T^* = 1.35$, 2.74 and 5.00 respectively

	$T^* = 1.35$		$T^* = 2.74$		$T^* = 5.00$	
ρ^*	β_T	$\beta\mu_{ex}$	β_T	$\beta\mu_{ex}$	β_T	$\beta\mu_{ex}$
0.1	0.705	−1.163	0.652	−0.882	0.603	−0.784
0.2	0.507	−1.589	0.402	−1.350	0.307	−1.295
0.3	0.369	−2.008	0.210	−1.788	0.064	−1.767
0.4	0.292	−2.402	0.070	−2.189	−0.133	−2.202
0.5	0.319	−2.699	0.026	−2.500	−0.244	−2.553
0.6	0.544	−2.790	0.171	−2.616	−0.171	−2.719
0.7	1.130	−2.505	0.674	−2.369	0.255	−2.527
0.8	2.346	−1.577	1.807	−1.483	1.312	−1.700
0.9	4.618	0.419	4.006	0.474	3.446	0.206
1	8.654	4.192	7.991	4.223	7.383	3.919

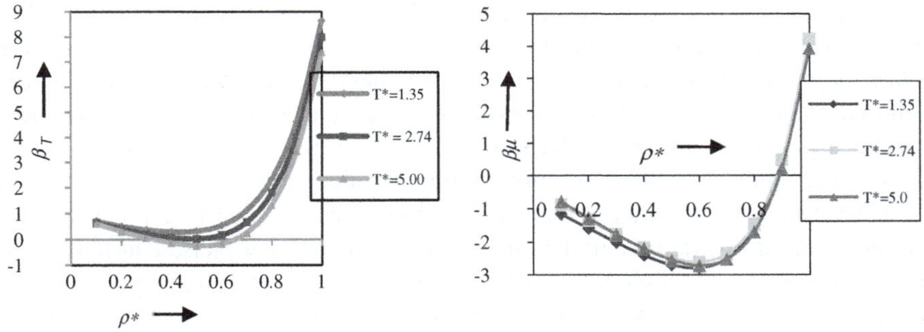

Fig. 39.3 Plot of isothermal compressibility (β_T) and chemical potential ($\beta\mu_{ex}$) values versus reduced density ranges from $\rho^* = 0.1$ to $\rho^* = 0.9$ for L-J fluid at three separate temperatures 1.35, 2.74, 5.00 respectively

39.2.2.1 Observation

The thermodynamic properties of interest like compressibility factor (β_T) and excess chemical potential ($\beta\mu_{ex}$) of L-J fluids have been calculated using series of equations from (39.19) to (39.23) and reported in Table 39.1. Figure 39.3 represents the change in these thermodynamic properties vs density ranges from $\rho^* = 0.1$ to $\rho^* = 0.9$ at three T^* values 1.35, 2.74 and 5.00. All the curves indicate that (β_T) and ($\beta\mu_{ex}$) show minimum value between $\rho^* = 0.5$–0.6. These curves in Fig. 39.3 have been compared with the result obtained by Charpentier and Jakse (p. 2290, Fig. 3 in Ref. [8]). It has been found that the result is quite agreeable at $T^* = 1.5$ and upto $\rho^* = 0.5$.

39.3 Discussion

The structural and thermodynamic properties of L-J fluid is, according to integral equation theory, depend on the bridge function that can be obtained from computer simulation. As per Sarkisov and Lomba [16] the computer simulation data still are not too accurate at either transition area or on the critical and triple point. The objective of present work is to calculate the bridge function for L-J fluid and investigate about the importance of this on thermodynamic properties of L-J fluid.

The bridge function data of L-J fluid calculated as mentioned in 2.1 shows a good agreement with Charpentier and Jakse [8] and that indicates the validity of present bridge function values. As it has been reported in Fig. 39.2 that $B(r)$ values do not vary so much outside the hard core, the dependency of $B(r)$ can be neglected in determination of thermodynamic properties of L-J fluid. Choudhury and Ghosh [7] reported that "the contribution of $B(r)$ on chemical potential is important inside the core ($r < \sigma$) and the contribution of long range part ($r \geq \sigma$) is not more than 2 %".

Thus it is an attempt to determine the isothermal compressibility and excess chemical potential of L-J fluid simply ignoring the bridge function. The result has been reported in Table 39.1 and Fig. 39.3. The isothermal compressibility curves show the minimum value at $\rho* = 0.5$ but the chemical potential curves show the minimum about $\rho* = 05$–0.6. However, the shape and nature of curves obtained has been matched with that of Sarkisov [17] which gives minimum at $\rho* = 0.5$. Charpentier and Jakse [8] also found minimum value at $\rho* = 0.5$.

Thus it may be considered that this method gives satisfied result at low temperature and at low density in determining the structure and phase equilibrium of L-J fluid. However, at high density there is a possibility to form a complex solution which may form a complex curve and require a suitable normalization procedure of calculation. In future, the behavior of L-J fluid at high density will be examined.

Once the chemical potential is being developed, the vapour-liquid phase diagram can be made at a given temperature and pressure. Thus further work regarding formation of vapour-liquid phase diagram for L-J fluid is in progress.

References

1. Ornstein LS, Zernike F (1914) Accidental deviations of density and opalescence at the critical point of a single substance. Proc Acad Sci Amsterdam 17:793–806
2. Martynov GA (1992) Fundamental theory of fluids: methods of distribution functions. Higher Bristol
3. Percus JK, Yevick GJ (1958) Analysis of classical statistical mechanics by means of collective coordinates. Phys Rev 110(1):1
4. Martynov GA, Sarkisov GN (1983) Exact equations and the theory of liquids. Mol Phys 49:1495–1504
5. Mortynov GA, Sarkisov GN, Vompe AG (1999) New closure for the Ornstein–Zernike equation. J Chem Phys 110:3961–3969

6. Lee LL (1997) The potential distribution-based closures to the integral equations for liquid structure: The Lennard-Jones fluid. J Chem Phys 107:7360–7370
7. Choudhury N, Ghosh SK (2002) Integral equation theory of Lennard-Jones fluids: A modified verlet bridge function approach. J Chem Phys 116:8517–8622
8. Charpentier I, Jakse N (2001) Exact numerical derivation of the pair-correlation function of simple liquids using the tangent linear method. J Chem Phys 114:2284–2292
9. Tikhonov DA, Sarkisov GN (2000) Russ J Phys Chem 74:470
10. Pollack GL (1991) Why gases dissolve in liquids. Science 251:1323–1330
11. Lee LL (1992) Chemical potentials based on the molecular distribution functions. An exact diagrammatical representation and the star function. J Chem Phys 97:8606–8616
12. Lomba F, Alsarez M, Lee LL, Almarza NG (1996) Phase stability of binary non-additive hard-sphere mixtures. A self-consistent integral equation study. J Chem Phys 104:4180–4188
13. Goin K, Mo KC, Starling KE (1977) Proc Okla Acad Sec 57:119
14. Verlet L, Weis JJ (1972) Equilibrium theory of simple liquids. Phys Rev A5:939–952
15. Carnahan NF, Starling KE (1969) Equation of states for nonattracting rigid spheres. J Chem Phys 51:635–636
16. Sarkisov G, Lomba E (2005) The gas-liquid phase-transition singularities in the framework of the liquid-state integral equation formalism. J Chem Phys 122(214504):1–6
17. Sarkisov G (2001) Approximate integral equation theory for classical fluids. J Chem Phys 114:9496–9505

Part VI
Cloud Computing and Algorithm

Chapter 40
A Group Decision Support System for Selecting an Open Source Tool for Social Media Integration

Arpan Kumar Kar

Abstract Today, traditional channels of communication needs to be supplemented with digital channels which can leverage upon the power of the internet. However, integration of the website with the social media platforms provide ample challenges in terms of choosing the best possible way for such integration. In this study, the application of a group decision support system has been presented for meeting this objective. The decision support approach prioritizes and aggregates multiple dimensions of few platforms, from the users' perspective by using the Delphi method, Fuzzy Set Theory and Analytic Hierarchy Process. A case study has been conducted on a web-based portal (Business Fundas) for selecting a suitable tool for social media integration.

Keywords Group decision making · Social media · Fuzzy sets · Analytic hierarchy process · Delphi method

40.1 Introduction

Today, increasingly, marketers are extending traditional channels of communication with digital channels to leverage upon the power of the social media. However, integration of the websites with the social media channels provide ample challenges. There are numerous tools and plugins for such a requirement which may be used. Also choosing a comprehensive yet context specific solution for a website is a complex problem. In this study, the application of a group decision support system has been presented for integration of a web portal with social

A. K. Kar (✉)
Indian Institute of Management Rohtak, Rohtak, Haryana, India
e-mail: arpan_kar@yahoo.co.in

S. Sengupta et al. (eds.), *Emerging Trends in Computing and Communication*, Lecture Notes in Electrical Engineering 298, DOI: 10.1007/978-81-322-1817-3_40,

media channels. First, the major evaluation dimensions are identified using the Delphi method. Then the group decision support approach prioritizes and aggregates multiple dimensions of evaluation, from the users' perspective by integrating Fuzzy Set Theory (FST) and Analytic Hierarchy Process (AHP) for group decision making. The case study has been conducted on a portal for selecting a solution for social media integration.

40.2 Review of Related Social Media Literature

Today, integrated marketing communications is the guiding principle organizations follow to communicate with their target markets [7]. This depends extensively on the effectiveness of the social media management strategies which have been adopted by the firm [14]. Today, significant resources are deployed in managing social media since it impacts organic search and assists acquisition of customers [17]. This would again result in higher returns on marketing from resource deployment. However, the challenge lies in identifying the suitable social media channel and using a suitable tool to integrate your communications with it. A strong relationship of the information type and the information channel acceptibility [5] indicates channels for communication needs to be chosen contextually. Now, the numerous social media channels can be classified under the following categories [14]: social networking sites (e.g. Facebook); creativity sharing sites (e.g. Flickr); intellectual property sites (Creative commons); user-sponsored sites (e.g. Cnet); company-sponsored websites (e.g. Vocalpoint); company-sponsored causes/help sites (click2quit.com); business networking sites (e.g. LinkedIn); collaborative websites (e.g. Wikipedia); e-commerce sites (e.g. eBay); podcasts (e.g. This American Life); news sites (BBC); educational sharing sites (e.g. MIT OpenCourseWare); open-source communities (e.g. linux.org); and social bookmarking (e.g. Digg). Within so many channels, the first step is to identify the relevant channels for a specific context. Subsequently, the next step is to select a suitable tool which will facilitate the integration with these channels. These objectives have been met by using the Delphi method, FST and AHP.

40.3 Contribution

A review of literature on social media highlights the need of integrated marketing communication for marketing and reaching out to the target segment. However not many studies focus on the decision support aspect of managing the social media channels or choosing the right way to manage such an initiative. In this study, an approach has been proposed which uses two well-developed methods and theories for group decision making for structuring a decision making process. The first stage of the computational process is selecting the important dimensions for evaluating the success of the outcome, i.e., the social media channels which should be considered for a given context. The next stage of the computational process is investigating the competencies of the listed available tools in the context of these evaluation dimensions, in a structured manner. The first stage has been addressed in the problem using the Delphi method. The second stage has been addressed by integrating FST and AHP for group decision making.

40.4 The Computational Approach

The study has taken the approach of using two methodologies for group decision making, namely the Delphi methodology and the AHP. The detailed description of the computational approach has been elaborated subsequently.

40.4.1 The Delphi Methodology

The Delphi methodology [6] has been widely used to obtain group consensus among domain experts. More specifically, it is a methodology for eliciting consensus in group decision from a panel of experts on a particular problem domain in an iterative manner, and experts are encouraged to revise their earlier decisions in light of the anonymous responses of other experts in the panel, in subsequent iterations [9, 13]. Most Delphi studies [8] have a sample size of 5–20 experts as respondents and 2 iterations are often sufficient to achieve consensus within the group. Consensus can be obtained in a number of ways including having over 80 % of the votes can be classified to two distinct categories [16]. In this study, the major social media channels which would be relevant in the given context has been identified using the Delphi methodology. After the identification of these channels, the channel specific competency would be evaluated for all the potential tools for the selection process.

40.4.2 The Fuzzy Extension of the Analytic Hierarchy Process

The AHP [1–3, 15] was developed for use in multi-hierarchical, multi-criteria decision making problems. AHP is extremely suitable for group decision making due to specific reasons. Firstly, AHP has robust theories to estimate consistency of priorities of decision makers. Secondly, there are systemic approaches to improve the consistency of priorities. Thirdly, it provides appropriate methods for the aggregation and consensus achievement of group preferences. To the best of our knowledge the application of these theories for group decision making is yet to be explored for tool selection for social media integration.

The prioritization of these channel categories obtained from Delphi and the prioritization of the tools is addressed through an integrated approach using FST and AHP. Inclusion of FST accommodates the subjectivity in the human decision making process for complex problems. In this approach, first the linguistic judgments of users are captured and mapped to quantifiable fuzzy judgments. Subsequently, these fuzzy linguistic judgments are converted to crisp priorities using AHP theory. Let $U = (u_1,\ldots u_n)$ be the set of *n users* having a relative importance of ψ_i such that $\psi = (\psi_1,\ldots\psi_n)$ is the weight vector of each user and $\sum\psi_i = 1$. Judgments $A = (a_{ij})_{k\times k}$ would be coded using FST extension of AHP [10, 15]. A triangular FST function [11] has been used for coding the judgments. The simple pairwise comparison approach [4, 12] for FST operations has been used for the fuzzy sets $\tilde{a}_i = \left(a_{i,1}, a_{i,2}, a_{i,3}\right)$ and $\tilde{a}_j = \left(a_{j,1}, a_{j,2}, a_{j,3}\right)$ as illustrated:

$$\tilde{a}_i \oplus \tilde{a}_j = \left(\left(\tilde{a}_{i,1} \oplus \tilde{a}_{j,1}\right), \left(\tilde{a}_{i,2} \oplus \tilde{a}_{j,2}\right), \left(\tilde{a}_{i,3} \oplus \tilde{a}_{j,3}\right)\right) \tag{40.1}$$

The individual priorities are obtained by solving the following system:

$$\min \sum\nolimits_{i=1}^{k} \sum\nolimits_{j>i}^{k} \left(\ln \tilde{a}_{i,j} - \left(\ln \tilde{w}_i - \ln \tilde{w}_j\right)^2\right) \text{ s.t. } \tilde{a}_{ij} \geq 0;\ \tilde{a}_{ij} \times \tilde{a}_{ij} = 1;\ \tilde{w}_i \geq 0, \sum \tilde{w}_i = 1. \tag{40.2}$$

The individual decision maker's priority vector is

$$\tilde{w}_i = \frac{\sqrt{[1/k]\prod_{j=1}^{k} \tilde{a}_{i,j}}}{\sum_{i=1}^{n} \sqrt{[1/k]\prod_{j=1}^{k} \tilde{a}_{i,j}}} \tag{40.3}$$

where $\tilde{w}_i$ is the priority of the decision criteria i such that $\tilde{W}_i = \{\tilde{w}_1, \tilde{w}_2, \ldots, w_7\}$ for *user i*. Collective preferences were evaluated using the aggregation of priorities.

$$\tilde{W}^{(c)} = \left(\tilde{w}_1^{(c)}, \tilde{w}_2^{(c)}, \ldots \tilde{w}_r^{(c)}\right) \quad \text{where} \quad \tilde{w}_i^{(c)} = \frac{\prod_1^n \left(w^{(k)}{}_i\right)^{\psi_i}}{\sum_1^r \prod_1^n \left(w^{(k)}{}_i\right)^{\psi_i}} \tag{40.4}$$

For crisp conversion of priority

$$|\tilde{w}_i| = w_{i,2} \cdot 0.25 + w_{i,2} \cdot 0.5 + w_{i,3} \cdot 0.25 \tag{40.5}$$

These crisp priorities would be mapped to criteria specific performance score of each tool to obtain the final score using a sum-product scalar multiplication approach. This would provide the overall competency based on aggregate scoring and is analogous to a weighted scoring approach for finding total capability score.

Priorities/Channel	C1	C2	C3	C4	C5	C6	GCI
Expert 1 Priority	0.300	0.300	0.100	0.100	0.100	0.100	0.133
Expert 2 Priority	0.450	0.222	0.082	0.082	0.068	0.098	0.105
Expert 3 Priority	0.155	0.479	0.134	0.068	0.082	0.082	0.098
Expert 4 Priority	0.299	0.299	0.100	0.120	0.083	0.100	0.160
Expert 5 Priority	0.268	0.068	0.178	0.079	0.194	0.214	-0.142
Aggregate Priority	0.303	0.250	0.124	0.096	0.106	0.121	

Fig. 40.1 Aggregated priorities and GCI of the 5 experts

Channels for Social Integration	C1	C2	C3	C4	C5	C6	PERFORMANCE	
Aggregate Priority	0.303	0.250	0.124	0.096	0.106	0.121	SoP Score	Rank
Digg Digg	5	3	4	5	3	5	4.162	1
WP Socializer	5	1	3	3	2	3	2.998	4
Calicotek Social Slider	4	1	2	2	2	4	2.597	5
Cevhershare Social Tool	4	2	2	3	2	5	3.064	3
WordPress Social Share Buttons	5	3	3	2	4	4	3.737	2

Fig. 40.2 Overall performance of the 5 tools for social integration

40.5 The Case Analysis and Results

The case study was conducted on a knowledge and news publishing portal (Business Fundas: `business-fundas.com`). First through a Delphi study, the important and relevant channel categories were identified. The participants of the Delphi study consisted of the five senior decision makers of Business Fundas, The participants were first exposed to all the thirteen possible channel categories for social media integration. The Delphi study highlighted that 6 channel categories, i.e., social networking sites (C1); user-sponsored blogs (C2); business networking sites (C3); collaborative websites (C4); podcasts (C5) and social bookmarking (C6) are more relevant. The consensus was achieved after 2 iterations with a minimum of 80 % of the participants voting for the selected channels and 100 % of the participants voting against the rejected ones. Then the relative priorities for the channels were completed (Fig. 40.1).

Subsequently, 5 open-source tools for social media integration were identified and evaluated for getting the final requirement specific score for each tool (Fig. 40.2).

As is evident, from among the 6 tools or plugins which were compared, based on the integration capabilities within the 6 social media channels, Digg Digg plugin performed the best, followed closely by the Wordpress Social Share Buttons.

40.6 Conclusion

The study highlights how the collective expertise of a group of decision makers can be jointly leveraged upon to select a suitable tool for social media integration based on a specific requirement. The use of an integrated approach of three methodologies on group decision making has been highlighted in this study, namely the Delphi method, FST and AHP. The approach highlights how selection of an appropriate tool is enriched by the collective decision making of a group of experts. Further, for the specific context of the Business Fundas website, the tool "Digg Digg" was estimated to be most suitable from among five other tools for social media integration. Since this is a context specific study, the results may not be generalizable, but the approach can be adopted while selecting a tool for social media integration, by taking the benefit of the collective expertise of a group of decision makers.

A limitation of the study is that like most case studies the findings are context specific and the outcome is thus less generalizable to other contexts, particularly if the requirements are different. However, the approach provided in this study should be easy to replicate with a different set of data which would be relevant for the new context, and thus arrive at the solution through a systematic approach.

References

1. Aguaron J, Escobar MT, Moreno-Jiménez JM (2003) Consistency stability intervals for a judgement in AHP decision support systems. Eur J Oper Res 145(2):382–393
2. Aguarón J, Moreno-Jiménez JM (2003) The geometric consistency index: approximated thresholds. Eur J Oper Res 147(1):137–145
3. Bolloju N (2001) Aggregation of analytic hierarchy process models based on similarities in decision makers' preferences. Eur J Oper Res 128(3):499–508
4. Buckley JJ (1985) Fuzzy hierarchical analysis. Fuzzy Sets Syst 17(3):233–247
5. Bystrom K (2002) Information and information sources in tasks of varying complexity. J Am Soc Inform Sci Technol 53(7):581–591
6. Dalkey N, Helmer O (1963) An experimental application of the Delphi method to the use of experts. Manage Sci 9(3):458–467
7. Eyrich N, Padman ML, Sweetser KD (2008) PR practitioners' use of social media tools and communication technology. Public Relat Rev 34(4):412–414
8. Hsu C, Sandford B (2007) The Delphi technique: making use of consensus. Pract Assess Res Eval 12(10):1–8
9. Hwang CL, Lin MJ (1987) Group decision making under multiple criteria: methods and applications. Springer, Berlin
10. Kar AK, Pani AK (2014) How can a group of procurement experts select suppliers? An approach for group decision support. J Enterp Inf Manage (In press)
11. Kar AK (2014) Revisiting the supplier selection problem: an integrated approach for group decision support. Expert Syst Appl (In press)
12. Kar AK, Rakshit A (2014) Pricing of cloud IaaS based on feature prioritization—a value based approach. Adv Intell Syst Comput 235:321–330

13. Khorramshahgol R, Moustakis VM (1988) Delphic hierarchy process (DHP): a methodology for priority setting derived from the Delphi method and analytical hierarchy process. Eur J Oper Res 137(3):347–354
14. Mangold WG, Faulds DJ (2009) Social media: the new hybrid element of the promotion mix. Bus Horiz 52(4):357–365
15. Saaty TL (1980) The analytic hierarchy process. McGraw Hill International, New York
16. Ulschak FL (1983) Human resource development: the theory and practice of need assessment. Reston Publishing Company, Reston
17. Xiang Z, Gretzel U (2010) Role of social media in online travel information search. Tourism Manage 31(2):179–188

Chapter 41
Revenue and Expense Optimization in a CRN Using DE Algorithm

Subhasree Bhattacharjee, Roukna Sengupta and Suman Bhattacharjee

Abstract Cognitive radio technology has been emerged to provide the solution of improvement of spectrum utilization. In this paper we consider a cognitive radio network in which there are primary users (PU) and a set of secondary users (SU). The spectrum is divided into channels using frequency division multiple access (FDMA). The channels are licensed to PUs. When PUs do not use the channels, they lease the vacant spectrum for monetary gain. SUs bid for the channels. PUs select the purchaser who provides highest bid value. Thus PUs can earn revenue by leasing the channels and SUs being the purchaser bid at a certain payoff. The main objective is to make both purchaser and the seller benefited. Here, using Differential evolution algorithm we solve both the objectives using a single objective function. The algorithm finds optimize value of the parameters.

Keywords Revenue · Expense · CRN · DE

41.1 Introduction

With rapid increases of wireless systems, scarcity of radio spectrum increases. To solve this problem, opportunistic users or unlicensed spectrum users are allowed to use the licensed spectrum when it is idle. This is known as dynamic spectrum

S. Bhattacharjee (✉)
Department of CA, Narula Institute of Technology, Kolkata, India
e-mail: bhattacharjeesubhasree@gmail.com

R. Sengupta (✉)
RCC Institute of Information Technology, Kolkata, India
e-mail: rouknasengupta@gmail.com

S. Bhattacharjee (✉)
IBM India Pvt. Ltd, Kolkata, India
e-mail: sumanbhattacharjee@in.ibm.com

S. Sengupta et al. (eds.), *Emerging Trends in Computing and Communication*, Lecture Notes in Electrical Engineering 298, DOI: 10.1007/978-81-322-1817-3_41,

access. Cognitive radio techniques are used to implement it [1]. So, cognitive radio (CR) is an autonomous unit in a communication environment that senses the environment and changes its operating parameters to achieve the dynamic spectrum access. Many research works are going on for dynamic spectrum sharing [2]. Considering economic aspect spectrum auction strategy brings a novel approach of spectrum sharing [3, 4]. Spectrum sharing has been modeled as a spectrum trading process where licensed users sell the spectrum to the unlicensed users. Game theory and pricing mechanism perform well in the auction framework. Spectrum auction has been proposed for unlicensed users in [5]. A multi-winner spectrum auction environment was discussed in [6]. Excellent research of spectrum sharing is going on cognitive radio [7]. In [8], double auction framework was proposed. In this paper, factors affecting bidding strategy are not considered. Our problem is based on spectrum trading where licensed users or primary users (PU) sell the vacant spectrum to the secondary users (SU) and earn revenue. SUs also try to minimize the expense which would have to pay to PUs. So, the problem has two objectives. One is to maximize the revenue and another is to minimize the expense. For multi objective optimization, many algorithms have been proposed. Most algorithms are in the field of evolutionary algorithms (EAs). Differential evolution (DE) is a particular evolutionary algorithm that has been used for multi objective optimization. DE is very simple but very much powerful tool, introduced by Price and Storn [9]. Due to page constraint, the algorithm is not described in this paper.

The rest of the paper is organized as follows. In Sect. 41.2, network model is discussed. In Sect. 41.3, experiments and results are discussed. We conclude the paper in Sect. 41.4.

41.2 Network Model

In this paper we consider multi cell CR network in which there are multiple PUs and SUs. Number of channels for each cell is equal to M. When the channels are vacant, PUs apply marketing strategy for auctioning the channels. SUs may know the bid value of each other. The auctioneer selects the purchaser who provides highest bid value. Thus PUs can earn revenue by leasing the channels and SUs being the purchaser bid at a certain payoff. The main objective is to make both purchaser and the seller benefitted.

Let, rev be an N X K channel assignment matrix where rev(PU, ch) = rev-PUch = r, if PU earns revenue r for channel ch. 0, if that channel is not assigned to PU. We can formulate the optimization problem of maximizing the revenue earned by all PUs in the system. The optimization problem is of the form:

$$arg\ max \sum_{PU=1}^{N} \sum_{ch=1}^{K} rev_{PU}^{ch} \tag{41.1}$$

Subject to the conditions that (1) Each PU requires at most one channel. (2) Each channel cannot be used by more than one PU. The conditions are as follows

$$c_{\min} \leq \sum_{ch=1}^{K} rev_{PU}^{ch} < c_{\max} \tag{41.2}$$

$$c_{\min} \leq \sum_{PU=1}^{N} rev_{PU}^{ch} < c_{\max} \tag{41.3}$$

where, $c_{max} = 2_* \, c_{min.}$

Here, c_{min} is equal to the minimum cost of a channel that a PU assigned and c_{max} is the maximum cost of a channel assigned by PU.

We have multiplied the term $(c_{max} - \sum_{ch=1}^{K} rev_{PU}^{ch})$ with λ_1 and add it with the objective function z. For the condition in (41.3), similarly get the third term λ_2 $(c_{max} - \sum_{PU=1}^{N} rev_{PU}^{ch})$. Now, the resultant constrained optimization problem is

$$\text{Max z1} = \sum_{PU=1}^{N} \sum_{ch=1}^{K} rev_{PU}^{ch} + \lambda_1 \left(c_{\max} - \sum_{ch=1}^{K} rev_{PU}^{ch} \right) + \lambda_2 \left(c_{\max} - \sum_{PU=1}^{N} rev_{PU}^{ch} \right) \tag{41.4}$$

If exp nse be an M X K channel assignment matrix where exp nse(SU, ch) = exp $nse_{SU}{}^{ch}$ = p, if SU bids the value p for channel ch. 0, if that SU not provides bid for that channel ch. We can formulate the optimization problem of minimizing the expense provided by all SUs in the system. The optimization problem is of the form:

$$arg\ min \sum_{SU=1}^{M} \sum_{ch=1}^{K} \text{expnse}_{SU}^{ch}$$

Subject to the conditions that total amount of bid provided by a single SU should not exceed a threshold value (b_{max})

$$0 \leq \sum_{ch=1}^{K} \text{expnse}_{SU}^{ch} \leq b_{\max} \tag{41.5}$$

We have multiplied the term $(b_{max} - \sum_{ch=1}^{K} \text{expnse}_{SU}^{ch})$ with λ_3 and add it with the objective function z1. Now, the resultant constrained optimization problem is

$$\text{Min z2} = \sum_{SU=1}^{M} \sum_{ch=1}^{K} \text{expnse}_{SU}^{ch} - \lambda_3 \left(b_{\max} - \sum_{ch=1}^{K} \exp\ nse_{SU}^{ch} \right) \tag{41.6}$$

λ_1, λ_2 and λ_3 are Lagrange coefficients.

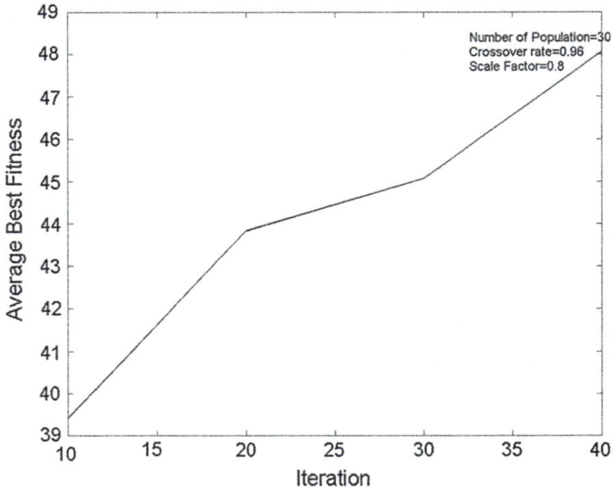

Fig. 41.1 Average fitness with iteration

Combining (41.4) and (41.6) we get,

$$\text{Max } z = \sum_{PU=1}^{N}\sum_{ch=1}^{K} rev_{PU}^{ch} + \lambda_1\left(c_{\max} - \sum_{ch=1}^{K} rev_{PU}^{ch}\right) + \lambda_2\left(c_{\max} - \sum_{PU=1}^{N} rev_{PU}^{ch}\right)$$
$$- \sum_{SU=1}^{M}\sum_{ch=1}^{K} \text{expnse}_{SU}^{ch} + \lambda_3\left(b_{\max} - \sum_{ch=1}^{K} \text{expnse}_{SU}^{ch}\right)$$

41.3 Result and Discussions

System Parameters for DE

Initial Population size POPSIZE = 30
Probability of Crossover = 0.96
Scale Factor = 0.8

We use DE algorithm. The structure of a chromosome in our system contains chrome [200], value of fitness and λ values. Where, chrome [200] contains real positive values. Here, number of PU is 10, number of SU is 10 and number of channels are = 10. rev matrix contains 100 revenue values and exp nse matrix contains 100 payoff values. Thus, chrome contains 200 elements in a row major form. We analyze the result of simulation of the proposed algorithm. Figure 41.1 shows the variation of average fitness with iteration. We vary the number of iteration from 10 to 60 and we see that average best fitness increases with the initial increase of iteration and then it decreases. Crossover ratio is kept at 0.96 and scale factor is fixed at 0.8. Figure 41.2 shows the variation of average payoff and

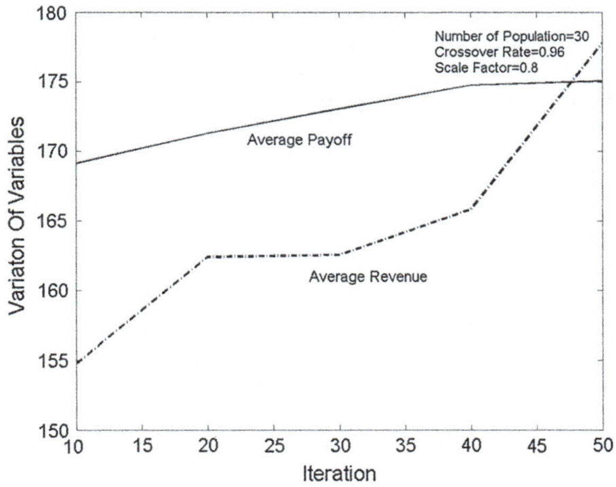

Fig. 41.2 Variation of average payoff and average revenue with iteration

revenue with the variation of iteration. We vary iteration from 10 to 50 and the average revenue and payoff value increase. Crossover rate is kept at 0.96 and number of population is fixed at 30.

41.4 Conclusions

In this paper, we consider a cognitive radio network where there are multiple PUs and SUs. The objective is to maximize the revenue earned by PUs and minimize the expense paid by SUs. The resultant objective function is constrained. We use DE algorithm to optimize the problem. The two decision variables revenue matrix and payoff matrix are optimized. The results are satisfactory.

References

1. Bhattacharjee S, Konar A, Bhattacharjee S (2011) Throughput maximization problem in a cognitive radio network. Int J Mach Learn Comput 1(4):332–336
2. Chang N, Liu M (2008) Competitive analysis of opportunistic spectrum access strategies. In: Proceedings of IEEE infocom, pp 1535–1542
3. Niyato D, Hossain E (2008) Spectrum trading in cognitive radio networks: a market-equilibrium-based approach. IEEE Wireless Commun 15(6):71–80
4. Beibei W, Yongle W (2008) Game theoretical mechanism design methods. IEEE Signal Process Mag 25(6):74–84
5. Huang J, Berry RA, Honig ML (2006) Auction-based spectrum sharing. Mob Netw Appl 11(3):418
6. Wu Y, Wang B, Liu K, Clancy TC (2008) A multi-winner cognitive spectrum auction framework with collusion-resistant mechanisms. In: Proceedings of 3rd IEEE symposium on new frontiers in dynamic spectrum access networks, pp 1–9

7. Chang HB, Chen KC (2010) Auction-based spectrum management of cognitive radio networks. IEEE Trans Veh Technol 59(4):1923–1935
8. Zhou X, Zheng H (2009) TRUST: a general framework for truthful double spectrum auctions. In: IEEE infocom'09, pp 999–1007
9. Price KV, Storn R (1997) Differential evolution—a simple evolution strategy for fast optimization. Dr. Dobb's J 22:18–24

Chapter 42
Implementation of an Algorithm for Minimum Spanning Tree in a Distributed Environment

Hara Prasad Rath, K. Sudipta Achary, Motahar Reza and Saroj K. Satpathy

Abstract High performance computing and its applications are innumerable and the fact that it has seen a tremendous change in the recent years has given it the chance of becoming the future of computing world. The growing need for computational speed has made parallel processing as a must for every sphere of computation. Java is seen as a suitable language for high performance computing due to its appealing features. In this paper we present a model which deploys an algorithm which is suitable for increasing the computational efficiency in calculating the Minimum Spanning Tree, using the features provided by remote method invocation in java. This model implements java's built-in features like Multi-Threading, Java RMI (Remote Method Invocation), and Object Serialization.

Keywords Minimum spanning tree · RMI · Distributed object model

42.1 Introduction

Java was designed to meet the real-world requirement of creating interactive, networked programs. To accomplish this, Java supports multithreaded programming, which allows us to write programs that do many things simultaneously. The

H. P. Rath · K. S. Achary · M. Reza (✉) · S. K. Satpathy
High Performance Computing Lab, School of Computer Science and Engineering, National Institute of Science and Technology, Berhampur, India
e-mail: reza@nist.edu

H. P. Rath
e-mail: hara.send@gmail.com

K. S. Achary
e-mail: sudipta.achary@gmail.com

S. K. Satpathy
e-mail: saroj@nist.edu

S. Sengupta et al. (eds.), *Emerging Trends in Computing and Communication*, Lecture Notes in Electrical Engineering 298, DOI: 10.1007/978-81-322-1817-3_42,

Java run-time system comes with an elegant yet sophisticated solution for multi-process synchronization that enables us to construct smoothly running interactive systems.

Java is designed for the distributed environment, as a fact it supports Remote Method Invocation (RMI) [1, 2]. RMI allows a Java object on one machine to invoke a method of a Java object on a different machine. And object may be supplied as an argument to that remote method. The sending machine serializes the object and transmits it. The receiving machine de-serializes it. This is an important feature, because it allows us to build distributed applications. Many libraries and middleware were developed to achieve high performance computing. Some of the libraries are JOMP [3], JCluster [4], JMPI [5], Java Fast Sockets JFS [6] and middleware like GridGain [7], and ProActive [8].

Using the above features described, we have implemented an algorithm to solve the problem of finding a Minimum Spanning Tree [9]. A spanning tree of a connected graph is its connected acyclic sub-graph (i.e., a tree) that contains all the vertices of the graph. A minimum spanning tree of a weighted connected graph is its spanning tree of the smallest weight, where the weight of a tree is defined as the sum of the weights on all its edges. The minimum spanning tree problem is the problem of finding a minimum spanning tree for a given weighted connected graph. Our discussion in Sect. 42.2 is solely concentrated on the related algorithms like Prims and Kruskal algorithm to find Minimum Spanning Tree. In the next Sect. 42.3 how java is used in high performance computing is discussed. Also the Remote Method Invocation and Multi-threaded concepts using the various classes of Java are discussed. Section 42.4 describes the proposed work, system specification and the working model for solving the problem while Sect. 42.5 describes the proposed algorithm. The result of performance analysis is discussed in Sect. 42.6.

42.2 Related Algorithms

42.2.1 Prim's Algorithm

Prim's algorithm [9] constructs a minimum spanning tree through a sequence of expanding sub-trees. The initial sub-tree in such a sequence consists of a single vertex selected arbitrarily from the set V of the graph's vertices. The current tree is expanded by simply attaching to it the nearest vertex not in that tree until all vertices are included.

The nature of Prim's algorithm makes it necessary to provide each vertex not in the current tree with the information about the shortest edge connecting the vertex to a tree vertex. We can provide such information by attaching two labels to a vertex: the name of the nearest tree vertex and the length (the weight) of the corresponding edge. Vertices that are not adjacent to any of the tree vertices can be given the label indicating their—infinite distance to the tree vertices a null label for the name of the nearest tree vertex. With such labels, finding the next vertex to be added to the current tree T = (VT, ET) become simple task of finding a vertex with the smallest distance label in the set V—VT. Ties can be broken arbitrarily.

After we have identified a vertex u* to be added to the tree, we need to perform two operations:

- Move u* from the set V—VT to the set of tree vertices VT.
- For each remaining vertex U in V—VT—that is connected to u* by a shorter edge than the u's current distance label, update its labels by u* and the weight of the edge between u* and u, respectively.

42.2.2 Kruskal's Algorithm

This is another greedy algorithm for the minimum spanning tree problem that also always yields an optimal solution. Kruskal's algorithm [9] looks at a minimum spanning tree for a weighted connected graph G = {V, E} as an acyclic sub-graph with |V| − 1 edges for which the sum of the edge weights is the smallest. Consequently, the algorithm constructs a minimum spanning tree as an expanding sequence of sub-graphs, which are always acyclic but are not necessarily connected on the intermediate stages of the algorithm.

- Initially we have a forest say 'V' set of trees, and every vertex in the graph is a tree
- Then the algorithm creates a set S containing all the edges in the graph
- While S is non-empty and V is not yet spanning

  1. Remove an edge with minimum weight from S
  2. If the edge links two different trees, then add it to the forest, combining two trees into a single tree
  3. Other-wise reject that edge (Fig. 42.1).

42.3 Remote Method Invocation in Java

Java RMI is a mechanism that allows one to invoke a method on an object that exists in another address space. The other address space could be on the same machine or a different one.

There are three processes that participate in supporting remote method invocation.

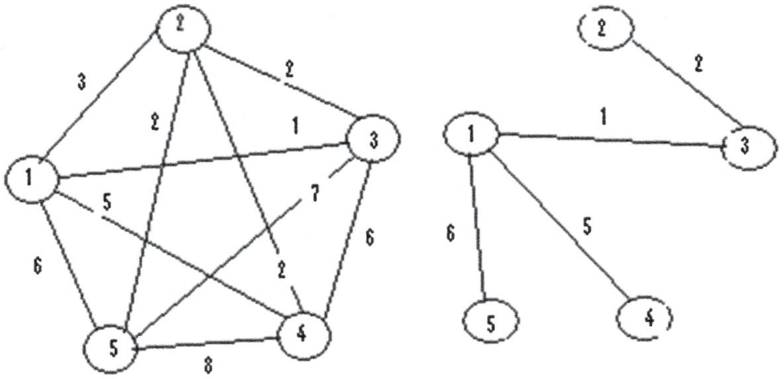

Fig. 42.1 Minimum spanning tree

Fig. 42.2 RMI architecture

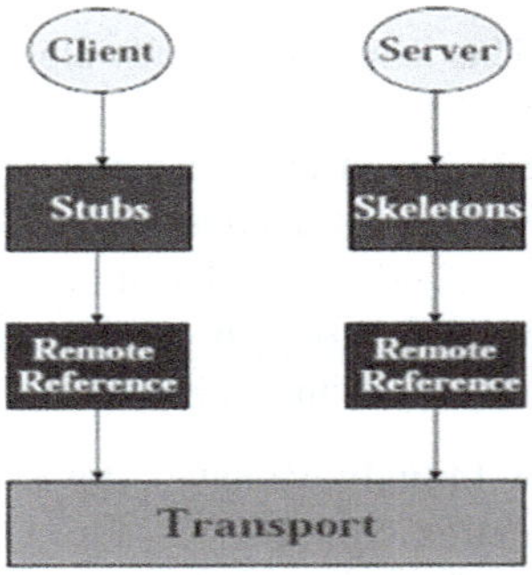

- The Client is the process that is invoking a method on a remote object.
- The Server is the process that owns the remote object. The remote object is an ordinary object in the address space of the server process.

The Object Registry is a name server that relates objects with names. Objects are registered with the Object Registry. Once an object has been registered, one can use the Object Registry to obtain access to a remote object using the name of the object. The main approach to use remote method invocation is its simplicity although it shows poor performance due to network delay and transmission (Fig. 42.2).

42.4 Proposed Work

42.4.1 Computational Setup

Task to be executed: A matrix of a certain order is taken as input which is an adjacency matrix. Now according to the proposed algorithm, we iterate through the whole matrix and in each row we find the minimum element in each row which is the edge from the ith node to jth node if i and j are the row index and column index respectively.

Now for a hugely large matrix it is really time consuming to iterate through the whole matrix in a serial execution environment. So, in the algorithm a parallel approach has been taken.

The process of finding the minimum from the rows of the matrix has been divided to a parallel environment in the system whose specification is mentioned as below. Each parallel running thread is assigned with equal number of nodes (rows) of the graph (adj. matrix).

42.4.2 System Specification

The experiment has been conducted in a Dell PowerEdge R610Rack Server with Dual Intel Xeon Quad core E5620 @2.93 GHz processors with a total 4 servers (64 Processors) 12 GB RAM/3 × 300 GB SAS HDD/RAID–5/RPS/Dual NIC.

42.4.3 The Algorithm

1. Given an Adjacency Matrix of a fully connected graph whose minimum spanning tree is to be found out. Nullify the lower half-mirror element of the matrix (because the distance from i to j is same as j to i).
2. For each row in the matrix we find the minimum element.
3. Simultaneously, we store the i and j index of the minimum found in the row.
4. At the same time we nullify that elements mirror element which is present in that matrix.
5. Then minimum element of each row is stored in a array. This array now contains the edges of the minimum spanning tree.
6. Adding up the contents we will get the total distance of the minimum spanning tree.

```
Adjacency Matrix M[n][n]
 N   ← Number Of Nodes In the Tree
 for  i ← 0 to N
      for  j ← 0 to N
          if  i = j
               M[i][j]=NULL
         else if j <  i
            M[i][j]=NULL
// Calculation of Minimum Spanning Tree
 Small_element [] ← Array of n elements
 element_index[]← Array of n elements

// Finding the Minimum edge from each node
for i ←   to  n
    Small_element[i]=M[i][0]
    elment_index[i]=0
    for j ←   to  n
     if((Small_element[i]>M[i][j]||Small_element[i]==0 )   && M[i][j]!=0)
  Small_element[i]=M[i][j]
              M[j][i]=NULL
//Storing the index of the edge (i----j)
elment_index[i]=j
  for i ←   to  n
  Total  MST  Distance ←  Total  MST  Distance +Small  element[i]
```

42.5 Computation Result

The computation was carried out in a Dell PowerEdge R610Rack Server with Dual Intel Xeon Quad core E5620 @2.93 GHz processors with a total 4 servers (64 Processors) 12 GB RAM/3 × 300 GB SAS HDD/RAID–5/RPS/Dual NIC.

First, the traditional algorithms of Prim and Kruskal were tested against different set of nodes (i.e. 1,000/2,000/5,000) in the graph. And the time for computation was noted for each algorithm.

Then, the Distributed algorithm was tested against the same number nodes. And the time for computation was noted. The result of the computation is as described by the table under:

Number of nodes in tree	Prim's serial algorithm (time in seconds)	Kruskal's serial algorithm (time in seconds)	Distributed algorithm (time in seconds)
1,000	265	4,560	26.16
2,000	1,957	25,187	28.21
5,000	8,124	984,303	57.65

Fig. 42.3 Comparison graph

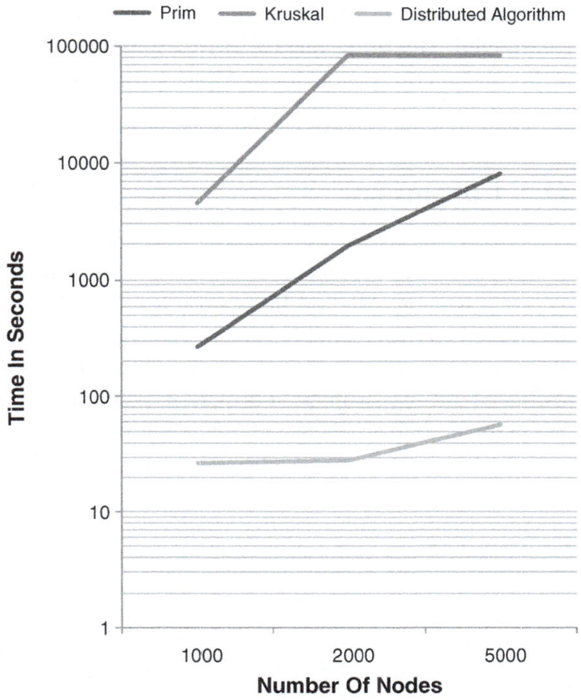

The experiment has been conducted in a Dell PowerEdge R610Rack Server with Dual Intel Xeon Quad core E5620 @2.93 GHz processors with a total 4 servers (64 Processors) 12 GB RAM/3 $\times$ 300 GB SAS HDD/RAID–5/RPS/Dual NIC.

42.6 Comparison Graph

The below graph shows a comparison between the different algorithms scrutinized in the discussion. The graph chart is drawn taking on the horizontal axis the no of nodes of the graph whose MST is to be found out. And on the vertical axis the time in terms of Seconds is taken on a log. Scale base-10.

As it can be observed from Fig. 42.3, Kruskal's algorithm takes a steep upward trend as the number of nodes in the tree increases. With 1,000 nodes in the tree the computation time marks 4,560 s, then it rises steeply to 25,187 s for 2,000 nodes and finally it concludes at a peak of 984,303 s for 5,000 number of nodes. In case of Prim's algorithm, it initiates with 265 s for 1,000 nodes in the tree then it rises gradually to 1,957 s for 2,000 nodes and finally it concludes at a high of 8,124 s for 5,000 number of nodes. Computation time for 1,000 nodes taken by the Distributed Algorithm marks 26.16 s initially, it gently rises to 28.21 s for 2,000 nodes and finally concludes at 57.65 s for 5,000 nodes.

42.7 Conclusion

In this paper a new algorithm has been explained which suitably uses the features of java's Remote Method Invocation to compute the minimum spanning tree in a distributed environment. This algorithm runs faster than the traditional algorithms of Prim and Kruskal both in a serial and parallel environment as it is confirmed by results of the experiment. Further, it facilitates for choosing the load for each processor node increasing the efficiency multifold.

References

1. Taboada GL, Teijeiro C, Tourino J (2007) High performance Java remote method invocation for parallel computing on cluster. In: Proceeding of the 12th IEEE symposium on computers and communications (ISCC'07), Aveiro, Portugal, pp 233–239
2. Taboada GL, Ramos SR, Exposito R, Tourino J, Doallo R (2013) Java in the high performance computing arena: research, practice and experience. Sci Comput Program 78:425–444
3. Kambites ME, Obdrzalek J, Bull JM (2001) An openMP like interface for parallel programming in Java. Concurrency Comput: Pract Experience 13(8–9):793–814
4. Zhang BY, Yang GW, Zheng WM (2006) Jcluster: an efficient Java parallel environment on a large scale heterogeneous cluster. Concurrency Comput: Pract Experience 18(12):1541–1557
5. Bang S, Ahn J (2007) Implementation and performance evaluation of sockets and RMI based Java message passing system. In: Proceedings of 5th ACIS international conference on software engineering research, management and applications, SERA'07, Busan, Korea, pp 153–159
6. Taboada GL, Tourino J, Doallo R (2008) Java fast sockets: enabling high-speed Java communications on high performance clusters. Comput Commun Arch 31(17):4049–4059
7. GridGain middleware. http://www.gridgain.com/online_reources.html
8. Amedro B, Caromel D, Huet F, Bodnartchouk V, Delbe C, Taboada GL (2009) ProActive: using a Java middleware for HPC design, implementation and benchmarks. Int J Comput Commun 3(3):49–57
9. Cormen TH, Leieron CE, Rivest RL, Stein C (2001) Introduction to algorithms. MIT press, Cambridge

Part VII
Poster Presentation

Chapter 43
An Analytical Approach to Study the Behavior of Defected Patch Structures

Ankan Bhattacharya

Abstract This paper introduces an analytical approach to study the behaviour of Defected Patch Structures. More specifically the effect of addition of a U-Shaped slot to a Microstrip Patch. Roger 4003 has been chosen as the substrate material for the proposed antenna which has electrical permittivity of 3.4. In the initial section a probe fed rectangular patch has been simulated, which is followed by the addition of the U-Shaped slot. The dimensions of the Patch as well as that of the Slot have been selectively chosen to generate optimum results.

Keywords Microstrip · Patch · Defected structures · Slot

43.1 Introduction

A microstrip antenna generally consists of a dielectric substrate sandwiched between a radiating patch on the top and a ground plane on the other side. The patch is generally made of conducting material such as copper or gold and can take any possible shape. The radiating patch and the feed lines are usually photo etched on the dielectric substrate. For simplicity of analysis, the patch is generally square, rectangular, circular, triangular, and elliptical or some other common shapes.

A. Bhattacharya (✉)
Department of Electronics and Communication Engineering,
Mallabhum Institute of Technology, Bishnupur, West Bengal, India
e-mail: bhattacharya.ankan1987@gmail.com

S. Sengupta et al. (eds.), *Emerging Trends in Computing and Communication*,
Lecture Notes in Electrical Engineering 298, DOI: 10.1007/978-81-322-1817-3_43,

Fig. 43.1 **a** Center-fed rectangular patch. **b** Modified rectangular patch

43.1.1 Patch Parameters

For a rectangular patch, the length L of the patch is usually in the range of 0.3333 $\lambda_o < L < 0.5\ \lambda_o$, where λ_o is the free space wavelength. The patch is selected to be very thin such that $t \ll \lambda_o$ (where *t* is the patch thickness). The height h of the substrate is usually $0.003\ \lambda_o \le h \le 0.05\ \lambda_o$. The dielectric constant of the substrate ϵ_r is typically in the range $2.2 \le \epsilon_r \le 12$. [1].

43.2 Designed Patches

Two rectangular patches have been designed as shown in the figures. Figure 43.1 shows a 'Center-Fed Rectangular Patch' structure. The patch shown in Fig. 43.1 is having a 'U-Shaped' slot.

Chosen parameters:

- Length, L: 36 mm
- Width, W: 26 mm
- Feed position: (0,0)
- Substrate: Roger 4003
- Substrate height, h: 5 mm
- Dielectric Constant, ϵ_r: 3.4
- Loss tangent: 0.002

43.3 Simulation Result and Discussion

$S_{11} = -10$ (VSWR = 1.9) is taken as the margin for resonant peaks. From Fig. 43.2 the maximum Return Loss observed is 6.2 dB ($S_{11} = -6.2$) at a frequency of 4.2 GHz, for the Rectangular patch. From Fig. 43.2 the maximum

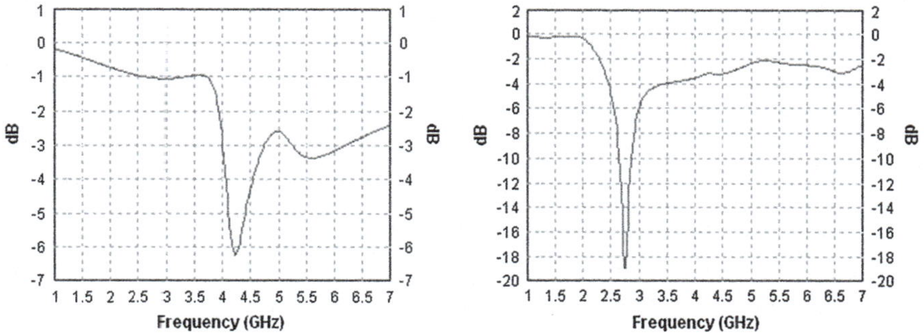

Fig. 43.2 **a** S_{11} versus frequency. **b** S_{11} versus frequency

Return Loss of 19 dB ($S_{11} = -19$) has been observed at a frequency of 2.7 GHz, for the Modified Rectangular Patch.

The frequency bandwidth obtained is given by,

$$BW = \frac{f2 - f1}{fc} \times 100\,\% \quad (43.1)$$

where, f_1, f_2 and f_c are the lower cut-off, upper cut-off and the center frequency respectively.

For the modified patch, $f_1 = 2.65$ GHz, $f_2 = 2.85$ GHz and $f_c = 2.75$ GHz as observed from Fig. 43.2. Therefore obtained BW = 7.27 %

43.4 Conclusion

It can be concluded that the Rectangular Patch antenna is practically inapplicable whereas the modified one shows promising results and is suitable for practical applications. The dimensions of the patch, the height of the substrate layer, the various properties of the substrate (e.g. the Dielectric Constant, Loss Tangent etc.), the geometrical shape as well as the dimensions of the cavity may be altered to generate more optimized results, which can be considered to be the future scope of this work.

Reference

1. Garg R, Bhartia P, Bahl I, Ittipiboon A (2001) Microstrip antenna design handbook. Artech House, Inc, Boston

Chapter 44
A New Hybrid Language Independent Keywords Extraction System

Vishal Gupta

Abstract Keywords extraction is identification of thematic words from a document which can depict the overall theme of the document. This paper concentrates on a new hybrid language independent keywords extraction system. Proposed hybrid keywords extraction system is hybrid of approach suggested by Bun and Ishizuka (Topic extraction from news archive using TF-PDF algorithm. In: Proceedings of third international conference on web information system engineering (WISE 02), pp 73–82 [2]) and Lee and Kim (News keyword extraction for topic tracking. In: Proceedings of fourth international conference on networked computing and advanced information management, pp 554–559 [3]). For identification of key terms from text we have used the NTF1-PSF and NTF2-PSF measures which are modified improved form of conventional TF-ISF measure. The first variant of TF is Normalized Term Frequency1 (NTF1) which is normalized by the maximum TF in a given sentence. The NTF2 is calculated by summing up the results of dividing the frequency of a given word appears in each sentence by the frequency of all words appear in each-sentence. Proportional-sentence-frequency (PSF) of a word in a given document is the exponential of the frequency of sentences containing the word j to the total sentence-frequency in the text document. Final keywords are obtained by taking intersection of keywords sets of NTF1-PSF and NTF2-PSF and union of title keywords set obtained by title keywords extraction. The efficiency of this language independent hybrid keywords extraction system is 84.39 %.

Keywords Language independent keywords · Keywords extraction · Hybrid keywords extraction · Information extraction · Text mining

V. Gupta (✉)
University Institute of Engineering and Technology, Panjab University,
Chandigarh, India
e-mail: vishal@pu.ac.invishal_gupta100@yahoo.co.in

S. Sengupta et al. (eds.), *Emerging Trends in Computing and Communication*, Lecture Notes in Electrical Engineering 298, DOI: 10.1007/978-81-322-1817-3_44,

44.1 Introduction to Keywords Extraction

Keywords extraction [1] is identification of thematic words from a document which can depict the overall theme of the document. This paper concentrates on a new hybrid language independent keywords extraction system. Proposed hybrid keywords extraction system is hybrid of approach suggested by Bun and Ishizuka [2] and Lee and Kim [3]. For identification of key terms from text we have used the NTF1-PSF and NTF2-PSF measures which are modified improved form of conventional TF-ISF [4, 5] measure.

44.2 Hybrid Language Independent Keywords Extraction

From input text, determine the boundary of individual words and sentences by identifying the presence of end markers of words and sentences. Stop words are those words which have very high-frequency. This is the language dependent component in our keywords extraction system. We have used the stop-words list for English language. More over We have used Porter's-stemmer [6] for converting our words into their stems. For extracting the keywords from a text document we are using the NTF1-PSF and NTF2-PSF [3] measures which are modified improved form of conventional TF-ISF [7] measure for extracting language independent keywords. NTF is normalized Term Frequency. PSF [8] is Proportional Sentence Frequency. TF means basically a frequency of a given word occurs in the whole sentence-collection of any text document. The TF is calculated by summing-up each count that a given word appears in each-sentence. But, TF-value can have some biases. So we require to normalize the value of TF. For this purpose, we have used two normalized-term-frequency equations (NTF1 and NTF2) [3]. The first bias of TF is that TF-value can be even larger than ISF-value. To eliminate this limitation, we have used the first version of normalized-TF that is normalized by the maximum TF in a given sentence collection. $\mathrm{TF_i} = \sum_{j=1}^{|S|} n_{i,j}$ $\mathrm{NTF1_i} = \mathrm{TF_i} / \mathrm{Max}\{\mathrm{TF_1},\ \mathrm{TF_2}\ldots\ldots\mathrm{TF_{|T|}}\}$ the next TF bias is that the words present in a long sentence may have larger-frequency and can be treated as more significant words. Hence, we wish to minimize such weight age of length of sentence, which leads to the new normalized-TF i.e. NTF. The NTF is calculated by summing up the values of dividing the frequency of a given word appears in each sentence by the frequency of all words appear in each-sentence. $\mathrm{NTF_i} = \sum_j \mathrm{n_{i,j}} / \sum_{k=1}^{|Tj|} n_{k,j}$ where j = 1 to |S| $\mathrm{PSF_j} = \exp(\mathrm{n_j}/\mathrm{N})$ where $\mathrm{n_j}$ = Number of sentences in a text document where word j is present. N = Total number of sentences in the document. Proportional-sentence-frequency (PSF) of a word in a text-document is $\exp(\mathrm{n_j}/\mathrm{N})$. PSF is the exponential of the frequency of sentences containing the word j to the total sentence-frequency in the text document. Words that occur in many sentences are more important than others. Words having more

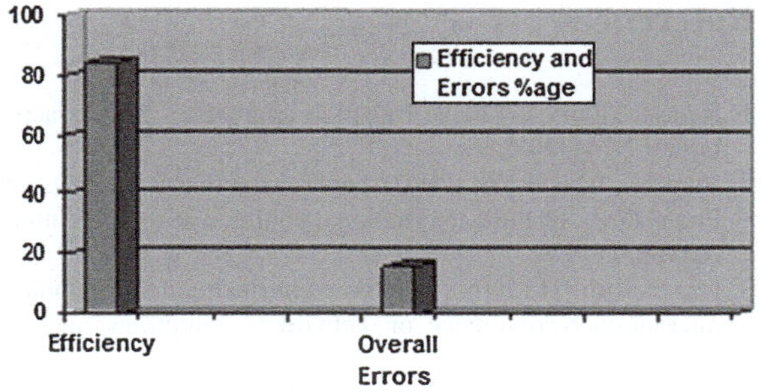

Fig. 44.1 Efficiency and errors %age for our keywords extraction system

value of NTF-PSF score are called keywords from this phase. For a given word i, we can calculate the NTF1-PSF score for that word as follows: $NTF1_i$-PSF_i = NTF1-Score of word i × PSF score of word i. Similarly we calculate NTF1-PSF score of each word in each sentence of that document. Top 20 % words (k_1) having maximum NTF1-PSF score are treated as keywords from this phase. Similarly for a given word i, we can also calculate the NTF2-PSF score for that word as follows: $NTF2_i$-PSF_i = NTF2-Score of word i × PSF score of word i. We can calculate NTF2-PSF score of each word in each sentence of that document. Top 20 % words (k_2) having maximum NTF2-PSF score are treated as keywords from this phase. Final keywords from this phase are the extracted by taking intersection of keywords set k_1 and k_2. $k = \{k_1 \cap k_2\}$ where k is keywords set obtained by taking intersection of keywords set k_1 and k_2. Where k_1 is set of top scored 20 % keywords obtained by applying NTF1-PSF measure and k_2 is set of top scored 20 % keywords obtained by applying NTF2-PSF. Next feature used in this keywords extraction system is Title keywords feature. Title-lines are the title-lines (head-lines) of text-documents. Those sentences containing title keywords [9] are important. Title keywords are obtained after removing stop words from title lines having headlines flag set to true. Stop word list is the language dependent component that we have added here. Set the Title_Keyword_Flag = True for those words which belong to title lines. Put all the title keywords in set k_3. Final keywords are extracted by taking union of keywords set k and k_3. $K_F = \{k \cup k_3\}$ where k is keywords set obtained by taking intersection of keywords set k_1 and k_2.

44.3 Results

The efficiency of our keywords extraction is 84.39 % (Fig. 44.1) which is ratio of correct keywords extracted by our system to the total number of keywords extracted by our system. This system has been tested by analyzing its results over 100 documents covering general articles, stories and news documents.

From the results we can conclude that this language independent keywords extraction system is performing reasonably well as compared to other language dependent keywords extraction systems.

References

1. Kaur J, Gupta V (2010) Effective approaches for extraction of keywords. Int J Comput Sci Issues 7:144–148
2. Bun KK, Ishizuka M (2002) Topic extraction from news archive using TF-PDF algorithm. In: Proceedings of third international conference on web information system engineering (WISE 02), pp 73–82
3. Lee S, Kim HJ (2008) News keyword extraction for topic tracking. In: Proceedings of fourth international conference on networked computing and advanced information management, pp 554–559
4. Neto JL et al (2000) Document clustering and text summarization In: Proceedings of 4th international conference practical applications of knowledge discovery and data mining, London, pp 41–55
5. Hulth A (2003) Improved automatic keyword extraction: given more linguistic knowledge. In: Proceedings of ACM EMNLP'03, Japan, pp 216–223
6. http://tartarus.org/martin/PorterStemmer/
7. Kian HH, Zahedi M (2011) An efficient approach for keyword selection; improving accessibility of web contents by general search engines. Int J Web Semant Technol 2:81–90
8. Gupta V, Lehal GS (2011) Automatic keywords extraction for Punjabi language. Int J Comput Sci Issues 8:327–331
9. Fattah MA, Ren F (2008) Automatic text summarization. Proc World Acad Sci Eng Technol 27:192–195

Chapter 45
A New System for Extracting Numeric Data from Punjabi Text

Vishal Gupta

Abstract In text documents, number data is always important. Extraction of numeric data is required for text summarization, Machine translation, keywords extraction, documents association etc. This paper concentrates on new system for extracting numeric data from Punjabi text. It covers rules for extracting dates, rules for extracting time, rules for extracting normal digits/decimal numbers/integers/roman numerals/Gurmukhi numerals, rules for extracting numeric data with specific prefix word, rules for extracting numbers as typical Punjabi words, rules for extracting numeric data with specific suffix words and rules for extracting numeric data in arithmetic expressions. The Precision, Recall and F-Score of Punjabi numeric data extraction system are 97.39, 93.547 and 95.43 % respectively.

Keywords Punjabi numeric data extraction · Number data extraction · Numeric data · Number data

45.1 Introduction

In text documents, number data [1] is always important. Extraction of numeric data is required for different text mining techniques like text summarization [3, 5], Machine translation, keywords extraction, documents association etc. This paper concentrates on new system for extracting numeric data from Punjabi text. This system covers rules for extracting dates, rules for extracting time, rules for

---

V. Gupta (✉)
University Institute of Engineering and Technology, Panjab University Chandigarh, Chandigarh, India
e-mail: vishal@pu.ac.in; vishal_gupta100@yahoo.co.in

S. Sengupta et al. (eds.), *Emerging Trends in Computing and Communication*, Lecture Notes in Electrical Engineering 298, DOI: 10.1007/978-81-322-1817-3_45,

extracting normal digits/decimal numbers/integers/roman numerals/Gurmukhi numerals, rules for extracting numeric data with specific prefix word, rules for extracting numbers as typical Punjabi words, rules for extracting numeric data with specific suffix words and rules for extracting numeric data in arithmetic expressions. Earlier a part of this system covering rules for extracting normal digits, roman numerals and Gurmukhi numerals had been implemented with Punjabi text summarization system [2].

45.2 Punjabi Numeric Data Extraction System

A number of rules specific for Punjabi language are formed to extract the language dependent features for numeric data.

Regarding Date rules for Extracting Dates from Punjabi text. We are using the same Date rules as defined by kaur and Gupta [4]. In Punjabi documents, dates are extracted in any format like: dd/mm/yyyy, dd-mm-yyyy or dd.mm.yyyy e.g. 13/12/2008, 13-12-2008 or 13.12.2008. Year can be in any format like yyyy or yy or yyyy-yy or yyyy-yyyy e.g 1995 or 95 or 1995-96 or 1995-1996.

Regarding rules for Time checking from Punjabi text, if current Punjabi word is ਏ.ਐਮ/ਏਐਮ or ਪੀ.ਐਮ/ਪੀਐਮ or ਵੱਜੇ/ਵਜੇ or ਘੰਟਾ/ਘੰਟੇ/ਘੰਟਿਆਂ or ਮਿਟ/ਮਿਟਾਂ or ਸਕਿਟ/ਸਕਿਟਾਂ and its previous word is between 1 and 24 then it is time and is extracted as such. Time can be in any format like hh:mm:ss or hh:mm like 05:57:30 or 05:57 i.e. if current word is between 1 and 24 followed by colon and then followed by minutes ranging between 1 and 60 followed by colon and then followed by seconds ranging between 1 and 60 then it is time and is extracted as such.

Regarding rules for extracting Digits/Decimal Numbers/Integers/Roman Numbers/Gurmukhi Numerals, we are processing Unicode based Punjabi text so normal digits are having their Unicode range between 48 and 57 for numbers 0–9, for Roman numerals (i, ii, iii etc.) the Unicode range is 8544–8575, for Gurmukhi numerals (੦, ੧, ੨, ੩, ੪, ੫, ੬, ੭, ੮, ੯) the Unicode range is 2662–2677. Decimal Numbers can be in two formats like .35 or 9.35. If current word is digit followed by decimal '.' and then followed by another digit. Then extract it as decimal number as whole. If current character is decimal followed by another digit then that is decimal number and extract it as such. If current digit is prefixed with '−' or '+' character then that number is integer and extract it as such.

Regarding rules for extracting numbers as typical Punjabi words, Some times numbers are written as Punjabi words like ਇੱਕ, ਦੋ, ਤਿਨ, ਚਾਰ, ਪੰਜ, etc. In Punjabi word ਇੱਕ is equivalent to 1, ਦੋ is equivalent to 2, ਤਿਨ is equivalent to 3 and so on. So Punjabi numbers ਇੱਕ, ਦੋ, ਤਿਨ, ਚਾਰ, ਪੰਜ, etc. are extracted as output of Punjabi number extraction system. Moreover some typical forms of Punjabi numbers are also covered like ਇਕਾਈ, ਦਹਾਈ,ਤਿਮਾਹੀ, ਛਿਮਾਹੀ, ਸਪਤਾਹੀ, ਪਹਿਲਾ, ਦੂਜਾ, ਤੀਜਾ, ਪੰਜਵਾਂ, ਛੇਵਾਂ, ਸਤਵਾਂ/ਸੱਤਵਾਂ, ਦੱਸਵਾਂ/ਦਸਵਾਂ and ਪੰਦਰਵਾਂ etc. ਇਕਾਈ is equivalent to 1, ਦਹਾਈ equivalent to 2, ਤਿਮਾਹੀ is equivalent to 3, ਪੰਜਵਾਂ is equivalent to 5, ਪੰਦਰਵਾਂ is equivalent to 15 and so on. A database for typical words has been made covering

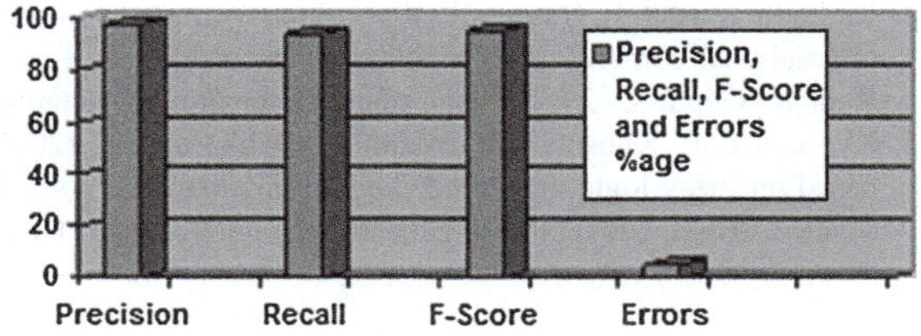

Fig. 45.1 Precision, recall, F-score and errors % age for Punjabi numeric data extraction system

different typical Punjabi words for numeric data and corresponding digit for that typical word is also stored in database for Punjabi typical words.

Regarding, extracting numeric data with specific prefix Word, If current Punjabi word is ਨੰ or ਨੰ: or ਨੰ- or ਨੰ/or ਨੰ@ or ਨੰ # or ਨੰਬਰ or ਨੰਬਰ: or ਨੰਬਰ. ਨੰਬਰ# or ਨੰਬਰ- or ਨੰਬਰ@ etc. Then next word is checked for numeric data in Unicode ranging between 48 and 57. Example: ਅਰਜੀ ਨੰ:88, ਟਕਿਟ ਨੰ # 74, ਮਕਾਨ ਨੰਬਰ.4 etc.

Regarding extracting numeric data with specific suffix word from Punjabi documents some times, numbers are followed by specific suffix words like ਰੁਪਏ, ਡਾਲਰ, ਯੂਰੋ, ਮੀਟਰ, ਸੈਂਟੀਮੀਟਰ, ਕੌਲਿ, ਗਰਾਮ, ਟਨ, ਦਰਾਮ, ਪਾਊਡ, ਸਾਲ'/ਸਾਲਾਂ or 'ਵਰ੍ਹਾ'/ਵਰ੍ਹੇ etc. These suffix words can include units of dimension or currency or some other measurements. A database has been developed covering these suffix Punjabi words. Word just previous to these suffix words are usually the numbers.

Regarding extracting numeric data in Arithmetic Expressions, some times number data is represented as arithmetic expressions in Punjabi documents like: (100 + 100), (100 − 60), (100/50), (100 × 100) etc. Punjabi Number extraction system checks whether current character is any arithmetic operator and if that arithmetic operator is surrounded by two number operands on its left and right then that numerical expression is extracted as such.

45.3 Results and Discussions

Punjabi numeric data extraction system has been tested by analyzing its results over fifty Punjabi news documents. The Precision, Recall and F-Score of Punjabi numeric data extraction system are 97.39, 93.547 and 95.43 % respectively (Fig 45.1). Errors of 4.57 % in the results of Punjabi numeric data extraction system, are due to absence of certain types of rules for extracting numeric data.

References

1. Fattah MA, Ren F (2008) Automatic text summarization. Proc World Acad Sci Eng Technol 27:192–195
2. Gupta V, Lehal GS (2012) Automatic Punjabi text extractive summarization system. In: Proceedings of international conference on computational linguistics, pp 191–198

3. Kaikhah K (2004) Automatic text summarization with neural networks. In: Proceedings of international conference on intelligent systems, IEEE, Texas, pp 40–44
4. Kaur K, Gupta V (2011) Topic tracking for Punjabi language. Comput Sci Eng Int J 1:37–49
5. Kyoomarsi F, Khosravi H, Eslami E, Dehkordy PK (2008) Optimizing text summarization based on fuzzy logic. In: Proceedings of international conference on computer and information science, IEEE, University of Shahid Bahonar Kerman, UK, pp 347–352

Chapter 46
A New Punjabi Keywords Extraction System

Vishal Gupta

Abstract Keywords are set of important thematic words that represent the whole document. This paper discusses a novel approach for determination of Punjabi keywords. Earlier Keywords extraction systems for Punjabi were not much efficient as those were using less number of features for extracting keywords. But the proposed Punjabi keywords extraction system is very efficient as it is using six features for extracting Punjabi keywords as compared to earlier systems like: TF-ISF feature, Noun frequency feature, font type feature (Bold font, Italics font and Underlined font), Cue phrase feature, Position feature and title keyword feature. The proposed approach also uses regression for assigning weights to these six features.

Keywords Punjabi keywords · Keywords extraction · Punjabi keywords detection

46.1 Introduction to Keywords Extraction

Keywords are set of important thematic words [1] that represent the whole document. This paper discusses a novel approach for determination of Punjabi keywords. For Punjab, Punjabi is its official language. Not much research has been done for Punjabi language because little amount of language-resources are present for Punjabi. Earlier Keywords extraction systems for Punjabi were not much efficient as those were using less number of features for extracting keywords. But the proposed Punjabi keywords extraction system is very efficient as it is using six features for extracting Punjabi keywords as compared to earlier systems like:

V. Gupta (✉)
University Institute of Engineering and Technology, Panjab University, Chandigarh, India
e-mail: vishal@pu.ac.in; vishal_gupta100@yahoo.co.in

S. Sengupta et al. (eds.), *Emerging Trends in Computing and Communication*,
Lecture Notes in Electrical Engineering 298, DOI: 10.1007/978-81-322-1817-3_46,

TF-ISF feature, Noun frequency feature, font type feature (Bold font, Italics font and Underlined font), Cue phrase feature, Position feature and title keyword feature. The proposed approach also uses regression for assigning weights to these six features.

46.2 Punjabi Keywords Extraction

From the Punjabi text, mark the sentences and words separately by presence of Punjabi words and sentences end markers. Punjabi stop words are highly frequent words with little importance. 615 unique Punjabi stop words are identified and deleted from the Punjabi text. Punjabi stemming is applied to convert the Punjabi words into their root forms. We have applied Punjabi noun/proper names stemming which is proposed by Gupta and Lehal [2]. This Punjabi stemmer applies 18 stemming rules. Various sub phases for Punjabi keywords extraction are:

For calculating first feature, we have counted the TF (Term Frequency) [3] of each noun in every sentence. Those nouns which are having highest value of this term frequency are given more importance and can be treated as key terms. Punjabi Noun words can be checked by their presence in Punjabi noun morph. Punjabi noun morph is having 37,297 Punjabi nouns.

For calculating second feature, we have calculated the scores of TF-ISF for each word. TF (Term-Frequency) is frequency of a word in the given sentence and ISF (Inverse-Sentence-Frequency) of a word is the frequency of sentences which contain the given word. It is clear that a word present in more sentences is more important and relevant than a word which is present many times in the same sentence. We have calculated TF-ISF score of each word as calculated by Neto et al. [4] and Gupta and Lehal [5, 6].

Our third feature is font feature for Punjabi words. Normally, words written in bold, italics or underlined fonts or having more font size are more important than the other words. In this phase, we identify the Punjabi words written in bold, italics or underlined fonts. Along with this we also identify the Punjabi words having more font size than the normal size used for regular text. Words identified from this face can be the candidates for keywords. Set Flag_Font_Feature = True for those Punjabi words which are in bold, italics or underlined fonts or having more font size.

Our Fourth feature is position feature. Punjabi words belonging to first or last sentence of first or last paragraphs are more important than the other sentences and these words are probable key terms. Usually first and last sentences of text carry more information about the topic of the text. Set the Position_flag = True for those Punjabi words which belong to first or last sentence of first or last paragraphs.

Our fifth feature is Title keywords feature. Title lines are the leading lines of any document. Title keywords [3, 5] are present in these title lines and these words are more relevant than others. Title keywords are identified after deleting

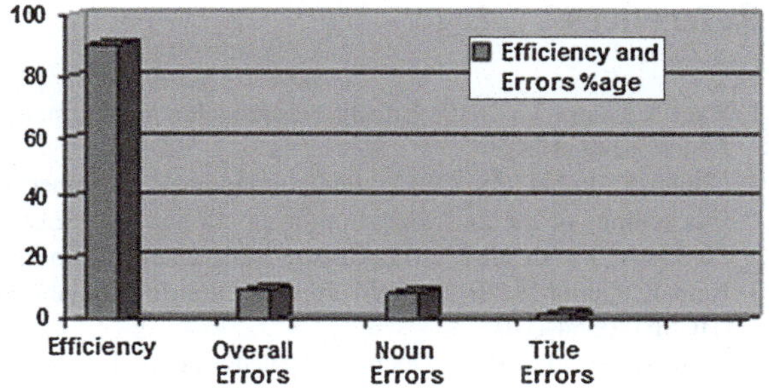

Fig. 46.1 Efficiency and errors percentage for Punjabi keywords extraction system

stop-words from title-lines having headlines flag set to true. The accuracy of Punjabi title keywords extraction is 97.48 %. Set the Title_Keyword_Flag = True for those Punjabi words which belong to title lines.

Our sixth feature is Cue-phrase feature. Cue Phrases [3] are type of key terms which are very important like ਉਦਾਹਰਣ, ਸਿੱਟਾ, ਅੰਤ ਵਿੱਚ, ਸ਼ੁਰੂਆਤ ਵਿੱਚ etc. Sentences which start with these terms or which possess these terms are more relevant. We have created Punjabi cue phrase list by consulting Punjabi corpus. Set the Cue Phrase_Keyword_Flag = True for those Punjabi words which belong to lines containing Punjabi cue phrases.

Regression has been applied to estimate the feature-weights as calculated by Fattah and Ren [7]. Learned weights for one to six features are 2.82, 0.65, 1.87, 0.85, 1.68 and 0.45 respectively. Final word-scores are calculated using feature-weight equation as: $w_1\text{Score }(f_1) + w_2\text{Score }(f_2) + \ldots w_6\text{Score }(f_6)$. Where Score (f_1), …, Score (f_6) are scores of six features and w_1, w_2, w_3, …, w_6 are assigned weights of these features.

46.3 Analysis of Results

The efficiency of Punjabi keywords extraction is 90.58 % (Fig. 46.1) which is ratio of correct Punjabi keywords extracted by our system to the total number of keywords extracted by our system. Punjabi keywords extraction system has been tested by analyzing its results over 50 Punjabi news documents.

Out of 9.48 % errors, 8.53 % errors are under the category of noun errors. Errors of 0.95 % are under the category of title keywords.

References

1. Kaur J, Gupta V (2010) Effective approaches for extraction of keywords. Int J Comput Sci Issues 7:144–148
2. Gupta V, Lehal GS (2011) Punjabi language stemmer for nouns and proper names. In: Proceedings of the 2nd workshop on South and Southeast Asian natural language processing (WSSANLP), IJCNLP 2011, Chiang Mai, Thailand, pp 35–39
3. Kaur K, Gupta V (2011) Keyword extraction for Punjabi language. Indian J Comput Sci Eng (IJCSE) 2:364–370
4. Neto JL et al (2000) Document clustering and text summarization. In: Proceedings of international conference on practical app of knowledge discovery & data mining, London, pp 41–55
5. Gupta V, Lehal GS (2011) Automatic keywords extraction for Punjabi language. Int J Comput Sci Issues 8:327–331
6. Gupta V, Lehal GS (2012) Automatic Punjabi text extractive summarization system. In: Proceedings of international conference on computational linguistics COLING '12, pp 191–198
7. Fattah MA, Ren F (2008) Automatic text summarization. Proc World Acad Sci Eng Technol 27:192–195

Chapter 47
Effect of Strain on the Band Line Up and Built in Electric Field of Strained AlGaN/GaN and InGaN/GaN Quantum Well

Sourav Dutta and Soumen Sen

Abstract One of the important parameters associated with the carrier confinement of hetero structures is their band line up. Here we have developed the relations to compute the band line up of materials $Al_xGa_{1-x}N/GaN$ and $In_xGa_{1-x}N/GaN$ and the effect of strain in the built in electric field. The band positions for the heterointerfaces of $InxGa_{1-x}N/GaN$ and $AlxGa_{1-x}N/GaN$ are calculated from equations developed, which are related with the position of the bands with the strain at the interface and are compared. Thus the strain is calculated and compared for both $In_xGa_{1-x}N/GaN$ and $Al_xGa_{1-x}N/GaN$ from the respective In and Ga mole fraction (Panda et al., International conference on computers and devices for communication, 2009) [1]. Due to the non-centrosymmetric structure of the hexagonal III-nitrides, spontaneous polarization exists in bulk films. In addition, when fabricating heterostructures of GaN/AlGaN or GaN/InGaN, the effect of piezoelectric polarization due to strain has to be taken into account. Both the spontaneous and piezoelectric polarization give rise to a large built-in field (Ng et al., J Electron Mater 30, 2001) [2].

Keywords Band lineup · Built-in electric field · Strain

47.1 Introduction

GaN based quantum wells have attracted much interest in the last few years since they have become the key material in commercial fabrication of longtime LED's and laser diodes. GaN has a large peak electron velocity, larger saturation velocity,

S. Dutta (✉) · S. Sen
Asansol Engineering College, Asansol, India
e-mail: souravdutta2012@gmail.com

S. Sen
e-mail: soumencccp1234@rediffmail.com

S. Sengupta et al. (eds.), *Emerging Trends in Computing and Communication*, Lecture Notes in Electrical Engineering 298, DOI: 10.1007/978-81-322-1817-3_47,

high breakdown voltage and thermal stability making it suitable to be used as a channel material in microwave power devices [3].

Thus we have considered strained AlGaN/GaN and InGaN/GaN quantum wells here, since they have been a subject of intense recent investigation and have emerged as attractive candidates for high voltage, high power operation and microwave frequency [3].

Another peculiarity results from the polar axis of the wurtzite crystal structure and the strong polarity of III-N bindings. All group-III nitrides in the wurtzite phase have a strong spontaneous macroscopic polarization and large piezoelectric coefficients. This has been found from ab initio calculations [4, 5].

47.2 Theoretical Details

For model calculations, we have considered an $In_xGa_{1-x}N$/GaN and $Al_xGa_{1-x}N$/GaN systems as the epitaxial layer where compositions of In and Al have been varied to change the strain. In and Al mole fraction(x) have been varied from 0 to 1.0. The strain in InGaN/GaN and AlGaN/GaN system have been calculated by the formulae (47.1) and (47.2) [1, 6],

$$\varepsilon = (d_e - d_s)/d_s \tag{47.1}$$

$$\varepsilon = (d_s - d_e)/d_e \tag{47.2}$$

where d_e and d_s are the lattice constants of the epitaxial layer and the substrate. The variation of strain with In and Al mole fraction is shown in Fig. 47.1. For $Al_xGa_{1-x}N$/GaN and for $In_xGa_{1-x}N$/GaN quantum wells, the equations deployed to determine the conduction and valence band positions are [1, 6].

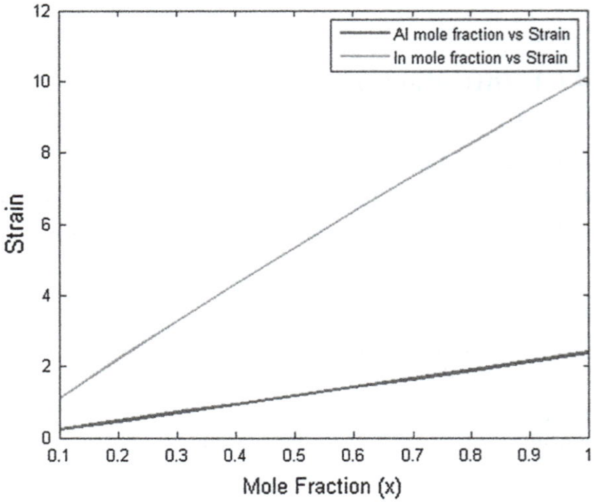

Fig. 47.1 Variation of the strain of AlGaN and InGaN with Al and In with mole fraction

The potential well, which has formed here to get quantum confinement is of asymmetric triangular shape. The built in electric field can be written as [7, 8]

$$E_p = -\left(\left(P_{SP}^{InGaN} + P_{PZ}^{InGaN} - P_{SP}^{GaN}\right)L^{GaN}\right)/\left(\varepsilon_0\left(2\varepsilon_r^{GaN}L^{InGaN} + \varepsilon_r^{InGaN}L^{GaN}\right)\right) \quad (47.3)$$

$$E_p = \left(P_{SP}^{AlGaN} + P_{PZ}^{AlGaN} - P_{SP}^{GaN} - P_{PZ}^{GaN}\right)/\left(\varepsilon^{GaN} + \varepsilon^{AlGaN}\left(L^{GaN}/L^{AlGaN}\right)\right) \quad (47.4)$$

For Eq. (47.3) and (47.4) ε is the static dielectric constant, L is the layer thickness, subscripts SP and PZ represent spontaneous and piezoelectric polarization, ε_r and ε_o are the dielectric constant of InGaN and permittivity of free space respectively, L^{InGaN} is the dot height and L^{GaN} is the barrier thickness respectively. The values of the parameters used in this paper are taken from Vurgaftman [9].

47.3 Results and Discussion

The variation of strain with Al mole fraction and In mole fraction are shown in Fig. 47.1. The strain varies almost linearly with the mole fraction of Al as well as In. The strain variation with respect to In mole fraction is always larger than the Al mole fraction Figures 47.2 and 47.3 show the variations of the conduction and valence band lineups with strain, respectively considering InN bandgap (Eg, InN = 1.95 eV) [1] for InGaN/GaN and (Eg, AlN = 6.13 eV) [6] for AlGaN/GaN quantum wells. Here we see that AlGaN/GaN quantum well is more linearly than InGaN/GaN quantum well. Figure 47.4 shows the variation of the energy band gap with the In and Al mole fraction for InGaN and AlGaN quantum well. Figures 47.5 and 47.6 show the variation of built in electric field for InGaN and AlGaN with In and Al mole fraction. Here we see that the built in electric field varies almost linearly with the mole fraction.

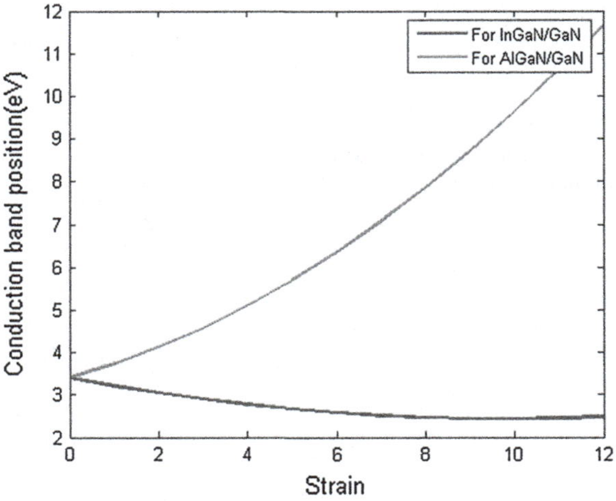

Fig. 47.2 Variation of conduction band position for AlGaN and InGaN with respect to the strain

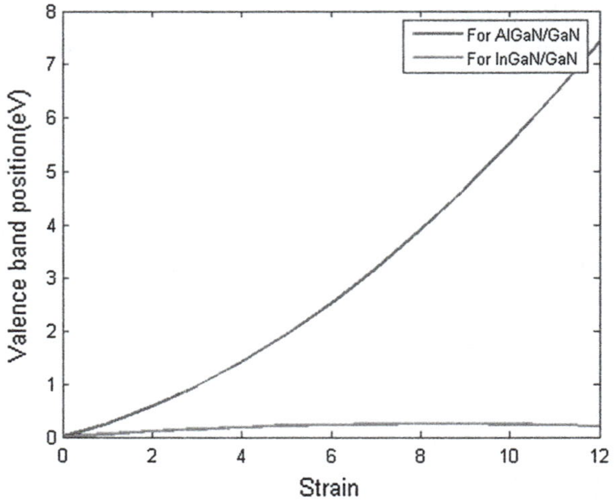

Fig. 47.3 Variation valence band position for AlGaN and InGaN with respect to strain

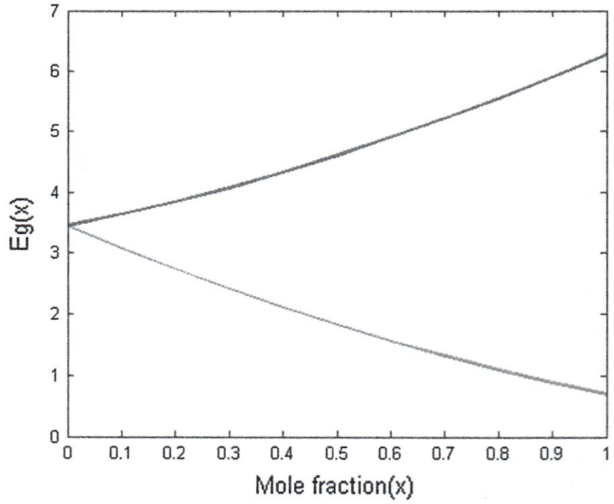

Fig. 47.4 Variation of energy band gap with mole fraction for AlGaN (*blue*) and InGaN (*green*)

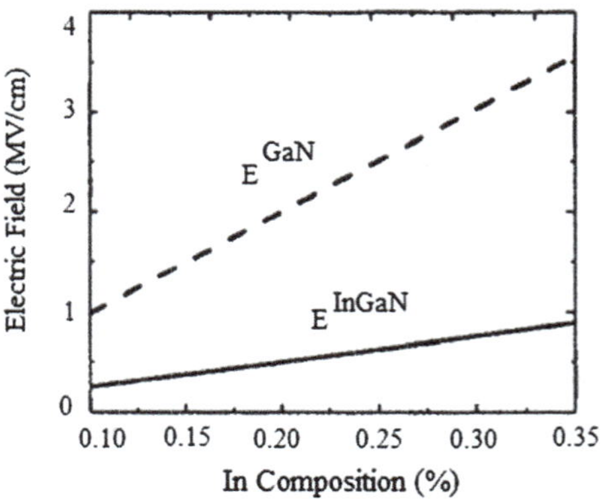

Fig. 47.5 Variation of built in electric field of InGaN with In mole fraction

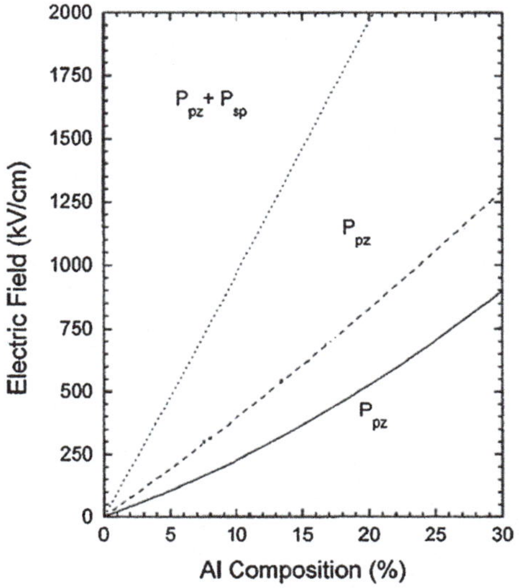

Fig. 47.6 Variation of built in electric field with Al mole fraction in AlGaN quantum well. Electric field due to pure piezoelectricity (P_{pz}) contribution is given as the *full line* [8] and the *dashed line* [5]

To summarize, in this paper, we have shown the effect of strain on the band line up and built in electric field of strained AlGaN/GaN and InGaN/GaN quantum well.

References

1. Panda S, Kabi S, Biswas D (2009) International conference on computers and devices for communication.
2. Ng HM, Harel R, Chu SNG, Cho AY (2001) J Electron Mater 30(3)
3. Ambacher O, Foutz B, Shealy JR, Wiemann NG et al (2000) J Appl Phys 87:334
4. Bernardini F, Fiorentini V, Vanderbilt D (1997) Phys Rev B 56:R10024
5. Shimada K, Sota T, Suzuki K (1998) J Appl Phys 84:4941
6. Das T, Kabi S, Biswas D (2009) J Appl Phys 105:046101
7. Bachir N, Hamdoune A, Chabane Sari NE, ISBN: 978-953-51-0549-7, InTech
8. Bykhovski AD, Kaminski VV, Shur MS, Chen QC, Khan MA (1996) J Appl Phys Lett 68:818
9. Vurgaftman I, Meyer JR (2003) J Appl Phys 94:3675

Chapter 48
A Decision Support System for Website Selection for Internet Based Advertising and Promotions

Arpan Kumar Kar

Abstract With the onset of the internet era, the focus shifted in marketing from Word of Mouth to the Word of Web. However, the size of the web increasing in leaps and bounds. A major challenge is to identify suitable websites for promotion of marketing campaigns. The operational issues in addressing the challenge can be classified into two parts. The first is to identify the critical factors for evaluating websites. The second is to evaluate the websites against these evaluation factors, and select them from a large pool of websites. Creating a schema for selecting such websites from a large list is a major challenge. This paper proposes a decision support system to select suitable partner websites from a list, by evaluating them against a set of context specific website-quality evaluation criteria. The integrated methodology for decision support uses Delphi, Analytic Hierarchy Process and Cuckoo Search.

Keywords Decision support system · Website selection · Internet marketing · Analytic hierarchy process · Cuckoo search · Delphi · Marketing analytics

48.1 Background Discussion and Focus of the Study

With the increasingly rising importance of the internet era, the focus shifted in traditional marketing from leveraging the Word of Mouth to the Word of Web. However, the size of the web is increasing in leaps and bounds. As of 31st October, 2013, "WorldWideWebSize" estimates that there are at least 3.59 billion nodes in this huge graph of websites connecting with each other. A major challenge is to identify suitable websites for the promotion of marketing campaigns

A. K. Kar (✉)
Indian Institute of Management Rohtak, Rohtak, Haryana, India
e-mail: arpan_kar@yahoo.co.in

S. Sengupta et al. (eds.), *Emerging Trends in Computing and Communication*, Lecture Notes in Electrical Engineering 298, DOI: 10.1007/978-81-322-1817-3_48,

through advertisements like sponsored posts, banner advertisements, text link advertisements and others. The operational issues in addressing the challenge can be categorized into two parts. The first is to identify the critical factors for evaluating web-sites. The second is to evaluate the websites against these evaluation factors, and select them from a large pool of websites. Creating a schema for selecting such websites from a large list is a major challenge. There has been no study which has attempted to structure the evaluation criteria in a multi-criteria multi-hierarchy decision making process and proposes a systematic approach for solving the problem. This paper proposes an integrated approach using Analytic Hierarchy Process (AHP) and Cuckoo Search Algorithm (CSA) decision support system which would also provide predictive decision support to select suitable partner websites from a list, by evaluating them against a set of context specific website-quality evaluation criteria. The approach has been validated through a case study conducted in association with an internet marketing consultancy firm, which identifies potential websites for promoting internet based advertising campaigns for its clients.

48.2 Research Methodology

48.2.1 Criteria for Evaluation of Websites

Literature focusing on the evaluation of websites have mostly focused on the end consumer's perspective for identifying factors that promotes adoption. Thus literature highlights the importance of factors like website content [1], ease of use [2], design [3], perceived usefulness [4], playfulness [5], price savings [4], complementary relationship [2], tangibility [6], security [6], interactivity [7], responsiveness [7], trustworthiness [6], privacy [8], user empowerment [8] and many other such factors. However, for a marketer, it is not is not possible to be evaluate the performance of individual websites against such factors simply because it would require collecting response of the target consumer segment for evaluating each possible website before selecting one for partnering. Thus marketers prefer to use other approaches for the internet marketing (Eyrich et al. [9]). Hence, there has been a rise of adoption of specific evaluation criteria, the performance data on which can be can be collected by using documented evidences from the public domain and the advertising rates commanded by the websites are often against their performances, against these criteria (Fig. 48.1).

The Delphi method [10] has been used in numerous group decision making literature to obtain group consensus among domain. Most Delphi studies [10] have a sample size of 5–20 domain experts as respondents and 2 iterations are often sufficient to achieve consensus in group decision making.

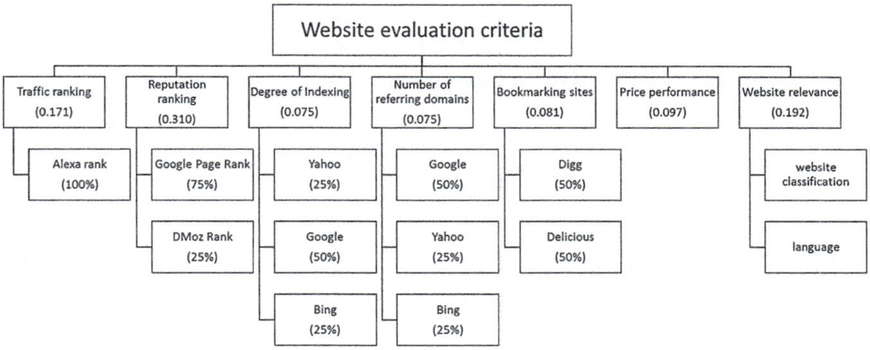

Fig. 48.1 Hierarchy of website evaluation criteria and their importance

48.2.2 Decision Support Approach for Evaluation and Selection

This study uses two methodologies for the identification of suitable websites for partnering in promotional campaigns. For the first phase, it uses the AHP for prioritizing the evaluation criteria. In the second phase, it uses the CSA for evaluating the performance of listed websites, against these prioritized evaluation criteria.

48.2.2.1 Analytic Hierarchy Process

The AHP [11, 12] was developed for use in multi-hierarchical, multi-criteria decision making problems. AHP has been extensively integrated with other theories for decision support in selection literature [13]. While prioritizing the evaluation factors using AHP, first the linguistic judgments of the decision maker is captured and mapped to quantified judgments [11]. Subsequently, these linguistic judgments are converted to crisp priorities using AHP theory. Let judgments $A = (a_{ij})_{k \times k}$ would be coded using Saaty's scale [11] for conversion of judgments. The individual priorities are obtained from the matrix of coded linguistic judgments by solving the following linear system:

$$\min \sum_{i=1}^{k} \sum_{j>i}^{k} \left(\ln \tilde{a}_{i,j} - \left(\ln \tilde{w}_i - \ln \tilde{w}_j \right)^2 \right) \text{s.t.}\, \tilde{a}_{ij} \geq 0; \tilde{a}_{ij} \times \tilde{a}_{ji} = 1; \quad \tilde{w}_i \geq 0, \sum \tilde{w}_i = 1. \tag{48.1}$$

The individual priority vector is obtained by

$$w_i = \frac{\sqrt{[1/k] \prod_{j=1}^{k} \tilde{a}_{i,j}}}{\sum_{i=1}^{n} \sqrt{[1/k] \prod_{j=1}^{k} \tilde{a}_{i,j}}} \tag{48.2}$$

In the subsequent step, the consistency of the priorities need to be estimated and if inconsistent, needs to be improved. This is achieved by estimating the Geometric Consistency Index (GCI) (Aguarón and Moreno-Jiménez [12].

$$GCI\left(\mathrm{A}^{(\mathrm{k})}\right) = \frac{2}{(r-1)(r-2)}\sum_{j>i}^{r}\left(\log|\tilde{a}_{i,j}| - \left(\log|\tilde{w}_i^{(k)}| - \log|\tilde{w}_j^{(k)}|\right)^2\right) \leq \overline{GCI_r} \tag{48.3}$$

The condition for consistency with seven evaluation criteria is GCI should not exceed 0.37. After the priorities are estimated, the next state is initiated, using the CSA.

48.2.2.2 Cuckoo Search Algorithm

CSA [14–16] is an optimization algorithm that can search in a smart manner from a list of potential solutions. A fitness function needs to be defined which will evaluate the competency of a potential solution, which in this case is a website. The fitness function is derived from the priority evaluated using AHP. The outcome is iteratively computed with each individual cuckoo C(i) as illustrated:

Defining the objective or fitness function:

$$\mathrm{F}\{\mathrm{Xi}\} = \sum_{1}^{7}(w_i \times S_{X_i}) \tag{48.4}$$

An initial population of n potential solutions are generated;

Start iterations While (t < MaxGeneration) or Maximum (F{Xi})

C(i) randomly replaces F{Xi} and evaluates

$$(\mathrm{F}\{\mathrm{Xi}\})\,\forall X_i \in (w_i \times S_{X_i}) \tag{48.5}$$

For maximization,

$$\mathrm{Fi} = \text{Maximum of } \mathrm{F}\{\mathrm{Xi}\}\})\,\forall X_i \in (w_i \times S_{X_i}) \tag{48.6}$$

Choose a nest among n (say, j) randomly;

$$\text{if (Fi > Fj), Fi = Fj, else discard Fj} \tag{48.7}$$

A fraction of the minimum (Fi) are abandoned, new ones are built while we keep Fi from each nest of solutions

$$\text{Rank the Fi, } F_i\,\forall\, i \in n \tag{48.8}$$

Save n_1 number of best solutions; Pass the best solutions to next iteration. End While iterations.

48.3 Conclusion

The proposed integrated approach provided decision support of identifying the 10 most suitable websites for partnering, based on a requirement specific context. Since the size of the web is fast expanding, such an approach can make it feasible to identify suitable websites for promotions and partnership, based on specific evaluation criteria. This integrated approach will therefore automate the selection process while taking into consideration individual priorities of the criteria.

A limitation of this paper is the lack of detailed illustration to illustrate the actual application of the algorithm in a step wise illustration. Further given the exhaustiveness of data across multiple iterations, detailed illustration of the systematic computation is beyond the current scope of the paper.

References

1. Agarwal R, Venkatesh V (2002) Assessing a firm's web presence: a heuristic evaluation procedure for the measurement of usability. Inf Syst Res 13(2):168–186
2. Loiacono ET, Chen D, Goodhue DL (2002) WebQual revisited: predicting the intent to reuse a web site. In American conference on information systems, 301–309
3. Barnes SJ, Vidgen R (2001) An evaluation of cyber-bookshops: the WebQual method. Int J Electron Commer 6:11–30
4. Devaraj S, Fan M, Kohli R (2002) Antecedents of B2C channel satisfaction and preference: validating e-commerce metrics. Inf Syst Res 13:316–333
5. Liu C, Arnett KP (2000) Exploring the factors associated with web site success in the context of electronic commerce. Inf Manage 38:23–33
6. Webb HW, Webb LA (2004) SiteQual: an integrated measure of web site quality. J Enterp Inf Manage 17:430–440
7. Palmer JW (2002) Web site usability, design, and performance metrics. Inf Syst Res 13(2):151–167
8. Wu F, Mahajan V, Balasubramanian S (2003) An analysis of e-business adoption and its impact on business performance. J Acad Mark Sci 31:425–447
9. Eyrich N, Padman ML, Sweetser KD (2008) PR practitioners' use of social media tools and communication technology. Public Relat Rev 34(4):412–414
10. Hsu CC, Sandford BA (2007) The Delphi technique: making sense of consensus. Pract Assess, Res Eval 12(10):1–8
11. Saaty TL (1980) The Analytic Hierarchy Process. McGraw Hill International, New York
12. Aguarón J, Moreno-Jiménez JM (2003) The geometric consistency index: approximated thresholds. Eur J Oper Res 147(1):137–145
13. Kar AK (2013) Revisiting the supplier selection problem: an integrated approach for group decision support. Expert systems with applications, (in Press)
14. Yang XS, Deb S (2009) Cuckoo search via Lévy flights. In: IEEE world congress on nature and biologically inspired computing, 210–214
15. Walton S, Hassan O, Morgan K, Brown MR (2011) Modified cuckoo search: a new gradient free optimisation algorithm. Chaos, Solitons Fractals 44(9):710–718
16. Gandomi AH, Yang XS, Alavi AH (2013) Cuckoo search algorithm: a metaheuristic approach to solve structural optimization problems. Eng Comput 29(1):17–35

Chapter 49
A Data-Aware Scheduling Framework for Parallel Applications in a Cloud Environment

B. Jaykishan, K. Hemant Kumar Reddy and Diptendu Sinha Roy

Abstract Cloud infrastructures are competent to providing massive processing capabilities of computational and data resources in virtualized environments. Introduction of big data analytics in many spheres of science, technology and business has led to the trend of employing data-parallel frameworks, like Hadoop for handling such massive data requirements. Since most Hadoop based systems make the two decisions of scheduling data and computation independently, it seems a promising prospective to map computations within cloud resources based on data blocks already distributed to them. This paper proposes a computation scheduling framework that adopts the strategy of improving computation and data co-allocation within a Hadoop cloud infrastructure based on knowledge of data blocks availability, hereafter referred to as Data Aware Scheduling (DAS) framework. The proposed DAS employs a dependency based grouping of data. Experiments have been conducted using standard map-reduce applications and results presented herein conclusively demonstrate the efficacy of the proposed framework.

Keywords Data-intensive computing · Hadoop · GridGain · Map-reduce

B. Jaykishan (✉) · K. Hemant Kumar Reddy (✉) · D. S. Roy (✉)
Department of CSE, National Institute of Science and Technology, Berhampur, India
e-mail: bairagijaykishan@gmail.com

K. Hemant Kumar Reddy
e-mail: khemant.reddy@gmail.com

D. S. Roy
e-mail: diptendu.sr@gmail.com

S. Sengupta et al. (eds.), *Emerging Trends in Computing and Communication*, Lecture Notes in Electrical Engineering 298, DOI: 10.1007/978-81-322-1817-3_49,

49.1 Introduction

Cloud computing has the promise of providing massive processing capabilities of computational and data resources in virtualized environments [1]. Hadoop is a data-parallel framework which when employed, provides massive data handling capability to applications. This paper proposes a scheduling (skew-map sequence) framework that adopts the strategy of improving computation and data co-allocation within a Hadoop cloud infrastructure based on knowledge of data blocks availability. The proposed framework has been referred to as Data Aware Scheduling (DAS) framework hereafter and uses dependency based data grouping technique. For demonstrating the efficacy of the proposed DAS framework, a pseudo-cloud infrastructure has been deployed using GridGain 5.2 [2] and Hadoop 2.0 [3].

49.2 GridGain Default Framework Versus Proposed DAS Framework

This section presents a brief overview of the working principle of Hadoop clusters when integrated within a GridGain setup versus proposed data aware scheduling framework using GridGain.

49.3 Experiments and Results

Figures 49.1 and 49.3 shows Hadoop's inbuilt framework and data placement strategy respectively; whereas Figs. 49.2 and 49.4 shows the same for the proposed DAS. Specifically the data placement in Fig 49.4 depends on a dependency based grouping scheme.

For this research, the standard Map-Reduce application chosen is an indexing operation on the genome data downloaded from [4] and ran the Bowtie indexing [5] Map-Reduce program. The MapReduce job running on the DAS's reorganized data has 60 % maps which benefits from having data locality, compared with 42 % from the randomly placed data. Figures 49.5 and 49.6 present the amount of data blocks moved during runtime of the aforementioned applications for Hadoop native distribution scheme and DAS' dependency based data distribution scheme respectively.

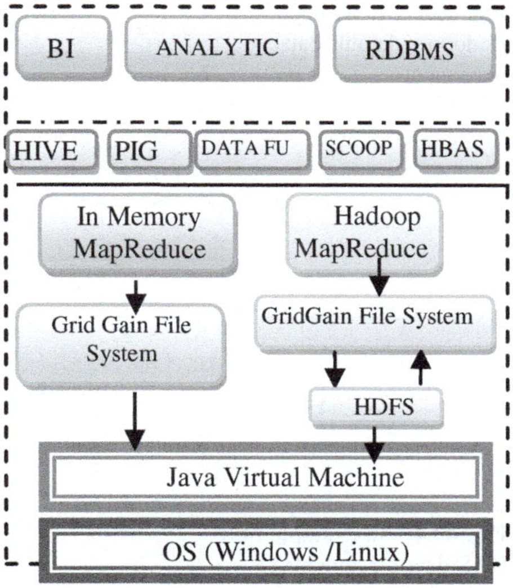

Fig. 49.1 GG Hadoop default framework

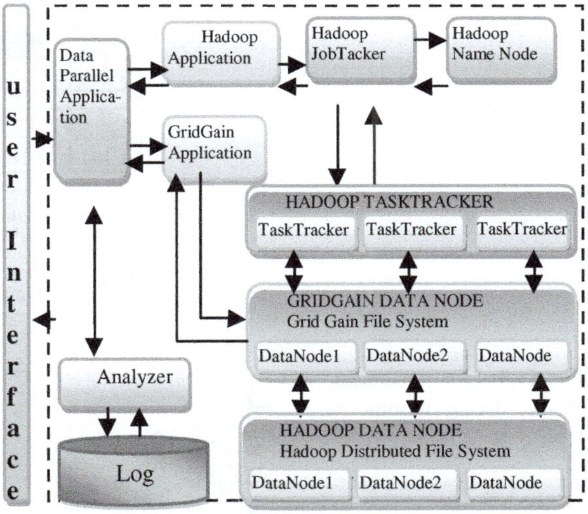

Fig. 49.2 Proposed DAS framework

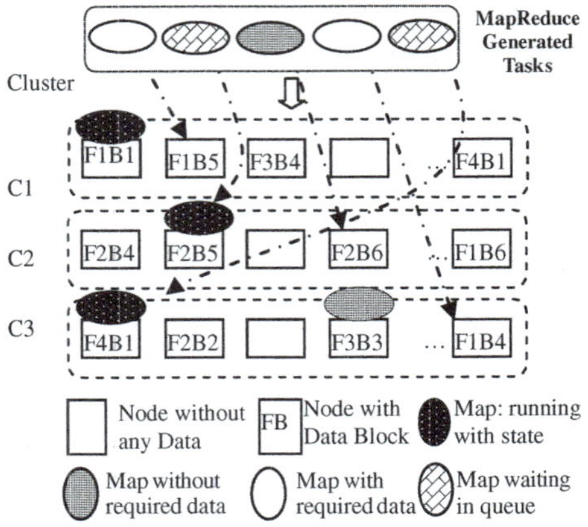

Fig. 49.3 An example of Hadoop's default data placement

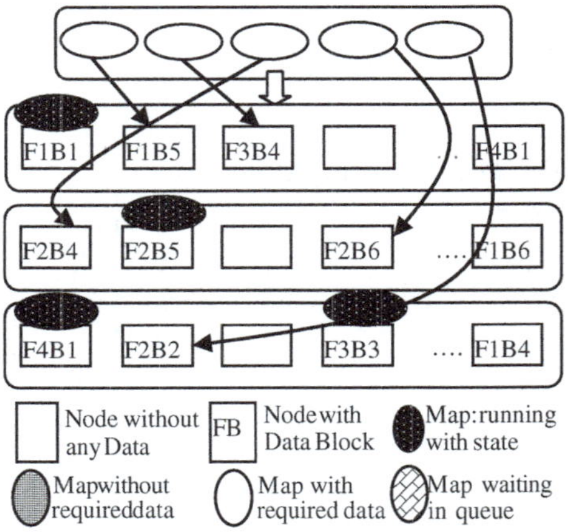

Fig. 49.4 An example of DAS data placement

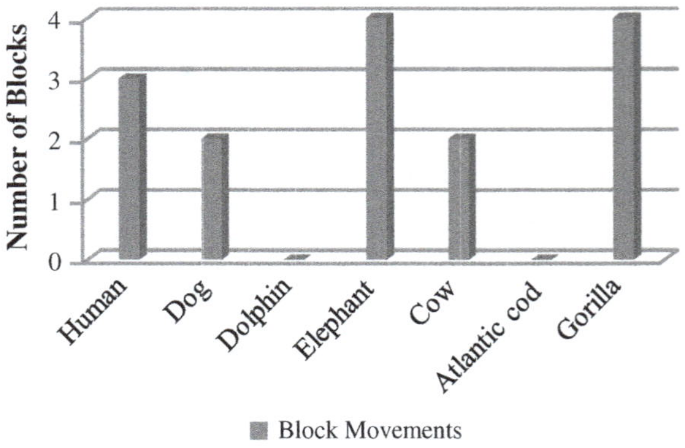

Fig. 49.5 Data movement enacted during run time for DAS

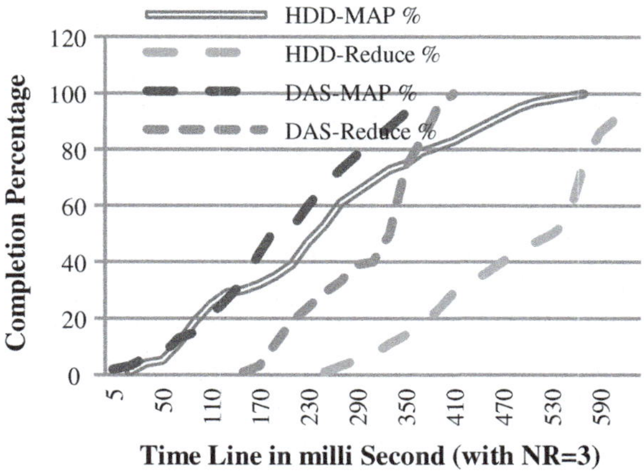

Fig. 49.6 Completion progress wrt time

49.4 Conclusion

This paper proposes a method for increased data-computation co-allocation in a Hadoop cloud based on a novel Data Aware Scheduling (DAS), where knowledge of data blocks' availability is considered for scheduling decisions. Experiments conducted using Bowtie indexing MapReduce programs on 40 GB genome data reveals that DAS shows around 60 % reduction in block movements with respect to Hadoop's default block distribution policy. This improvement can be attributed to the dependency based grouping and subsequent data placement among nodes employed in the proposed DAS framework.

Acknowledgments This work has been carried out at the Data Sciences Lab, Department of Computer Science and Engineering, National Institute of Science and Technology, Berhampur.

References

1. Yuan D, Yang Y, Liu X, Chen J (2010) A data placement strategy in scientific cloud workflows. Future Gener Comput Syst 26:1200–1214
2. www.gridgain.com. Last accessed 5 Dec 2013
3. http://hadoop.apache.org/. Accessed 15 Nov 2013
4. http://genome.ucsc.edu/
5. http://bowtiebio.sourceforge.net/index.shtml

About the Editors

Sabnam Sengupta has completed her Ph.D. in Computer Science and Engineering in the field of Software Engineering from Jadavpur University, Kolkata, India in 2008. She has completed her Post Graduation from University of Calcutta in 1998 and has completed her bachelors from Presidency College, Kolkata in 1995. She has published one book, three book chapters, along with several publications in the International journals and conference proceedings. She is serving as a member of editorial boards of a few International journals. Dr. Sengupta has 10 years of experience in academics and 3 years of experience in the software industry, both in India and USA. She is currently working as an Associate Professor and Head, Department of Information Technology, B. P. Poddar Institute of Management and Technology, Kolkata. Her biography has been published in Marquis Who's Who in Science and Technology, 28th Edition, 2010. She is a member of IEEE. Her research interests include formal modelling, software requirements, verification and testing and software architecture.

S. Sengupta et al. (eds.), *Emerging Trends in Computing and Communication*, Lecture Notes in Electrical Engineering 298, DOI: 10.1007/978-81-322-1817-3,

Kunal Das has received the Bachelor of Science degree in Electronics Science and Bachelor of Technology degree in Information Technology from University of Calcutta, Kolkata, India, in 2001 and 2004 respectively. He has completed his Master of Technology in Information Technology from the same university in 2006. He is currently pursuing his Ph.D. He has authored 7 Int. journals papers and several conference papers. He is Co-author of book titled "Quantum Dots and Quantum Cellular Automata: Recent Trends and Applications", NOVA Science publishers, USA. He is a member of IEEE and an invited member of American Nano Society, USA. Mr. Das is currently working as an Assistant Professor, Department of Information Technology, B. P. Poddar Institute of Management and Technology; Kolkata. His research interests include Quantum Dot Cellular Automata, VLSI architectures, and hardware.

Gitosree Khan received the Bachelor of Engineering degree in Computer Science and Engineering from Berhampur University, Orissa, India, in 2005 and Master of Technology in Computer Science and Application from University of Calcutta in 2010. She is currently pursuing Ph.D. She has authored an International journals paper and published a few papers in conference proceedings. She has reviewed several Int. journals papers and conference papers. She is a member of IEEE. She is currently working as an Assistant Professor, Department of Information Technology, B. P. Poddar Institute of Management and Technology; Kolkata. Her research interests include Cloud Computing and Software engineering.

Author Index

S. Sengupta et al. (eds.), *Emerging Trends in Computing and Communication*,
Lecture Notes in Electrical Engineering 298, DOI: 10.1007/978-81-322-1817-3,

The manufacturer's authorised representative in the EU is Springer Nature Customer Service Centre GmbH, Europaplatz 3, 69115 Heidelberg, Germany. If you have any concerns regarding our products, please contact ProductSafety@springernature.com

Printed and bound by CPI Group (UK) Ltd, Croydon, CR0 4YY
30/09/2026
02225447-0016